ACS SYMPOSIUM SERIES **579**

Polymeric Materials for Microelectronic Applications

Science and Technology

Hiroshi Ito, EDITOR
IBM Corporation

Seiichi Tagawa, EDITOR
Osaka University

Kazuyuki Horie, EDITOR
University of Tokyo

Developed from a symposium sponsored
by the Polymers for Microelectronics and Photonics Group
of the Society of Polymer Science, Japan, and the
Division of Polymeric Materials: Science and Engineering, Inc.,
of the American Chemical Society,
at the Polymers for Microelectronics 1993 Meeting,
Kawasaki, Japan,
November 15–19, 1993

American Chemical Society, Washington, DC 1994

Library of Congress Cataloging-in-Publication Data

Polymeric materials for microelectronic applications: science and technology / Hiroshi Ito, editor, Seiichi Tagawa, editor, Kazuyuki Horie, editor.

p.   cm.—(ACS symposium series, ISSN 0097–6156; 579)

"Developed from a symposium sponsored by the American Chemical Society Division of Polymeric Materials: Science and Engineering, Inc., and the Polymers for Microelectronics and Photonics Group of the Society of Polymer Science, Japan, at the Polymers for Microelectronics 1993 Meeting, Kawasaki, Japan, November 15–19, 1993."

Includes bibliographical references and indexes.

ISBN 0–8412–3055–2

1. Microelectronics—Materials—Congresses. 2. Polymers—Congresses.

I. Ito, Hiroshi. II. Tagawa, Seiichi. III. Horie, Kazuyuki, 1941– . IV. American Chemical Society. Division of Polymeric Materials: Science and Engineering, Inc. V. Polymers for Microelectronics Meeting (1993: Kawasaki, Japan) VI. Series.

TK7871.15.P6P59   1994
621.381—dc20   94–38922
                                                                      CIP

The paper used in this publication meets the minimum requirements of American National Standard for Information Sciences—Permanence of Paper for Printed Library Materials, ANSI Z39.48–1984. ∞

PRINTED IN THE UNITED STATES OF AMERICA

TK 7871.15.P6P59.1994.CHEM

1994 Advisory Board

ACS Symposium Series

M. Joan Comstock, *Series Editor*

Foreword

THE ACS SYMPOSIUM SERIES was first published in 1974 to provide a mechanism for publishing symposia quickly in book form. The purpose of this series is to publish comprehensive books developed from symposia, which are usually "snapshots in time" of the current research being done on a topic, plus some review material on the topic. For this reason, it is necessary that the papers be published as quickly as possible.

Before a symposium-based book is put under contract, the proposed table of contents is reviewed for appropriateness to the topic and for comprehensiveness of the collection. Some papers are excluded at this point, and others are added to round out the scope of the volume. In addition, a draft of each paper is peer-reviewed prior to final acceptance or rejection. This anonymous review process is supervised by the organizer(s) of the symposium, who become the editor(s) of the book. The authors then revise their papers according to the recommendations of both the reviewers and the editors, prepare camera-ready copy, and submit the final papers to the editors, who check that all necessary revisions have been made.

As a rule, only original research papers and original review papers are included in the volumes. Verbatim reproductions of previously published papers are not accepted.

M. Joan Comstock
Series Editor

Contents

INSULATING POLYMERS

Preface

POLYMERIC MATERIALS HAVE FOUND WIDESPREAD USE in the manufacture of electronic devices and have been employed in increasingly diverse areas of electronics. In addition to their high processability and ease of fabrication, a wide range of chemical and physical properties provided by molecular engineering makes polymeric materials highly versatile in the electronics industry.

Polymers have been used as lithographic resists and dielectrics in the fabrication of semiconductor devices and as passivation and insulating materials in electronics packaging. In addition to these "passive" applications, "active" applications of conductive, photoresponsive, and optoelectronic polymers have emerged, are growing, and provide exciting research opportunities to polymer scientists and engineers. The unique optical, electronic, and lithographic properties of silicon-containing polymers have also spawned extensive research activities.

Recognizing the worldwide research efforts on these important polymers, the American Chemical Society Division of Polymeric Materials: Science and Engineering, Inc. and the Society of Polymer Science, Japan, jointly organized the International Symposium on Polymers for Microelectronics (PME) in Toyko in 1989, which was followed by the second Symposium in San Francisco in 1992. This book is based upon the third PME Symposium, which was held in Kawasaki, Japan, in 1993. The volume should help the reader appreciate the diversity and the rapid progress of the research activities on advanced polymer materials for use in microelectronics.

Acknowledgments

We thank the authors for their efforts in preparing their manuscripts for this volume and the referees for reviewing each paper. Our special thanks also go to Anne Wilson and the production staff of the ACS Books Department for their efforts in assembling this volume.

HIROSHI ITO
IBM Almaden
  Research Center
650 Harry Road
San Jose, CA 95120

September 13, 1994

SEIICHI TAGAWA
Osaka University
8-1 Mihogaoka
Ibaraki, Osaka 567
Japan

KAZUYUKI HORIE
Faculty of Engineering
University of Tokyo
7-3-1 Hongo
Bunkyo-ku, Tokyo 113
Japan

CHEMISTRY AND PHYSICS
OF POLYMER IRRADIATION

Chapter 1

Photophysics, Photochemistry, and Photo-optical Effects in Polymer Solids

Kazuyuki Horie

Faculty of Engineering, University of Tokyo, 7–3–1 Hongo, Bunkyo-ku, Tokyo 113, Japan

Photophysics, photochemistry, and optical phenomena of polymers for microelectronics are closely related to one another. By using polyimides as typical examples, it is discussed how photophysics such as charge transfer process affects photochemistry such as photosensitivity of polyimides. Photochemistry such as photoisomerization induces the change in refractive indices providing the phase modulation in optical devices. Microstructure of a polymer liquid crystal is also elucidated by fluorescence spectroscopy.

The interaction of light with materials is one of the fundamental subjects in natural science and technology. Phenomena related to light are called in Japanese "hikari" by using a single word, but there are one word and two prefixes concerned with light in English; "light", "photo-" and "opto-". Photochemistry and photophysics deal with the change in materials by light which usually chemists are interested in, and optics deal with the change in light by materials which has been developed mainly by physicists (1). Both aspects of the interaction between light and materials as well as their hybrid phenomena will become more and more important in the field of polymers for microelectronics and photonics. Typical examples of the materials with hybrid phenomena are photorefractive polymers where non-linear optical effects are induced by the space charge generated by the photoirradiation of photoconductive polymers (2) and a command surface where the alignment of nematic liquid crystals are controlled by photoisomerization of azobenzene layer (3). Several examples of photo- and opto-functional materials are summarized in Table I.

Polyimides (PI) are well-known because of their excellent thermal stability, and they have become important as high temperature insulation materials in the microelectronics field. They are also expected to be potential materials for various thermostable microelectronics and photonics devices. In the present paper, we use polyimides as typical materials for studying photophysics, photochemistry, optical phenomena and their relations in polymer solids, and report several recent results of the charge-transfer (CT) fluorescence study of aromatic polyimides, the influence of CT state on the photochemistry of photosensitive polyimides, third-order non-linear optical properties of aromatic polyisoimides, and photochemically-induced optical effects in dye-doped polymer films. Charge-transfer and dimer fluorescence studies have also been applied to polymer liquid crystals.

0097–6156/94/0579–0002$08.00/0

Table I. Typical examples of photo- and opto- functional materials

Functions	Example
Photophysical Functions	Organic Photo-Conductors
	Electro-Luminescence
	Excited Energy Transfer
	Scintillators
	Fluorescent Probes
Photochemical Functions	Photoresists
	Photochromism
	Photochemical Hole Burning (PHB)
	Photoresponsive Polymers
	Photocatalysts
	Photo-Energy Conversion
Optical Functions	Optical Fibers, Wave guides
	Non-Linear Optical Materials
	Phase & Frequency Modulation Devices
	Liquid Crystals

Hikari phenomena ⋯ The Interaction of Light with Materials

Hikari ─┬─ Light
        ├─ Photo- ─┬─ Photochemistry ⎱ ⋯⋯ Changes of Materials
        │   └─ Photophysics   ⎰   by Materials
        └─ Opto-   optics ⋯⋯⋯⋯⋯⋯ Changes of Light
                                           by Materials

Charge Transfer Fluorescence of Aromatic Polyimides

Aromatic polyimides (PI) have outstanding thermal stability. Their excellent physical properties are correlated with both the rigidity of the main chains and also strong intermolecular interaction caused by the stacking of aromatic rings. Another important factor affecting the physical properties of aromatic polyimides is the state of aggregation of polymer chains. Typical examples are the high-modulus and high-strength polyimide films developed using thermal imidization of cold drawn polyamic acid. It is especially important therefore to control the molecular aggregation (4).

Although the microscopic structures for PIs have been observed by using wide-angle X-ray diffraction (WAXD) and small-angle X-ray scattering (SAXS), these methods are limited to elucidation of crystalline structure and large-scale ordered structure, respectively. Fluorescence spectroscopy is known to be very useful in the investigation of dynamics and structure of solid polymers (5) and is complementary to the X-ray methods.

Several aromatic polyimide films are known to show broad and structureless fluorescence spectra with peak wavelengths in 450-600 nm region, considerably longer than those of usual aromatic molecules. Typical fluorescence spectra are shown in Figure 1 (6). As the emission spectra become red-shifted with increasing electron-withdrawing ability of aromatic imide moiety and electron-donating ability of aromatic moiety from diamine, the emission is assigned to be due to an excited-state charge transfer (CT) interaction between imide moiety and aromatic moiety from diamine. When a polyimide, PI(BPDA/PDA), from biphenyltetracarboxylic dianhydride (BPDA) and *p*-phenylenediamine (PDA) (No. 4 in Figure 1) prepared by stepwise imidization up to 250°C (2hr) was annealed at 330°C for 2hr, as is shown in Figure 2, fluorescence intensity at 525 nm excited by 350 nm irradiation increased 4 times without shift of peak wavelength, showing the intermolecular nature of this CT fluorescence, and a strong new fluorescence at 540 nm appeared with 465 nm excitation due to the formation of also a ground state intermolecular CT complex (7). A linear increase in the intensity of CT fluorescence with the increase in density of polyimide films due to increase in imidization temperature observed for both PI(BPDA/PDA) and PI(BPDA/ODA), where ODA is oxydianiline, also supports the intermolecular mechanism of the CT complex (7).

The study of model bisimides in solution (Figure 3) (8) shows also the intramolecular CT formation in the excited states. Strong fluorescence at 400 nm is observed for the model bisimide with cyclohexylamine, M(BPDA/CHA), (No.4 in Figure 3) which is emitted from the direct excited state of monomer because the sample has no possibility to make either intra- or inter- molecular CT interaction. The model bisimide of BPDA with aniline, M(BPDA/A), is non-fluorescent, suggesting that effective deactivation occurs *via* strong intermolecular CT formation of the excited state, but M(BPDA/mEA) (No. 1 in Figure 3) shows weak CT fluorescence. The fluorescence spectra of M(BPDA/o-Tol) (No. 2) and M(BPDA/i-PrA) (No. 3) show CT character in excited states although they cannot form coplaner structure. Spectroscopic studies of pyromellitic polyimides (No.1 in Figure 1) and their model compounds discussing the intramolecular CT absorption or fluorescence have also been done by a few groups (9-11).

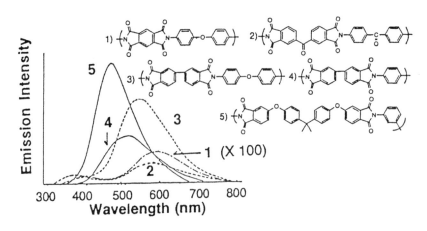

Figure 1. Fluorescence Spectra (300 nm excitation) for Several Polyimide Films (6).

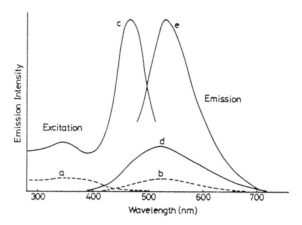

Figure 2. Effects of annealing (330°, 2hr) on the emission and excitation spectra of PI(BPDA/PDA). --- (a) excitation (monitored at 530 nm) and (b) emission (excited at 350 nm) spectra of PI without annealing; --- (c) excitation spectrum (monitored at 530 nm) and emission spectra of PI excited at 350 nm (d) and 465 nm (e) with annealing (Reproduced with permission from Ref. 6).

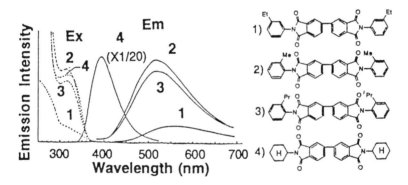

Figure 3. Excitation and fluorescence spectra of several model compounds in CH_2Cl_2 solution (8).

Photochemistry of Benzophenone-type Photosensitive Polyimides

Benzophenone-containing polyimide prepared from benzophenonetetracarboxylic dianhydride (BTDA) and bis(4-amino-3-ethylphenyl)methane (DEDPM) is known to have excellent negative photoresist properties as well as thermal stability. Photophysical processes for the photocrosslinking reaction of PI(BTDA/DEDPM) are shown in Figure 4 (12). An excited triplet state benzophenone moiety is formed *via* singlet state by the absorption of a photon. A part of the excited state is deactivated back to the ground state. The excited triplet state benzophenone moiety abstracts a hydrogen atom from an ethyl group of a neighboring polymer chain, resulting in two radicals; a ketyl radical and a benzyl radial. Some of them disproportionate and others combine with each other to form a crosslink. The quantum yields for the photocrosslinking of PI(BTDA/DEDPM), ϕ_x, are obtained by measuring the change in their molecular weights during photoirradiation by using GPC, and ϕ_x under air at room temperature was obtained to be 0.001.

Table II shows the quantum yields for photoreaction of model compound, M(BTDA/DMA), and benzophenone, BP, in ethylbenzene solution, and the corresponding kinetic data. The quantum yields for photoreaction for M(BTDA/DMA) under air, ϕ_{air}, and in vacuum, ϕ_{vac}, are an order of magnitude smaller than those for BP. The lowest excited state of M(BTDA/DMA) is considered to be made up by the superposition of the electronic configuration in the excited state of the CT and that in the benzophenone group. The rate constant for hydrogen abstraction from ethylbenzene, k_a^{EB}, for M(BTDA/DMA) is smaller than that of BP because the n-π^* nature in the excited state of M(BTDA/DMA) should decrease owing to the increase in CT nature. On the other hand, the sum of the rate constants for deactivation, $k_d + k_d^{EB}[EB]$, of M(BTDA/DMA) is greater than that of BP, probably due to the intermolecular CT formation with ethylbenzene. The ϕ_x is expected to be improved by restricting CT formation in solid state PI.

In order to prepare new types of photosensitive polyimides which have much higher photocrosslinking efficiency than PI(BTDA/DEDPM), two methods were carried out in the author's laboratory. One is the introduction of alicyclic diamines(*13*) which prevents the occurrence of charge transfer (CT) leading to the increase of ϕ_x up to the level of model compounds. The other is the use of carbene or nitrene species as reactive intermediates for photocrosslinking of polyimide. A solvent-soluble polyimide with alicyclic diamine PI(BTDA/DMDHM) was prepared, which showed ϕ_x = 0.004 suggesting a 4 times improvement of photocrosslinking efficiency compared to PI(BTDA/DEDPM)(*14*) (Table III). The introduction of alicyclic groups instead of phenylene groups did not decrease the physical thermostability, i.e., T_g of the corresponding polyimides. The hydrogen abstraction reaction, ϕ_{ab} for M(BTDA/nBA) is 22 times higher than ϕ_{ab} for M(BTDA/oEA). The quantum yield of photocrosslinking, ϕ_{cr}, for the alicyclic polyimide is just 2-4 times higher than ϕ_{cr} for the aromatic polyimide. The difference in these ratios can be attributed to the difference in the reactivity of disproportionation between benzophenone ketyl radical and alkyl radical (*14*).

The introduction of diazo-group in polyimide main chain, PI(DZDA/DEDPM) gives the very high value of photocrosslinking efficiency (ϕ_x=0.13) (*15*) (Table IV). The photocrosslinking of PI(DZDA/DEDPM) occurs by insertion of carbene which is produced from the singlet state of diazo-moiety and does not loose its reactivity before the occurrence of reaction. The decomposition reaction of diazo groups is not influenced by oxygen.

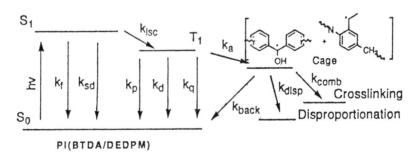

Figure 4. Photophysical processes for PI(BTDA/DEDPM) (*12*).

Table II. Various kinetic parameters for the photoreactions of M(BTDA/DMA) and benzophenone in ethylbenzene (*12*)

	Φvac	Φair	k_a^{EB} $(M^{-1}s^{-1})$	$k_d+k_d^{EB}[EB]$ (s^{-1})
Benzophenone	0.88	0.29	1.0×10^6	1.2×10^6
M(BTDA/DMA)	0.066	0.025	9.3×10^4	1.1×10^7

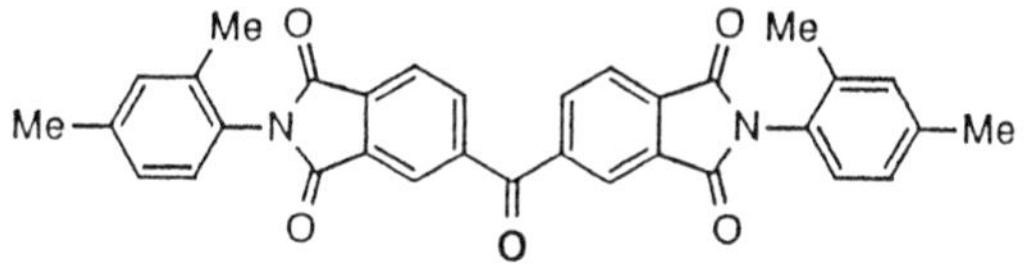

M(BTDA/DMA)

Table III. Quantum yields for hydrogen photoabstraction, Φ_{ab}, and photocrosslinking, Φ_{cr}, of aromatic and alicyclic polyimides and their model compounds (*14*)

		Φ_{ab}		Φ_{cr} (film)	
Compound	Solvent	Air	Vac	Air	Vac
PI(BTDA/DMDHM)				0.004	0.004
PI(BTDA/DEDPM)				0.001	0.002
M(BTDA/n-BA)	THF	0.56	0.56		
	MTHF	0.34			
M(BTDA/o-EA)	THF	0.054			
	Ethylbenzene	0.025	0.066		
Benzophenone	Ethylbenzene	0.3	0.88		

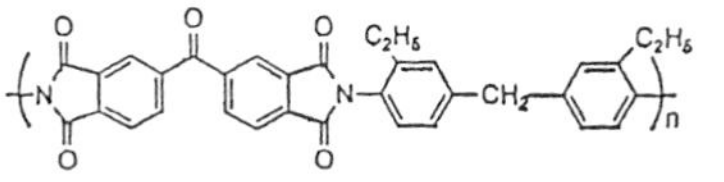

PI(BTDA/DEDPM)

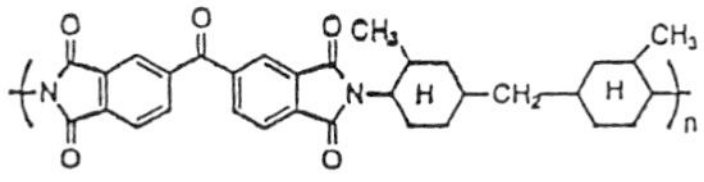

PI(BTDA/DMDHM)

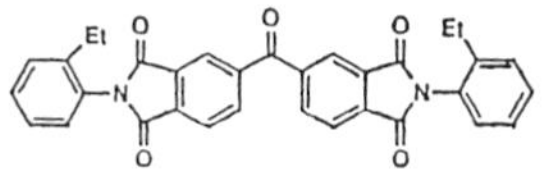

M(BTDA/o-EA)

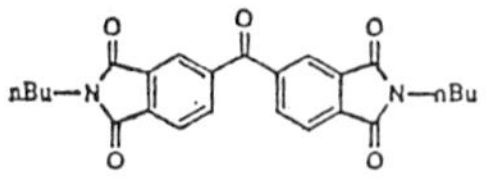

M(BTDA/nBA)

Table IV. Quantum yields for photodecomposition, Φ_{dec}, and photocrosslinking, Φ_{cr}, of polyimides and their models (15)

Compound	Φ_{dec}		Φ_{cr}
	Air	Vac	
PI(BTDA/DEDPM)	—	—	0.002
M(BTDA/o-EA)	0.025	0.066	
M(BTDA/n-BA)	0.36		
BP	0.3	0.88	
PI(DZDA/DEDPM)	0.06	—	0.01, 0.13
M(DZDA/o-EA)	0.15	0.14	
M(DZDA/n-BA)	0.20		
DDM	1.1		

PI(DZDA/DEDPM)

M(DZDA/o-EA)   DDM

Charge Transfer Fluorescence from Polymer Liquid Crystals

The first observation of charge-transfer fluorescence from polymer liquid crystals has been reported recently (16). Rigid-rod aromatic polyester with long alkyl side-chains from 1,4-dialkylesters of pyromellitic acid and 4,4'-biphenol (B-C16) forms layered structures which exhibit mesomorphic properties(17). X-ray diffraction pattern showed that its aromatic main chains are in an extended form and packed into a layered structure with the average lateral packing distance of 4Å between main chain within a layer. Though X-ray diffraction cannot reveal the intermolecular arrangement between pyromellitic group and biphenyl group within a layer in the liquid crystalline phase, fluorescence spectroscopy shows charge-transfer (CT) complex formation between these groups (Figure 5) (16), concluding that the biphenyl moieties lie adjacent to the pyromellitic groups in a layer. The most reliable driving forces responsible for the organization of these rigid-rod polyesters with flexible side chains into layered liquid crystalline structures are usually thought to be the difference in polarity and geometric shape between the aromatic main chains and aliphatic side chains. However, the formation of interchain charge transfer complex in the aromatic main chains is also suggested to play a significant role in the stability of layered segregation structure. Fluorescence due to ground-state complex was observed for quenched films of thermotropic liquid crystalline polyesters such as poly(ethylene terephthalate-oxybenzoate) and poly(ethylene naphthenate-oxybenzoate) (18).

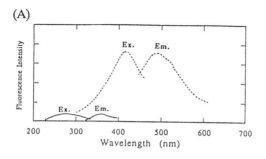

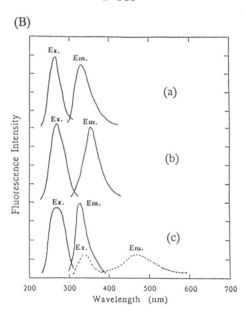

Figure 5.   (A) The fluorescence and its excitation spectra for the frozen-in layered mesophase of B-C16, which was prepared between quartz slides by quenching the mesophase at 100°C to room temperature. ——: monomer fluorescence, ---: fluorescence from the ground-state CT complex. (B) The fluorescence and its excitation spectra for (a) 4,4'-diacetoxy biphenyl, (b) 1,4-hexadecyl pyromellitic acid, and (c) their equimolar mixture in chloroform solution of 10^{-1} M (Reproduced with permission from Ref. *16*). ---: fluorescence due to the ground-state CT complex.

Non-Linear Optical and Photo-Optical Effects in Polyimides and Other Polymers

Aromatic polyimides have been used in recent years as a new class of thermostable optical materials in waveguide application (19). Several attempts have been done for including and poling non-linear optical (NLO) molecules or groups in polyimides in order to obtain thermally-stable electro-optic response (20). Aromatic polyisoimides (PiI), isomers of polyimides, have conjugated structure along their main chains, and hence is supposed to show intrinsic third-order NLO properties. The third-order nonlinear susceptibility of several polyisoimide films was estimated to be in the order of 10^{-12} esu by THG measurements (Figure 6) with the highest values of 6.7×10^{-12} esu for PiI (PMDA/PDA) (21).

Pi I(R1/R2)	R1	R2	isoimide contents	λmax. /nm	absorption edge /nm	$\chi^{(3)}$ /10^{12} (e.s.u.)
Pi I(PMDA/PDA)	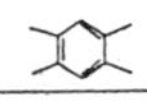	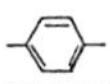	85%	432	520	6.7
Pi I(PMDA/ PDA;DEDPM)	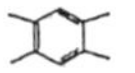	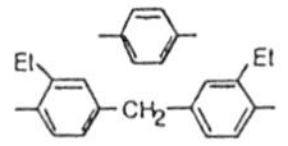	94%	420	505	5.0
Pi I(BPDA/PDA)	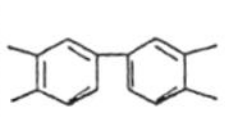	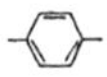	97%	388 297	470	3.2
Pi I(BTDA/PDA)	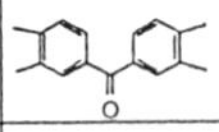	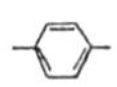	98%	395	470	1.3
Pi I(BTDA/TDA)	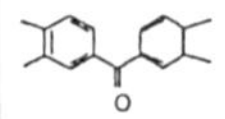	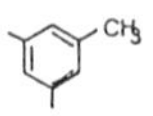	98%	360	460	1.2
PI (PMDA/PDA)	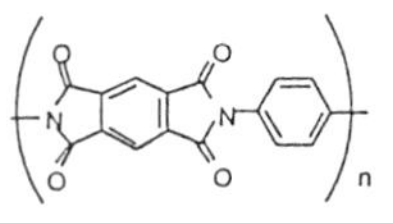		0%	276 330	350	1.4

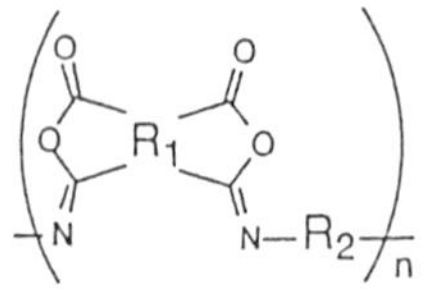

Pi I(R1/R2)

Figure 6. The chemical structure of synthesized polyisoimides and their absorption peak (λmax.), absorption edge, $\chi^{(3)}$, and isoimide contents (21).

The optical switching devices in optical computing systems usually takes advantage of electro-optic, acousto-optic and nonlinear optical effects. Photochemical reactions have potential applicability to photo-optical switching devices, because they change the electronic structure of the molecules and the density of the system (22-25). The change in refractive indices and the anisotropy of PMMA (poly(methyl methacrylate)) films containing DAAB (*p*-dimethylaminoazobenzene) were measured during polarized-light-induced photoisomerization, and the possibility as a photo-optical controlling material was examined (*26*). DAAB was photoisomerized from trans-form to cis-form by irradiating 410 nm non-polarized or polarized light. Refractive indices were measured by a prism-film coupling method (*27*) using a He-Ne laser (632.8 nm) after the photoisomerization reached to the steady state. The conversion of photoisomerization was determined to be 70 %. Refractive indices decreased by photoisomerization of DAAB and the difference from the initial value, dn, amounted to -3×10^{-3} for PMMA films containing 5 wt% of DAAB. The absorption dichroism during polarized-light-induced photoisomerizaion with Hg-lamp and polarizer increases with irradiation time and then levels off. The observed birefringence shown in Figure 7 (a) during the polarized-light-induced photoisomerization of 2.5 % DAAB in PMMA is attributed as the first approximation to the difference in trans-form concentrations sensitive to TE and TM modes which decreased at a certain time. Figure 7 (b) gives the changes in refractive indices dn for both TE and TM modes during polarized-light irradiation with the abscissa reduced to the same trans-form concentrations, showing that the difference in dn between two curves still exists due to the much-higher cis-form concentration for PM mode than that for PE mode at a same trans-form concentration. This suggests the possibility of selective control of orientation of dye molecules in films by light irradiation.

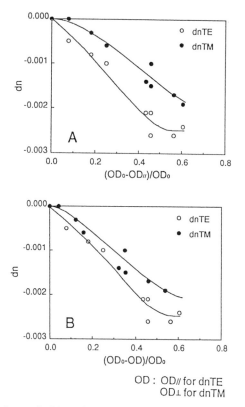

Figure 7. The change in TE and TM mode refractive indices, dn, of a PMMA film with 2.5 % trans-DAAB during photoisomerization with polarized light (*26*).

Acknowledgment

The author expresses his sincere gratitude to Prof. emeritus I. Mita, Dr. T. Yamashita, Dr. M. Hasegawa, and all his collaborators who have made it possible to carried out the works cited in this article.

Reference

1. Horie, K.; Ushiki, H. *Science of Photofunctional Molecules*: Kodansha; Tokyo, 1992.
2. Ducharme, S.; Scott, J.; Twieg, R.; Moerner, W. E. *Phys. Rev. Lett.* **1991**, *66*, p.1846.
3. Ichimura, K.; Suzuki, Y.; Seki, T.; Hosoki, A.; Aoki, K. *Langmuir* **1988**, *4*, 1214.
4. Kochi, M.; Uruji, T.; Iizuka, T.; Mita, I.; Yokota, R. *J. Polym. Sci.* **1987**, Part C *25*, 441; Yokota, R.; Horiuchi, R.; Kochi, M.; Soma, H.; Mita, I. *J. Polym. Sci.* **1988**, Part C *26*, 215.
5. *Photophysical and Photochemical Tools in Polymer Sience*; Winnik, M. A., Ed.; Reidel: Dordrecht, 1986.
6. Hasegawa, M.; Kochi, M.; Mita, I.; Yokota, R. *Polym. Prepr. Jpn.* **1987**, *36*, 1158, 3554; *J. Polym. Sci.* **1989**, C *27*, 263.
7. Hasegawa, M.; Kochi, M.; Mita, I.; Yokota, R. *Eur. Polym. J.* **1989**, *25*, 349.
8. Hasegawa, M.; Shindo, Y.; Sugimura, T.; Ohshima, S.;Horie, K.; Kochi, M.; Yokota, R.; Mita, I. *J. Polym. Sci.*; *Polym. Phys. Ed.* **1993**, *31*, 1617.
9. Wachsman, E. D.; Frank, C. W. *Polymer* **1988**, *29*, 1191.
10. LaFemina, J. P.; Arjavalingam, G.; Hougham, G. *J. Chem. Phys.* **1989**, *90*, 5154; *Polymer* **1990**, *31*, 840.
11. Ishida, H.; Wellinghoff, S. T.; Baer, E.; Koenig, J. L. *Macromolecules* **1980**, *13*, 826.
12. Higuchi, H.; Yamashita, T.; Horie, K.; Mita, I. *Chem. Mater.* **1991**, *3*, 188.
13. Jin, Q.; Yamashita, T.; Horie, K.; Yokota, R.;Mita, I. *J. Polym. Sci.*; *Polym. Chem. Ed.* **1993**, *31*, 2345.
14. Jin, Q.; Yamashita, T.; Horie, K. *J. Polym. Sci.*; *Polym. Chem. Ed.* **1993**, *31*, in press.
15. Yamashita, T.; Horie, K. *ACS PMSE Prepr.* **1992**, *66*, 245; *Chem. Mater.* to be published.
16. Sone, M.; Harkness, B. R.; Watanabe, J.; Yamashita, T.; Torii, T.; Horie, K. *Polym. J.* **1993**, *25*, 997.
17. Harkness, B. R.; Watanabe, J. *Macromol.* **1991**, *24*, 6759.
18. Hashimoto, H.; Hasegawa, M.; Horie, K.; Yamashita, T.; Ushiki, H.; Mita, I. *J. Polym. Sci.*; *Polym. Phys.* **1993**, *31*, 1187.
19. Reuter, R.; Franke, H.; Feger, C. *Appl. Opt.* **1988**, *27*, 4565; **1993**, *32*, 2927.
20. Wu, J. W.; Valley, J. F.; Ermer, S.; Binkley, E. S.; Kenney, J. T.; Lipscomb, G. F; Lytel, R. *Appl. Phys. Lett.* **1991**, *58*, 225.
21. Morino, S.; Horie, K.; Yamashita, T.; Wada, T.; Sasabe, H. *Appl. Phys. Lett.* to be submitted.
22. Todorov, T.; Tomova, N.; Nikolova, L. *Opt. Commun.* **1983**, *47*, 123.
23. Todorov, T.; Nikolova, L.; Tomova, N. *Appl. Opt.* **1984**, *23*, 4309, 4588.
24. Lückemeyer, Th.; Franke, H. *Polym. Eng. Sci.* **1991**, *31*, 912.
25. Xie, S.; Natansohn, A.; Rochon, P. *Chem. Mater.* **1993**, *5*, 403.
26. Morino, S.; Horie, K.; Yamashita, T. *Polym. Prepr. Jpn.* **1993**, *43*, 668.
27. Tien, P. K. *Appl. Opt.* **1971**, *10*, 2395.

RECEIVED September 13, 1994

Chapter 2

Photochemistry of Liquid-Crystalline Polymers

David Creed[1], Richard A. Cozad[1], Anselm C. Griffin[1], Charles E. Hoyle[2],
Lixin Jin[1], Petharnan Subramanian[1], Sangya S. Varma[1], and
Krishnan Venkataram[1]

[1]Department of Chemistry and Biochemistry and [2]Department of Polymer
Science, University of Southern Mississippi, Hattiesburg, MS 39406

Several aspects of the photophysics and photochemistry of liquid
crystalline polymers containing aryl cinnamate and stilbene
chromophores are reported. UV-VIS spectra of these materials in
various phases in thin films are perturbed relative to spectra in solution.
These perturbations are attributed to chromophore aggregation that is
dependent on the phase type. Fluorescence and fluorescence excitation
spectra of a stilbene polyester show the presence of emissive
aggregates. The photochemical reactions of these materials are also
dependent on phase. In polyarylcinnamates the ratio of 'dimer' or
'dimer like' to photo-Fries photoproducts depends both on temperature
(phase type) and wavelength of irradiation. 'Dimer' or 'dimer like'
products are formed preferentially under conditions where
chromophore association and/or excitation is maximized. In the
stilbenes a photocycloaddition is observed that is photoreversible at
elevated temperature. Parallel studies are reported on small molecule
model compounds.

The photochemistry of liquid crystalline (LC) materials is of interest both as a means
of modifying the properties of these potentially useful materials and because they have
the unique combination of one or two dimensional order and translational mobility.
There have been many reports of the photochemistry of small molecule LC materials
(1) but comparatively few reports (2-14) of the photochemistry of LC polymers. We
have already reported the synthesis and some aspects of the photochemistry and
photophysics of nematic (2,3,4) and smectic (5,6) LC polycinnamates such as 1 and 2,
and preliminary observations (7) of the photochemistry of nematic stilbene polyesters
such as 3, that were originally synthesized by Jackson and Morris (15). We now
report further observations on the photophysics and photochemistry of these
polymeric LC materials, paying particularly attention to ground state chromophore
association and its unique photophysical and photochemical consequences.

0097–6156/94/0579–0013$08.00/0

1a, x = 9
1b, x = 10

Experimental

Polymers and small molecule model compounds were synthesized and characterized as described previously (*2,5,8,15*). The keto-coumarin triplet sensitizer **8** was synthesized as described elsewhere (*16*). UV-VIS spectra were obtained using a Perkin Elmer Lambda 6 spectrophotometer with a variable temperature cell as described elsewhere (*4*). Corrected fluorescence and fluorescence excitation spectra were obtained using a Spex Fluorolog spectrophotometer. Polymer films were cast from chloroform, methylene chloride, or tetrachloroethane on to quartz plates or cuvettes using an EC101D series photoresist spinner P/N 8-13242 from Headway Research Inc. In variable temperature experiments, cuvettes coated on one outside face with the polymer film were filled with an optically transparent, high boiling point fluid, such as ethylene glycol, to facilitate heat transfer from the variable temperature cell to the film. Most irradiations were conducted in the spectrophotometer cavity using a Bausch and Lomb SP200 high pressure mercury lamp and monochromator (6.4 nm bandpass) together with appropriate cut-off filters. In some cases irradiations were conducted at room temperature with a Hanovia 450W medium pressure mercury lamp and cut-off filters.

Results and Discussion

Chromophore Aggregation in LC Polymers. We have observed perturbations of the UV-VIS spectra that we attribute to chromophore association or aggregation in thin (ca. 0.5-2.0 μm) films of <u>all</u> the LC polymers we have studied to date. These perturbations are present in all phases of films of the polymers. Thus the spectrum of an 'as cast' film of a main-chain LC (MCLC) polyarylcinnamate, **1b**, is broader than the spectrum in solution (Fig. 1). Much more pronounced effects are observed on the UV-VIS spectra of the LC phases and especially in the crystalline phase of the side-chain LC (SCLC) polyarylcinnamate, **2**. Spectra of LC films of both MCLC and SCLC polyarylcinnamates show enhanced blue-shifted absorption and weak enhancement of absorption to the red of that of the 'isolated' chromophore (*2,5*), ie., the polymer in a good solvent or a small molecule model compound in solution. Spectral effects attributed to formation of these aggregates are quite dependent on the phase type, annealing, and the molecular weight of the sample. Data for low ($M_n \approx$ ca. 10^4) and higher ($M_n \approx$ ca. 5 x 10^4) molecular weight glassy nematic samples of **1a** prepared by cooling of the isotropic melt are shown in Figure 2. Spectral perturbations attributed to chromophore association are most dramatic (*5,6*) for the side-chain substituted polymer, **2**, particularly in 'as cast', partially crystalline films (Fig. 3) and in smectic B films, and are less apparent in less ordered nematic and isotropic films. The spectra of 'as cast' films strongly resemble those of 50% dispersions of the model compound **4** (*vide infra*) in poly(methyl methacrylate), (PMMA). In the case of thin films of MCLC stilbene polyesters such as **3**, temperature dependent chromophore association can also be observed (*7*). UV-VIS spectra of 'as cast' films of **3** at room temperature show enhanced absorption both to the blue and to the red of the spectra of the same polymer in a good solvent such as chloroform or a model compound in solution. Perturbed spectra analogous to those of the polymer films can be obtained using the small molecule model compound, **5**,

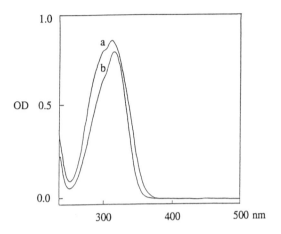

Figure 1. UV-VIS Spectra of Polymer **1b**, (a) as an 'As Cast' Thin Film, and (b) in Chloroform.

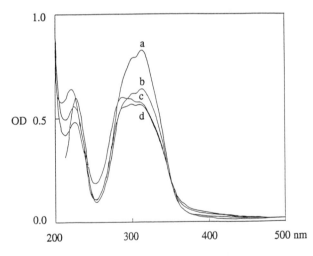

Figure 2. UV-VIS Spectra of Thin Films of MCLC Polymer **1a** Cooled from the Isotropic Melt (HMW = high molecular weight, LMW = low molecular weight), (a) 'As Cast' Film at Room Temperature, (b) HMW Film, Fast Cool, (c) HMW Film, Slow Cool, (d) LMW Film, Fast Cool.

dispersed in PMMA (*vide infra*) where aggregation readily occurs. Perturbed spectra of all three polymers, **1**, **2**, and **3**, can also be obtained by first dissolving the polymers in 'good' solvents such as chloroform, methylene chloride, or ethanol and then adding 'poor' solvents such as cyclohexane or water. An example is shown (Fig. 4) of the spectrum of a low molecular weight sample of **1a** in chloroform, ethanol, and 50 % aqueous ethanol at very low concentrations. Presumably the hydrophobic effect in aqueous non-solvents leads to inter-chromophore associations that are absent when the polymers are dissolved in good solvents. Others (*9,11*) have published UV-VIS spectra of cinnamate containing polymers in LC phases. Keller (*8*) reported the strongly perturbed spectrum of a film of an aryl cinnamate substituted SCLC polysiloxane but did not comment on the perturbation or its origin. Noonan and Caccamo (*11*) attributed perturbed spectra of a PMMA substituted with a 4-methoxycinnamate chromophore in its side chain, to alterations of the conformations of the chromophores as a result of enhanced side chain packing.

We have begun to explore the fate of singlet excitation energy using the fluorescence of stilbene polyesters such as **3**. Such experiments are not possible with the polyarylcinnamates **1** and **2** since they are weakly fluorescent and by the time the emission spectrum is recorded (using narrow excitation and broad emission slits) the material has undergone significant photolysis. On the other hand, the model stilbene and stilbene polyesters such as **3** are highly fluorescent. The structured fluorescence spectra (Fig. 5) and fluorescence excitation spectra of **3** in dilute solution in good solvents resemble those of model compounds such as **5** in dilute solution. However, the fluorescence spectra of films of **3** display the broad, red shifted, structureless emission characteristic of excimers or excited aggregates (Fig. 5), with little or no contribution from emission of 'isolated' stilbene chromophores. Fluorescence excitation spectra (Fig. 6) of films of **3** are quite different from fluorescence excitation spectra of solutions of **3**, but are very similar to absorption spectra of films of **3**. The implications of these observations for the photochemistry of **3** and related compounds are currently under investigation.

Chromophore Aggregation in Model Compounds. In order to improve our understanding of the aggregation process in the polymers and its photophysical and photochemical consequences, we have investigated the behavior of small molecule model compounds such as 4-pentyloxyphenyl-4'-pentyloxycinnamate, **4**, and diethyl stilbene dicarboxylate, **5**, that serve as models for the chromophores in the cinnamate and stilbene polymers, respectively. We do not observe aggregate spectra in concentrated solutions (up to 0.1 M) of these model compounds in organic solvents. However, most of the spectral changes attributed to chromophore association in the polymers can be duplicated using dispersions of these model compounds in PMMA. For example, UV-VIS spectra of 50 % dispersions of **4** in PMMA (Fig. 7), which are partially opaque and therefore almost certainly contain microcrystalline **4**, strongly resemble spectra (Fig. 3) of 'as cast' thin films of partially crystalline SCLC polymer **2**. Spectra at intermediate concentrations (eg. 25 % of **4**) resemble spectra of the films of MCLC polymer **1** (eg. Fig. 1), although the polymer spectra are not as structured as those of the model compound. It is also intriguing to note that the shape of the intermediate (25 % of **4**) spectrum can be duplicated almost perfectly (Fig. 8) by

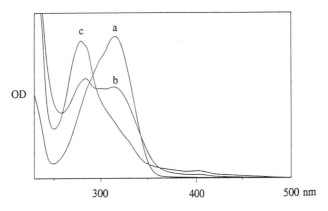

Figure 3. Normalized UV-VIS Spectra of SCLC Polymer, **2**, (a) in Methylene Chloride, (b) as a Smectic B Film at 70 °C, and (c) as an 'As Cast' Film at Room Temperature.

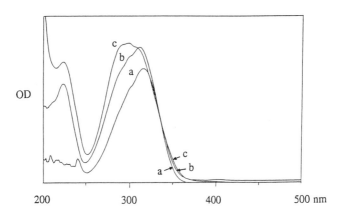

Figure 4. UV-VIS Spectra of Low Molecular Weight Polymer **1a** in (a) Chloroform, (b) Ethanol (b), and (c) 1:1 Ethanol:Water at Room Temperature.

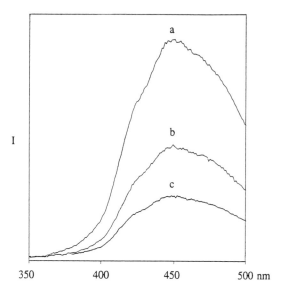

Figure 5. Fluorescence Spectra of an 'As Cast' Film of MCLC Polymer **3** Excited at (a) 330, (b) 290, and (c) 370 nm.

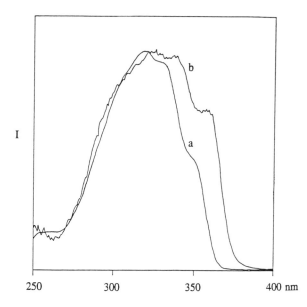

Figure 6. Fluorescence Excitation Spectra of Polymer **3**, (a) in Solution in Chloroform and (b) as an 'As Cast' Film at Room Temperature.

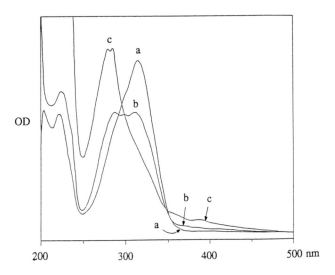

Figure 7. UV-VIS Spectra of (a) 5, (b) 25, and (c) 50 Weight Percent of Model Compound **4** Dispersed in PMMA.

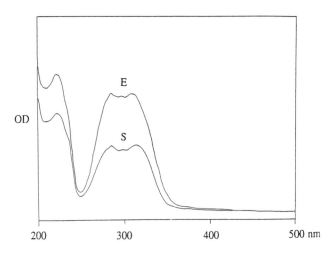

Figure 8. Experimental (E) and Simulated (S) Spectra of a 25 % Dispersion of **4** in PMMA.

addition of the spectra of the concentrated (50 %) and dilute (5 %) dispersions in a 2:1 ratio. This strongly suggests that only two chromophores, corresponding to 'isolated' and 'aggregated' species, are present in dispersions of the model compound. We have also compared spectra of SCLC polymer **2** with spectra of mixtures of the 5 and 50 % dispersions of its methacrylate monomer. We hope ultimately to make reasonable estimates of the fraction of light absorbed by isolated and aggregated chromophores in different phases of the polymer. Obtaining this information is a necessary pre-requisite to obtaining quantum yields of reaction of both isolated and aggregated chromophores.

Photochemistry of Polymers and Model Compounds. Long term irradiation (313 nm) of 'as cast' films of polyarylcinnamates **1** and **2** leads to insolubilization and spectroscopic changes attributed to saturation of the cinnamate double bond and to photo-Fries rearrangement (*2,5*). Solid state ^{13}C nmr shows that saturation of the double bond involves 2 + 2 photocycloaddition in both **1** and **2**, although other reactions leading to saturation of the cinnamate double bond may also occur (*17*). The photochemical behavior is strongly dependent on phase type. In the case of low molecular weight samples of fluid nematic **1**, irradiation leads initially to hyperchromism and a change of the spectral shape indicative of generation of 'isolated' chromophores and loss of chromophore association in the initial phase of the reaction (*2*). However, this initial hyperchromism is not observed in the higher molecular weight sample, suggesting it is dependent on the viscosity of the film (Fig. 9). Since there is a large difference in absolute optical densities at 313 and 390 nm, the OD data in Figure 9 are shown as percentages of the maximum OD obtained for each sample. Interestingly, a small hyperchromic effect is observed in the glassy nematic phase of high molecular weight samples of **1**. In the SCLC polymer **2**, hyperchromism and spectral changes are most dramatic in highly organized smectic B films but greatly diminished in the less organized smectic A and nematic phases (*5,6*). These effects are most likely due to photocycloadduct formation since they occur upon excitation at wavelengths (366 or >380 nm) where aggregates preferentially absorb and little or no photo-Fries rearrangement occurs and upon triplet sensitization when photo-Fries rearrangement does not occur (*vide infra*). Spectral changes that occur after this initial hyperchromism are due to saturation of the cinnamate double bond and to photo-Fries rearrangement (*2,5*). Aggregation also leads to wavelength dependent photoproduct formation (*18*) from polyarylcinnamates such as **1** and **2**. Thus, when aggregates are specifically excited at long wavelengths (eg. above 380 nm), spectral changes attributed to [2 + 2] photocycloaddition are observed, whereas, when both aggregates and unassociated chromophores are excited at shorter wavelengths (eg. 313 nm), spectral changes attributed to both photocycloaddition and photo-Fries rearrangement are observed.

The ratio of 'dimers' (saturated products) to photo-Fries rearrangement products upon 313 nm irradiation of low molecular weight samples of **1** has been investigated as a function of phase type at different temperatures at both low (11-13 %) and high (>90 %) conversion (Table I). This ratio was estimated from the simple assumption that only three types of chromophores contribute to the UV-VIS spectrum of the irradiated polymer: the aryl cinnamate type (represented by **4**), a 'dimer' type (represented by **6**), and the photo-Fries type (represented by **7**). In both

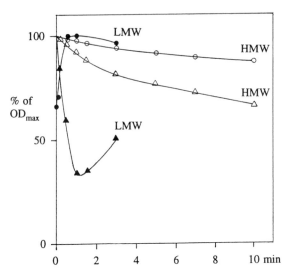

Figure 9. Optical Density (OD) Changes at 390 (Triangles) and 313 nm (Circles) Upon Irradiation (313 nm) of Fluid Nematic Films of Low (LMW) and High (HMW) Molecular Weight Samples of **1a**.

6

7

8

the amorphous and glassy nematic phases at 25 °C there is a strong preference for 'dimer' over 'Fries' type products at low conversions. This preference is greatly diminished in the isotropic melt at 128 °C, presumably because of both reduced chromophore aggregation and increased fluidity of this phase. Data for high

Table I. Dimer:Fries Ratios (D/F) from Irradiation (313 nm) of Thin Films of Low Molecular Weight 1b at Low (L) and High (H) Conversions

T, °C	Phase	Conversion	D/F
25	Amorphous	L	>2.5
25	Amorphous	H	1.8
25	Nematic glass	L	>4.5
25	Nematic glass	H	2.0
88	Nematic	L	-[a]
88	Nematic	H	0.5
128	Isotropic	L	0.55
128	Isotropic	H	0.85

[a]cannot be obtained because of initial hyperchromism of irradiated fluid nematic samples.

conversions indicate a preference for 'photo-Fries' products (low D/F ratios). Presumably, aggregated chromophores are consumed in the early stages of the reaction to give 'dimer' products. The remaining unaggregated chromophores are more likely to undergo the unimolecular photo-Fries rearrangement leading to the lower D/F ratios observed at high conversions. To confirm this possibility we exhaustively irradiated films of 1 at 366 nm to convert all aggregates into 'dimers'. We then changed the irradiation wavelength to 313 or 334 nm and observed spectral changes consistent with conversion of the remaining chromophores to photo-Fries products. Absolute quantum yields for these reactions and those of the stilbene polyesters 3 cannot be obtained because of the heterogeneity of chromophores in the absorbing material.

Triplet quenching and sensitization experiments on both 1 and model compounds suggest that the formation of photo-Fries rearrangement products from aryl cinnamates occurs from the singlet or possibly an upper triplet state (19). Thus we are able to suppress photo-Fries product formation and formation of radical fragments [observed by flash photolysis (19)] from 4 by triplet sensitization (4,19). We have used ketocoumarin triplet sensitizers (20) in this work because they have a convenient 'window' in their UV-VIS absorption spectra between ca. 230 and 320 nm that enables us to observe cinnamate photochemistry without interference from the absorbance of the sensitizer or its photolysis products. However, we had to synthesize a modified ketocoumarin sensitizer, 8, with a long alkoxy 'tail' because the commercially available ketocoumarins phase-separate (4) from the glassy nematic and nematic phases of 1. Preliminary results using 8 are most promising and suggest there is extensive triplet energy migration in films of polyarylcinnamates such as 1. We have not yet begun to study triplet sensitization of the reactions of the SCLC polyarylcinnamate 2 or stilbene polyesters such as 3.

Irradiation (313 nm) of the stilbene polyester, **3**, in solution or in cast films leads to disappearance of the long wavelength absorption due to the *trans*-stilbene chromophore and appearance of a new absorption band at ca. 250 nm (*6*). We believe the major products of this reaction are cyclobutane dimers (eg. **9**). In the case of the film irradiations, the shape of the UV-VIS spectrum changes as reaction proceeds until it more closely resembles the solution spectrum. This suggests that aggregates are preferentially consumed in the early stages of reaction. When an exhaustively 313 nm irradiated solution (methylene chloride) or film at 65 °C is irradiated at 254 nm there is recovery of absorption above 300 nm but with a broader, blue-shifted absorption consistent with regeneration of a mixture of *trans*- and *cis*-stilbenes by photochemically allowed cleavage of the cyclobutane dimers, such as **9**, believed to be generated by longer wavelength irradiation. The two isomers presumably arise from the two possible cleavage pathways A and B (Scheme) for **9**. The 313 nm disappearance/254 nm recovery behavior exhibited by films irradiated at 65 °C is not observed at room temperature. This observation, together with changes in the UV-VIS spectra of films that occur at about 40 °C (*vide supra*) suggest the possibility of a T_g for **3** at about 40 °C, although, to date, we have not been able to observe such a T_g by DSC. This type of photoreversible photoreaction does not occur with the cinnamate polymers **1** and **2** either in films or in solution.

The photochemistry of dispersions of model compounds in PMMA differs appreciably from that of the polymer films. In general, dispersions of model compounds at high loadings at room temperature (below the T_g of PMMA), where there is extensive aggregation, are relatively unreactive compared to the polymer films. We estimate quantum yields of reaction are at least an order of magnitude less for the model compounds relative to polymer films of comparable optical density. We also attempted to obtain cyclobutane dimers of the model compounds by irradiations (>300 nm) of powdered crystals of **4** and **5**. However these materials are also quite unreactive under these conditions which suggests that the intermolecular distance and/or orientations in crystals of these model compounds do not permit dimerization to occur. This would be an example of the well-known phenomenon of topochemical control (*21*) of a solid state dimerization reaction. In dilute dispersions the situation is quite different. A 5 % dispersion of the cinnamate model compound **4** in PMMA when irradiated at 313 or 366 nm undergoes a clean reaction, as evidenced by the UV-VIS spectrum (Fig. 10), to afford a spectrum characteristic of the photo-Fries rearrangement product, **7**. A similar effect is observed upon irradiation of **4** in solution. We consider the observation of an isosbestic point upon irradiation in PMMA to be significant to the question of whether the polymers undergo any *trans-cis* isomerization upon irradiation. The isosbestic point indicates that, for **4**, photo-Fries rearrangement is the only unimolecular photochemical reaction occurring in the glassy polymer medium. We believe that this observation and the absence of bands due to the *cis*-isomer in irradiated films of **1** and **2** indicate *trans-cis* isomerizations are not major photochemical reactions of either the polyarylcinnamates or the model compounds that we have studied. In contrast to the relatively clean reaction of **4**, the stilbene **5**, in dilute dispersions in PMMA, undergoes irreversible photolysis under 313 nm irradiation perhaps due to ester fragmentation and consequent saturation of the stilbene double bond. Isomerization, as in the case of the cinnamates, does not

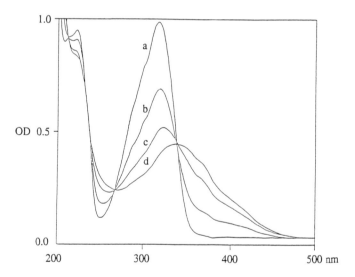

Scheme. Photochemical Cleavage of Cyclobutane Dimer, **9**.

Figure 10. OD Changes During Irradiation (313 nm) of a Dilute (5 %) Dispersion of Model Compound **4** in PMMA for (a) 0, (b) 30, (c) 90, and (d) 170 min.

occur. We have not attempted any further characterization of the products of irradiation of dilute dispersions of **5**.

Acknowledgments.

This work was supported by the National Science Foundation (EPSCoR program), the State of Mississippi, and the University of Southern Mississippi. We thank Drs. Richard Weiss and Joachim Stumpe for helpful discussions.

Literature Cited.

1. Weiss, R. G. *Tetrahedron.*, **1988**, *44*, 3413-3475.
2. Creed, D.; Griffin, A. C.; Gross, J. R. D.; Hoyle, C. E.; Venkataram, K. *Mol. Cryst. Liq. Cryst.*, **1988**, *155*, 57-71.
3. Haddleton, D. M.; Creed, D.; Griffin, A. C.; Hoyle, C. E.; Venkataram, K. *Makromol. Chem. Rapid Commun.*, **1989**, *10*, 391-396.
4. Subramanian, P.; Creed, D.; Hoyle C. E.; Venkataram, K. *Proc. SPIE - Int. Soc. Opt. Eng. (Photopolym. Device Phys. Chem. Appl. 2)*, **1991**, *1559*, 101-109.
5. Singh, S.; Creed, D.; Hoyle, C. E. *Polymer Preprints*, **1993**, *34*, 743-744.
6. Singh, S.; Creed, D.; Hoyle, C. E. *Proc. SPIE - Int. Soc. Opt. Eng.*, **1993**, *1774*, 2-11.
7. Creed, D.; Cozad, R. A.; Hoyle, C. E.; Morris, J. C.; Jackson, Jr., W. J. *Proc. SPIE -Int. Soc. Opt. Eng.*, **1993**, *1774*, 69-73.
8. Koch, T.; Ritter, H.; Buchholz, N.; Knochel, F. *Makromol. Chem.*, **1989**, *190*, 1369-1377.
9. Keller, P. *Chem. Mater.*, **1990**, *2*, 3-4.
10. Ikeda, T.; Lee, C. H.; Sasaki, T.; Lee, B.; Tazuke, S. *Macromolecules*, **1990**, *23*, 1691-1695.
11. Noonan, J. M.: Caccamo, A. F. In *Liquid Crystalline Polymers*; Editors, Weiss, R. A.; Ober C. K. American Chemical Society Symposium Series, American Chemical Society, Washington, DC, 1990, Vol. 435, 144-157.
12. Stumpe, J.; Muller, L.; Kreysig, D.; Hauck, G.; Koswig, D.; Ruhmann, R.; Rubner, J. *Makromol. Chem. Rapid Commun.*, **1991**, *12*, 81-87.
13. Whitcombe, M. J.; Gilbert, A.; Mitchell, G. R.; *J. Polym. Sci., (A), Polym. Chem.*, **1992**, *30*, 1681-1691.
14. Gangadhara, Kishore, K. *Macromolecules*, **1993**, *26*, 2995-3003.
15. Jackson, Jr., W. J.; Morris, J.C. *J. Appl. Polym. Sci: Appl. Polym. Symp.*, **1985**, *41*, 307-326.
16. Woods, L. C.; Fooladi, M. *J. Chem. Eng. Data*, **1963**, *2*, 624-627.
17. Egerton, P. L.; Pitts, E.; Reiser, A. *Macromolecules*, **1981**, *14*, 95-100.
18. Creed, D.; Griffin, A. C.; Hoyle, C. E.; Venkataram, K. *J. Amer. Chem. Soc.*, **1990**, *112*, 4049-4050.
19. Subramanian, P.; Creed, D.; Griffin, A. C.; Hoyle, C. E.; Venkataram, K. *J. Photochem. Photobiol. A: Chem.*, **1991**, *61*, 317-327.
20. Specht, D. P.; Martic, P. A.; Farid, S. *Tetrahedron*, **1982**, *38*, 1203-1211.
21. Schmidt, G. M. J. *Pure Appl. Chem.*, **1971**, *27*, 647-657.

RECEIVED July 12, 1994

Chapter 3

Dissociative Electron Capture of Polymers with Bridging Acid Anhydrides
Matrix Isolation Electron Spin Resonance Study

Paul H. Kasai

Research Division, IBM Almaden Research Center, 650 Harry Road, San Jose, CA 95120–6099

It is postulated that high electron-beam sensitivity of polymers embodying bridging acid-anhydride sectors toward chain scission is due to a dissociative electron capture sequence of the following scheme.

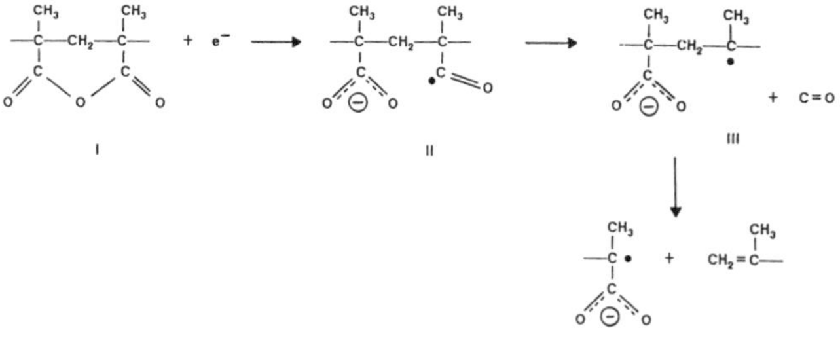

Using the matrix isolation ESR technique, consequences of electron capture by succinic anhydride and glutaric anhydride molecules were examined. In support of the postulate, the study has revealed that (1) both anhydrides readily capture low energy electrons and isomerize to yield the acyl radicals, as II in the postulated scheme, and (2) the resulting acyl radicals, on exposure to mild radiation ($\lambda = 600$ nm) at 4 K, decarbonylate to yield the alkyl radicals corresponding to III in the postulated scheme.

PMMA, poly(methyl methacrylate), is a classical one-component positive resist that has been widely used in electron beam, x-ray, and ion-beam microlithography, especially when the highest possible resolution is sought (1). Excellent film-forming characteristics, resistance to chemical etchants, absence of swelling or distortion are the attributes that have

0097–6156/94/0579–0027$08.00/0

led to its utility in these applications. The sensitivity of PMMA toward these radiations, however, is low, and many chemical modifications have been made to enhance its sensitivity.

One of the exemplary and more successful such modifications is that of "Terpolymer" reported by Moreau et al. (2). They formed copolymer of methyl methacrylate and methacrylic acid, and generated bridging anhydride structures between adjoining methacrylic acid units (scheme 1).

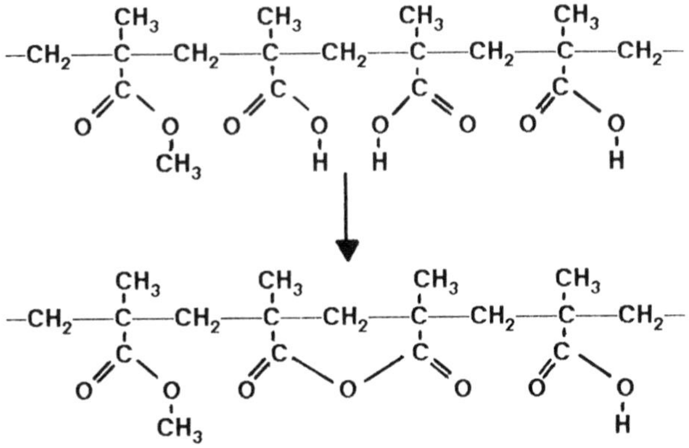

$$
\begin{array}{l}
\text{CH}_3 \quad\quad \text{CH}_3 \quad\quad \text{CH}_3 \quad\quad \text{CH}_3 \\
-\text{CH}_2-\text{C}-\text{CH}_2-\text{C}-\text{CH}_2-\text{C}-\text{CH}_2-\text{C}-\text{CH}_2- \\
\end{array}
$$

Scheme 1

The G_s value (the number of chain scission produced per 100 eV of energy absorbed) of neat PMMA in e-beam irradiation is ~ 1 and that of poly(methacrylic acid) is ~ 2. The terpolymer of scheme 1 (with the mol % composition of 70/15/15 for methyl methacrylate, methacrylic acid, and methacrylic anhydride) has been found to have the e-beam sensitivity with G_s of 4.5 (1,2). The sensitivity increase is undoubtedly due to the cyclic anhydride sectors.

The exposure mechanism of poly(methacrylic anhydride) under electron beam was speculated earlier based on the product analysis, and ESR data observed at 77 K (3). Homolytic cleavage of a bond in the anhydride sector, *on impact with high energy electron*, with subsequent decarboxylation has been suggested.

When a beam of high energy electrons traverses through a material, a swarm of thermal energy electrons are generated through cascading ionization processes. When the energy of the primary beam is ~ 10 KeV, for example, the total number of secondary electrons produced has been estimated to be 10^3 times that of the primary electrons (4). Resists of the epoxylated poly(methacrylate) type (5), and the poly(butene sulfone) type (6) have been especially developed as high-sensitivity e-beam resists, the former being a negative type while the latter is a positive resist. We have shown recently that the high e-beam sensitivity of these resists is due to dissociative electron capture (of the abundant secondary electrons) by the respective electrophilic sectors, the epoxy ring, and the sulfone moiety (7,8).

Taking cognizance of the high stability of carboxylate anions, we surmised that the exposure process of the terpolymer under electron beam might also be that initiated by dissociative electron capture. The most likely sequence leading to main-chain scission would hence be as follows (scheme 2).

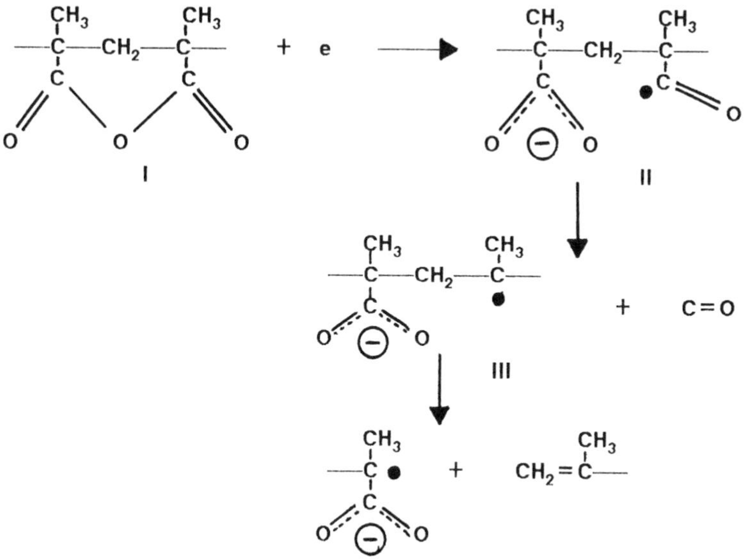

Scheme 2

It has been generally observed that poly(olefin) chains having two adjoining back-bone carbons both with hydrogen atom(s) attached would undergo cross-linking when irradiated with high energy radiation (x-ray and e-beam) (9). In contradiction to the rule, Pohl et al. (10) found that copolymers of maleic anhydride underwent chain scission predominantly on exposure to electron beams. Chain scission occurred with G_s of 2.5 for copolymer of methyl methacrylate and maleic anhydride, for example.

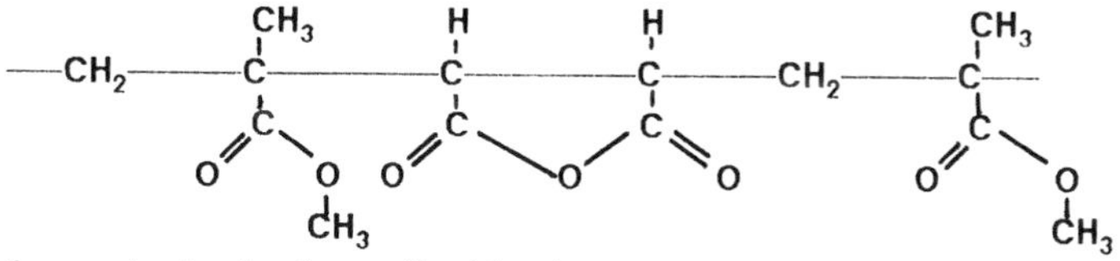

No reaction mechanism has been offered for the process. We surmised that chain scission observed for copolymers of maleic anhydride was also the result of the sequence initiated by dissociative electron capture in a manner completely analogous to scheme 2.

In order to substantiate these postulated sequences, we have examined, using the matrix isolation ESR technique, the consequence of electron capture by succinic anhydride, a model system for the bridging anhydride sectors in maleic anhydride copolymers, and glutaric anhydride, a model system for the bridging anhydride sectors of Terpolymer. The result is presented in this paper. The study revealed that both cyclic anhydrides readily captured (low energy) electrons and spontaneously isomerized (ring opening) to yield the acyl radicals (as II in scheme 2). The resulting acyl radicals, when irradiated with mild radiation (λ = 600 nm) at 4 K, readily decarbonylated to yield the expected alkyl radicals (as III in scheme 2).

Experimental Section

The technique of codepositing Na atoms and molecules of some electron affinity in an argon matrix at ~ 4 °K, and effecting an electron transfer between them by mild radiation (λ = 590 nm corresponding to Na 3s → 3p transition) was described some time ago (11). The resulting anions are isolated from the Na cations, and are usually unaffected by the ensuing radiation. In the present experiment, succinic anhydride and glutaric anhydride were sublimed from glass vessel heated to 120 °C, and were trapped in argon matrices together with Na atoms generated from a resistively heated stainless steel tube (250 °C). The concentrations of Na atoms and the anhydride molecules were estimated to be ~ 0.1 mol % each.

The ESR spectrometer used was an X-band system (IBM Instruments, Model 200D). All the spectra were obtained at 4 °K, and the frequency of the system locked to the loaded sample cavity was 9.420 GHz. For photoirradiation of the matrix, a light beam from a high pressure xenon-mercury lamp (Oriel, 1 Kw unit) was passed throguh a water filter and a broad band interference filter of choice, and was focussed on the cold finger 40 cm away.

Succinic anhydride, perdeutero succinic anhydride, and glutaric anhydride were all obtained from Aldrich Chemical, and were purified by sublimation prior to deposition.

Spectra and Analyses

Succinic anhydride: The ESR spectrum, observed prior to photoirradiation, of an argon matrix in which succinic anhydride and Na atoms had been codeposited is shown in Fig. 1. The ESR spectrum of Na atoms ($3s^1$) isolated in an argon matrix was first studied by Jen, et al. (12); the spectrum comprises a sharp, widely spaced quartet due to a large isotropic hfc (hyperfine coupling) interaction with the ^{23}Na nucleus ($I = 3/2, A_{iso} = 330\,G$). Individual components are often split further by the multiple trapping site effect. The quartet indicated by letter A in Fig. 1 is that of isolated Na atoms. Codeposition of Na atoms with succinic anhydride results in two additional patterns, the second quartet B (with broader line width, and smaller spacings in comparison to A), and a broad signal C with partially resolved structures at the position corresponding to $g = 2.00$.

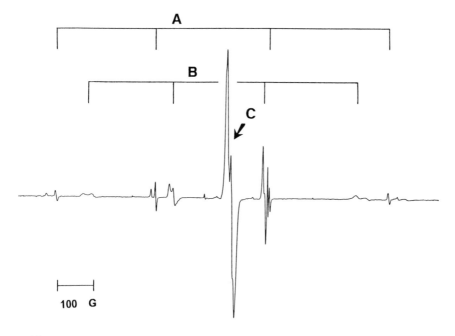

Figure 1. ESR spectrum of the Na/succinic anhydride/Ar system observed prior to photoirradiation. The quartet due to isolated Na atoms, A, the quartet due to Na:(succinic anhydride) complex, B, and the signal due to succinic anhydride anions, C, are recognized as indicated.

We have shown recently that Na atoms and ethers (e.g. diethyl ethers, tetrahydrofuran, etc.), when encountered in an argon matrix, form a weak complex spontaneously (7). The complex is formed by dative interaction of the oxygen lone-pair electrons into a vacant s-p hybridized orbital of Na, and the unpaired electron resides in an s-p hybridized orbital of Na pointing away from the oxygen. The Na hfc constant of the complex is hence $20 \sim 30\%$ smaller than that of isolated Na atoms. We have also found that a similar complex is formed between Na atoms and acetone (13). The quartet B observed here (Fig. 1) is attributed to a similar complex formed between Na and succinic anhydride; which oxygen(s) is involved in its formation is not known presently, however.

We have also observed that, in the matrix isolated Na-to-molecule electron transfer experiment, if the electron affinity of the acceptor molecule is sufficiently high, electron transfer occurs spontaneously during collision in space (as the two beams merge above the cold finger) or in a fluid surface layer of the matrix during deposition (8). The broad signal C is attributed to the succinic anhydride anions produced by such spontaneous electron transfer.

Fig. 2a shows, in an expanded scale, the central section of Fig. 1 encompassing the inner two components of quartet A. Fig. 2b shows the spectrum of the same matrix (the same section) observed after the matrix had been irradiated with "red light" ($\lambda = 700 \pm 50 \, nm$) for 8 min. The lowest electronic transition of Na atoms (3s → 3p) is at 590 nm. The corresponding transition in the Na:(succinic anhydride) complex should be red-shifted. In accord with these energetics, on exposure to red light, quartet B disappears completely with concurrent increase of signal C, while quartet A remains unaffected. The reaction realized here is thus:

$$\text{Na:(succinic anhydride)} \xrightarrow[\text{700 nm}]{h\nu} \text{Na}^+ + \text{(succinic anhydride)}^-$$

Fig. 2c shows the spectrum of the same matrix (the same section) observed after the matrix was subsequently exposed to "yellow light" ($\lambda = 600 \pm 50 \, nm$) for 8 min. Consistently with the energetics presented above, the Na atom signals (quartet A) disappear on exposure to yellow light. It is attributed to the electron transfer from isolated Na atoms to succinic anhydride. Most interestingly signal C also disappears on exposure to yellow light, and in its stead, a complex well-resolved pattern D appears centered about the position of $g = 2.00$.

For the succinic anhydride anion, the following two structures are plausible.

II-ALT   II

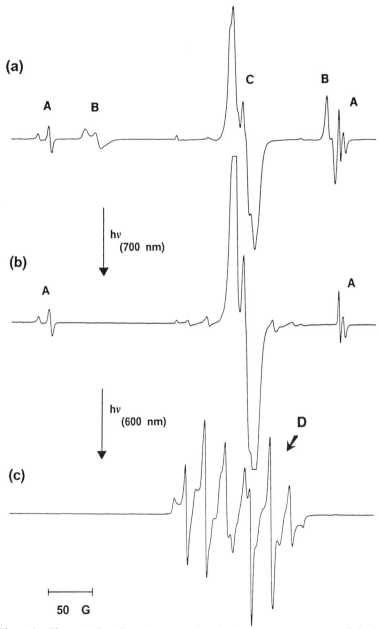

Figure 2. The central sections (encompassing the inner two components of the Na atom quartet) of the ESR spectra of the Na/succinic anhydride/ Ar system: (a) Observed prior to photoirradiation, (b) observed after irradiation with red light (700 ± 50 nm) for 8 min, and (c) observed after subsequent irradiation with yellow light (600 ± 50 nm) for 8 min. The letters A, B, and C indicate isolated Na atoms, Na:(succinic anhydride) complex, and succinic anhydride anions, as in Fig. 1. The multiplet pattern D is ascribed to alkyl radical III (see text).

In structure II-ALT, the negative charge is localized in the particular oxygen, and the anhydride bridge remains intact. In structure II, the extra electron (the negative charge) is accommodated in the resonance stabilized carbonate orbital at the expense of a C-O bond.

It has been well-established that, for an alkyl radical, the hfc constant of a β proton is essentially isotropic and is given by eq. 1 (14).

$$A(H_\beta)_{\text{in gauss}} = 4 + 50\,cos^2\theta \qquad (1)$$

where θ is the dihedral angle between the p orbital of the unpaired electron and the C_β—H_β bond. In structure II-ALT, the unpaired electron is localized in the carbon p_π orbital perpendicular to the molecular plane, and the dihedral angles of both the β protons would be ~30°. The hfc interaction of ~42 G from each of the two β protons is thus expected. In structure II, the unpaired electron orbital lies in the C–C=O plane and points away from the β protons. Fig. 3a shows, in an expanded scale, the ESR spectrum C of the succinic anhydride anion observed after irradiation with red light (signal C in Fig. 2b). The overall spread of the signal (30 G) is significantly smaller than that expected from structure II-ALT. The overall spread is compatible with structure II, an acyl radical.

Fig. 3b shows the ESR spectrum C observed when the Na/succinic anhydride/argon matrix was prepared using perdeuterio succinic anhydride and exposed to red light. The asymmetric spectral pattern revealed here is that due to an orthorhombic g tensor. The principal g values have been determined as given in the figure. Fig. 3c shows the simulated spectrum based on them. The anisotropy of the g tensor determined here is unusually large for an organic radical. The g-tensor of the formyl radical, HC=O, has been determined by Adrian et al. (15); the principal g values are $g_1 = 2.0042$, $g_2 = 2.0027$, and $g_3 = 1.9960$. The proximity of the g-tensor of the succinic anhydride anion determined presently and that of the formyl radical is intriguing, and is a strong support of the acyl radical form, II, of the anion.

The spectral pattern D (of Fig. 2c) is shown expanded in Fig. 4a. The corresponding D spectrum observed from the Na/perdeuterio succinic anhydride/argon system after irradiation with yellow light is shown in Fig. 4b. The multiplet structure of D is thus entirely due to hfc interactions with protons. The pattern was recognized as a doublet-of-doublet-of-triplet as indicated and was assigned to the alkyl radical III resulting from decarbonylation of the acyl radical II.

The anisotropic triplet part with separations from the central component of 20~30 G is that expected from two α protons that are exchanging their positions either by rotation or tunnelling (16). The essentially isotropic doublet-of-doublet part with the respective

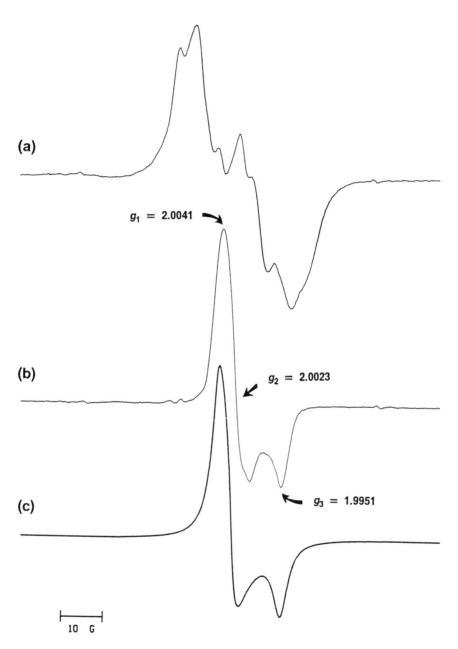

Figure 3. The ESR spectra, signals C, of the succinic anhydride anions: (a) observed form the Na/succinic anhydride/Ar system, (b) observed from the Na/perdeutero succinic anhydride/Ar system, and (c) simulated based on the g values determined in (b).

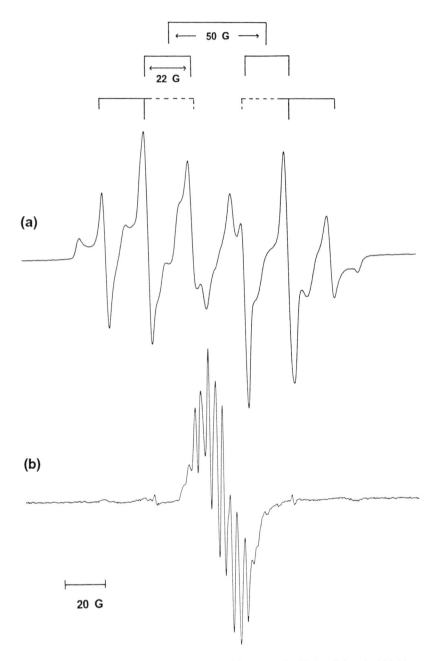

Figure 4. The ESR spectra, signals D, observed from: (a) the Na/succinic anhydride/Ar system, (b) the Na/perdeutero succinic anhydride/Ar system. The doublet-of-doublet-of-triplet pattern was recognized in (a) as indicated.

spacings of 50 G and 22 G is then ascribed to the β protons. More detailed analysis of spectrum D is presented in the discussion section.

Glutaric Anhydride: The ESR spectrum, observed prior to photoirradiation, of an argon matrix in which glutaric anhydride and Na atoms had been cocondensed also showed (1) signals A of isolated Na atoms, (2) signals B due to weak Na:(glutaric anhydride) complex, and (3) signal C due to glutaric anhydride anion of the acyl form. In a manner exactly analogous to the sequence observed with succinic anhydride, exposure to red light caused signals B to disappear with concurrent increase in signal C, and subsequent exposure to yellow light resulted in disappearance of A signals and complete conversion of the acyl radical signal C to the alkyl radical pattern D.

Fig. 5a and 5b show, respectively, the ESR spectra of the acyl radical IV, and the alkyl radical V thus observed from the Na/glutaric anhydride/argon system.

As expected, these spectra are indeed very similar to those of the acyl and alkyl radicals, II and III, observed from the Na/succinic anhydride/argon system. The doublet-of-doublet spacings of the pattern D of radical V are also 50 G and 22 G, respectively. The spectra of radicals III and V, however, differ in many small details. They are discussed in the following section.

Discussion

The extra line-width (with partially resolved structures) of the ESR signal of acyl radical II derived from normal succinic anhydride in comparison to the signal of deuterated species (Fig. 3a vs. 3b) must be ascribed to proton hfc interactions. From the overall spread of the signal, it was concluded that $\sum_i A_{iso}(H_i) \cong 15$ G. Further refinement of the proton hfc interactions observed here was not pursued because of marginal resolution, and complications expected from non-equivalence of the two β protons, the anisotropy of the hfc tensors comparable to the isotropic terms, and non-coincidence of the principal axes of the g tensor and those of the hfc tensors.

Examination of the propionyl radical, $CH_3-CH_2-\overset{\bullet}{C}=O$, by the MNDO molecular orbital method (17) revealed the energy minima at both the cis and trans planar conformations of the C-C-C=O link, and predicted that $\sum_i A_{iso}(H_i) \cong 18$ and 5 G for the cis and trans cases, respectively. The calculation also showed that the barrier for rotation of the carbonyl group out of the C-C-C plane would be extremely low (< 0.5 Kcal). For the acyl radicals II and IV generated from succinic anhydride and glutaric anhydride,

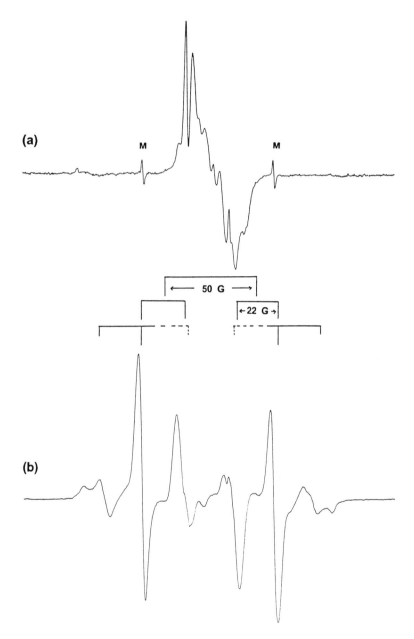

Figure 5. ESR spectra observed from the Na/glutaric anhydride/Ar system. (a) Spectrum due to acyl radical IV observed after irradiation with red light (700 ± 50 nm). The weak, sharp signals (indicated by M) are due to inadvertently formed methyl radicals (the outer components of the quartet). (b) Spectrum due to alkyl radical V observed after irradiation with yellow light (600 ± 50 nm). The doublet-of-doublet-of-triplet pattern was recognized in (b) as indicated.

respectively, because of the interaction with the carbonate group, the $C\text{-}C\text{-}C=O$ link would surely be nonplanar and the two β protons would not be equivalent.

The ESR spectra of alkyl radicals III and V are determined by the hfc interactions of the two α protons and two β protons. The hfc tensor of an α proton of an alkyl radical is well understood; it is highly anisotropic and typically assumes the following form (18,19):

$$A_{\parallel} \cong 12 \text{ G}, \quad A_{\perp,\parallel} \cong 22 \text{ G}, \quad A_{\perp,\perp} \cong 35 \text{ G}. \tag{2}$$

Here $A_{\perp,\parallel}$, for example, is the principal element of the tensor along the direction perpendicular to the C-H bond and parallel to the p_{π} orbital of the α carbon. Thus, for an alkyl radical having two α protons, as the orientations of the hfc tensors of the two protons differ by $\sim 120°$ about the p_{π} orbital, the central component of the triplet would not have the prominence normally observed from the hfc interaction with two equivalent protons. This effect of differing orientations would diminish if the two protons are rapidly exchanged either through rotation or by tunnelling. The central component of the triplet discerned within the spectral pattern D of the Na/succinic anhydride/Ar system (Fig. 4a) has some prominence indicating the presence of such exchange process. For the alkyl radical V generated in the Na/glutaric anhydride/Ar system (Fig. 5b), the central component of the triplet is even more prominent indicating a faster rate of exchange.

As stated earlier, the doublet-of-doublet patterns (with the respective spacings of 50 G and 22 G) discerned within the spectral patterns D of alkyl radicals III and V (Fig. 4a and 5b) are ascribed to the hfc interactions with the β protons. As stated already, the hfc interaction of a β proton of an alkyl radical is essentially isotropic and its magnitude depends on the dihedral angle of the $C_{\beta}\text{-}H_{\beta}$ bond relative to the p_{π} orbital of the unpaired electron as given in eq. 1. As the dihedral angles of the two β protons always differ by $\sim 120°$, it is not unexpected that the two β protons of radical III (or V) have different coupling constants. The observed H_{β} coupling constants of 50 G and 22 G, however, are not compatible with eq. 1 with a single, static conformation.

Most revealingly, in both Fig. 4a and 5b, the four major components constituting the doublet-of-doublet pattern are not of all equal prominence. In both cases, the inner two components are broader and/or weaker than the outer two components. A situation was hence assumed whereby the α CH_2 plane oscillates rapidly $\pm \eta$ degrees centered about a certain orientation. The hfc constant of one of the β protons would be the average of the two values given by the two dihedral angles $\theta \pm \eta$, while the hfc constant of the second β proton would be the average of the two values given by the dihedral angles $\theta + 120° \pm \eta$. Here θ would be "the average dihedral angle" of the first $C_{\beta}\text{-}H_{\beta}$ bond. Analysis of the observed hfc constants (50 G and 22 G) by this scheme yields $\theta = 6°$, and $\eta = 17°$. The Newman diagram of the conformation thus revealed is shown in Fig. 6. Similar conformational and dynamical features have been concluded earlier for n-propyl radicals by Adrian et al. (20), and more recently for trimethylene oxide anions (of the ring-ruptured form) generated in argon matrices (7).

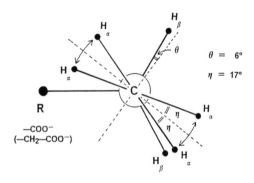

Figure 6. The Newman diagram showing the conformation of alkyl radicals III and V. The α CH$_2$ plane oscillates $\pm \eta$ degrees as indicated.

A computer program that would simulate the ESR powder pattern of a radical having multiple sets of hfc tensors of differing orientations has been described sometime ago (19). The program computes and accumulates the spectral pattern for an ensemble of randomly oriented radicals. The effect of exchanging the pair of α protons or oscillation of the α CH_2 plane can be dealt with by applying the modified Bloch-equation process at each orientations (21,22). Using the program so modified, we simulated the ESR patterns expected for radicals III and V of the conformation (and dynamics) given in Fig. 6.

Fig. 7a shows the simulated spectrum of radical III. The simulation was done based on (1) the model α proton coupling tensor of eq. 2, (2) the model β proton coupling constants given by eq. 1, (3) exchange of the pair of α protons at 30 MHz, and (4) oscillation of the α CH_2 plane at 300 MHz. For each "individual signal" of static conformation, a Lorentzian line shape of 1.0 G width was assumed. Fig. 7b shows the spectrum of radical V simulated by increasing the α proton exchange rate to 300 MHz. All other parameters were the same as those used for Fig. 7a. In these simulations the g tensor was assumed isotropic, and surely nontrivial dipolar part of the β proton hfc tensor was completely neglected. In view of these and other simplifying assumptions made, the proximity of the observed and simulated spectra (Fig. 4a vs. Fig. 7a; and Fig. 5b vs. Fig. 7b) is remarkable, and provide added credence to the assignment. The slower α proton exchange rate determined for radical III must be due to closeness of the charge-bearing carbonate group to the radical site.

For comparison the spectrum was computed for the case of extremely slow α proton exchange (3 MHz), and equally slow oscillation of of the α CH_2 plane (3 MHz), and also for the case of extremely rapid α proton exchange (900 MHz), but slow oscillation of the the α CH_2 plane (3 MHz). The results are shown in Fig. 8. The sensitivity of the ESR powder pattern of these radicals on their conformational and dynamical aspects in both the gross and small subtle features is intriguing.

The present study has clearly shown (1) the propensity of cyclic acid anhydrides to capture low energy electrons and to spontaneously isomerize (ring rupture) to yield the acyl radicals, and (2) the facility with which the resulting acyl radicals to decarbonylate to yield the alkyl radical of the main sector. In the present matrix environment at 4 K, the latter, decarbonylation step required irradiation of the matrix with yellow light (λ = 600 nm, or $\sim$ 48 Kcal/mol). At this stage we are not certain whether the process is truly a photo-excited one, or that induced thermally. Further examination of the propionyl radical, $CH_3-CH_2-\overset{\bullet}{C}=O$, by the MNDO molecular orbital method showed that the C_α $-C_\beta$ bond of the radical is unusually weak ($\sim$ 30 Kcal/mol), suggesting the possibility of a thermal pathway at ambient conditions. In either case, the dissociative electron capture sequences demonstrated here for the succinic anhydride and glutaric anhydride molecules are in strong support of the mechanism, scheme 2, postulated for the electron beam exposure process of the "Terpolymer" resist and maleic anhydride copolymers.

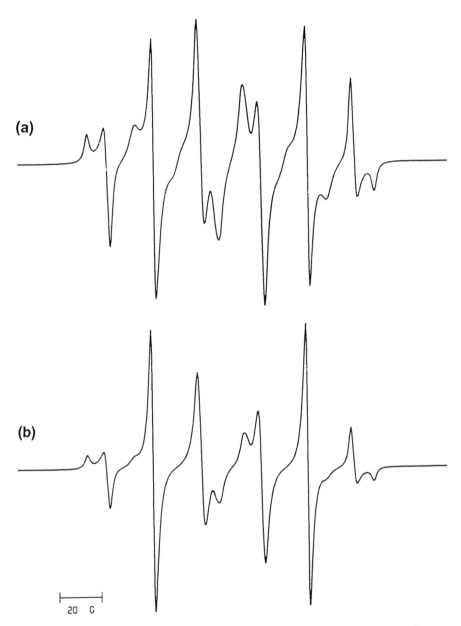

Figure 7. ESR powder patterns simulated based on the model α and β proton coupling tensors and the conformation of Fig. 6. (a) For radical III assuming the α proton exchange rate of 30 MHz, and the α CH$_2$ plane oscillation at 300 MHz. (b) For radical V assuming the α proton exchange rate of 300 MHz, and the α CH$_2$ plane oscillation at 300 MHz.

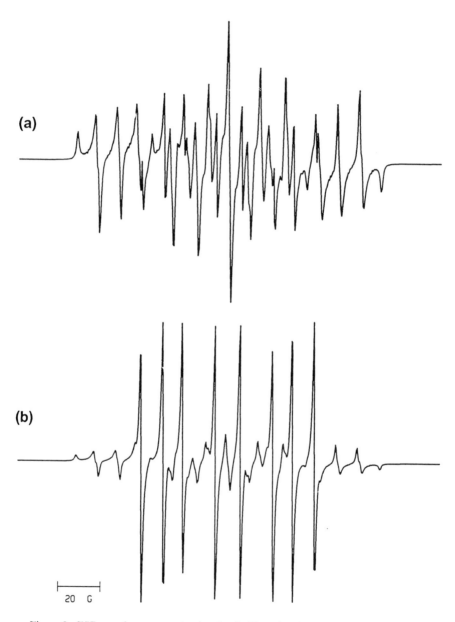

Figure 8. ESR powder patterns simulated as in Fig. 7 but for extreme cases of: (a) the α proton exchange rate of 3 MHz, and the α CH_2 plane oscillation at 3 MHz; and (b) the α proton exchange rate of 900 MHz, and the α CH_2 plane oscillation at 3 MHz.

References

1. See, for example, a chapter by Bowden, M. J. in *Materials for Microlithography*; Thompson, L. F.; Willson, C. G.; Fréchet, J. M. J. Eds.; American Chemical Society: Washington, D.C., 1984: pp 39-117.

2. Moreau, W.; Merritt, D.; Moyer, W.; Hatzakis, M.; Johnson, D.; Pederson, L. *J. Vac. Sci. Tech.* **1979**, *16*, 1989.

3. Hiraoka, H. *IBM J. Res. Dev.* **1977**, *21*, 121.

4. Parikh. M. *J. Chem. Phys.* **1980**, *73*, 93.

5. Thompson, L. F.; Feit, E. D.; Heidenreich, R. D. *Polymer Eng. Sci.* **1974**, *14*, 529.

6. Thompson, L. F.; Bowden M. J. *J. Electrochem. Soc.* **1973**, *120*, 1722.

7. Kasai, P. H. *J. Am. Chem. Soc.* **1990**, *112*, 4313.

8. Kasai. P. H. *J. Am. Chem. Soc.* **1991**, *113*, 3317.

9. Dole, M. *The Radiation Chemistry of Macromolecules* ; Academic Press: New York, 1973; Vol. II, p 98.

10. Pohl, K. U.; Rodriguez, F.; Namaste, Y. M. N.; Obendorf, S. K. in *Materials for Microlithography* ; Thompson, L. F.; Willson, C. G.; Fréchet, J. M. J. Eds.; American Chemical Society: Washington, DC, 1984: p 323.

11. Kasai, P. H. *Acc. Chem. Res.* **1971**, *4*, 329.

12. Jen, C. K.; Bowers, E. L.; Cochran, E. L.; Foner, S. N. *Phys. Rev.* **1962**, *126*, 1749.

13. Kasai, P. H. *unpublished result.*

14. Morton J. R. *Chem. Rev.* **1964**, *64*, 453.

15. Adrian, F. J.; Cochran, E. L.; Bowers, V. A. *J. Chem. Phys.* **1962**, *36*, 1661.

16. The excange of α protons by rotation is not to be confused with "free rotation" which would lead to equal hfc constants of the β protons. See also ref. 7.

17. For the MNDO calculation, MOPAC (V.5) by J. P. Stewart of Frank J. Seiler Research Laboratory (U.S. Air Force Academy, Colorado Springs, CO) was used.

18. McConnell, H. M.; Strathdee, J. *Mol. Phys.* **1959**, *2*, 129.

19. Kasai, P. H. *J. Am. Chem. Soc.* **1972**, *94*, 5950.

20. Adrian, F. J.; Cochran, E. L.; Bowers, V. A. *J. Chem. Phys.* **1973**, *59*, 3946.

21. See, for example, Martin, M. L.; Martin, G. J.; Delpuech, J. *Practical NMR Spectroscopy*; Heyden: London, 1980: pp 293-311.

22. An example of the modified Bloch equation method applied to an ESR powder pattern (due to g-tensor anisotropy) is treated in: Schlick, S.; Kevan, L. *J. Am Chem. Soc.* **1980**, *102*, 4622.

RECEIVED September 13, 1994

Chapter 4

Luminescence Study of Ion-Irradiated Aromatic Polymers

Y. Aoki, H. Namba, F. Hosoi, and S. Nagai

Japan Atomic Energy Research Institute, Takasaki Radiation Chemistry Research Establishment, Watanuki-machi, Takasaki, Gunma 370–12, Japan

Luminescence spectra of solid polymer films such as polystyrene, poly(2-vinyl naphthalene) and poly(N-vinyl carbazole) were measured during ion irradiation with 200 keV He^+ ions. For each polymer, the specific feature of luminescence spectrum due to the excimer of the pendant aromatic group was observed. This excimer fluorescence was decreased in intensity with the irradiation dose over the fluence range of 1-2 x 10^{13} ions/cm^2, while a new luminescence was observed to grow at the longer wavelength, for polystyrene and poly(2-vinyl naphthalene), but not for poly(N-vinyl carbazole). The decreasing rate of the excimer fluorescence corresponded to the increasing rate of the new luminescence. Thus, a new kind of luminescent species are produced by ion irradiation, accompanied by the disappearance of the excimer fluorescence. Its formation process might be associated with the overlapping effect of ion tracks occurred in ion-irradiated polymers.

Ion irradiation effects on polymer materials have attracted interests of many scientists in the field of materials research, especially of polymer modification with ion beam(*1*). The ion irradiation effects of polymers could be roughly classified into two kinds of effects at present. One is the intra-track effect which occurs within an individual ion track and it is dependent on the distribution of energy deposition by only one ion and it is specified by the energy and the kind of the ion. The other one is the dose effect which can be observed when more than two ion tracks influence the same part of polymer. Then intra-track effects seem to occur predominantly at low dose while the dose effects are expected in a higher fluence region because the former is a local effect. According to previous reports(*1-3*), at low ion fluence the changes in the chemical bonding, chain scission and/or crosslinking, of polymers are obtained, and at high fluence most of polymers are decomposed nearly completely to inorganic hydrogenated carbonaceous materials which show good electric conducting property.

0097–6156/94/0579–0045$08.00/0

However, there are few reports describing the effects, in the intermediate fluence region, on the structures of the polymers, the chemical reactions, etc. An ion pulse radiolysis study of polystyrene(4) shows excimer fluorescence and un-identifed luminescence in the luminescence spectrum. In this report, the un-identified luminescence from ion-irradiated polystyrene and poly(2-vinyl naphthalene) will be shown, which can be considered to result from overlapping of ion tracks and to give some information about ion radiolysis of polymers in the intermediate region.

Experimental

Polystyrene (578, Scientific Polymer Products), poly(2-vinyl naphthalene)(V-284, General Science Corporation), and poly(N-vinyl carbazole)(093C, Scientific Polymer Products) were used as received. The polymer films were prepared by casting on Si wafers or quartz plates. The casting solvents were toluene for polystyrene and poly(2-vinyl naphthalene), and dichloromethane for poly(N-vinyl carbazole).

Ion irradiation was done with 200 keV He^+ beam from a 200 keV ion implanter at JAERI-TAKASAKI. The ion beam was scanned in vertical and horizontal directions on a 10mm-diameter aperture to obtain uniform ion doses. The current density was estimated from the ion current through the aperture and the aperture size. The apparatuses for luminescence measurement were composed of a monochrometer/ spectrograph (HR250, Jovin-yvon), a multichannel photodetector with image intensifier(OMH-42, ATAGO BUSSAN) and a photomultiplier tube (R2758, HAMAMATSU) in photon counting mode. Luminescence spectra were measured repeatedly during irradiation under the current density of 12.7 nA/cm². The dose for one spectrum was 3.2 x 10^{12} ions/cm². In the measurement of the fluence dependence of luminescence by photon counting system, the rise time of the photomultiplier output was ca. 2.7 ns and the channel advance time was set to be 500 ms.

Molecular products emitted from polymer samples during ion irradiation were also detected by a quadrupole mass spectrometer (QMG-420C, Balzers) as a complemental product analysis. In this measurement, the beam current density (51.0 nA/cm²) was higher than that for luminescence measurement in order to obtain appropriate yields of emitted products.

Results and discussion

Figure 1(a), (b) and (c) show the luminescence spectra of polystyrene, poly(2-vinyl naphthalene) and poly(N-vinyl carbazole) induced by 200 keV He^+ irradiation. The fluorescence of sandwich-type excimer was observed around 330 nm for polystyrene(5), and around 420 nm for poly(2-vinyl naphthalene)(6,7) and poly(N-vinyl carbazole)(8-10). The so-called second excimer fluorescence was also observed around 370 nm for both poly(2-vinyl naphthalene) and poly(N-vinyl carbazole). There was no monomer fluorescence in the spectra for all polymers investigated here. In addition, a new kind of luminescence was seen in the spectra of polystyrene and poly(2-vinyl naphthalene) whose spectral region was 450-650 nm and 600-750 nm,

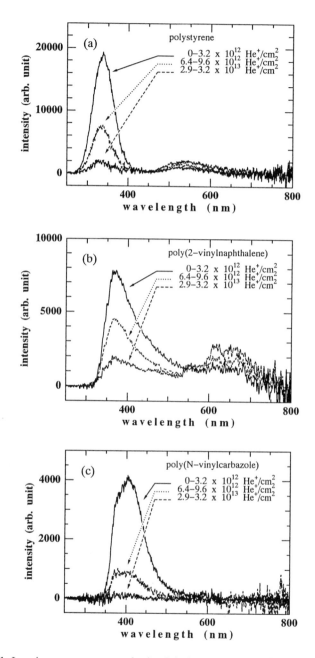

Figure 1 Luminescence spectra obtained during 200 keV He⁺ irradiation.
(a) polystyrene, (b) poly(2vinyl naphthalene) and (c) poly(N-vinyl carbazole).

respectively. The intensities of the two kinds of excimer fluorescence decreased and that of the new luminescence increased in the same fluence region. And after a break of irradiation for several minutes, the excimer fluorescence and the growing luminescence showed the same intensities and the same spectra as those before the break. Thus, the changes in intensity observed during irradiation may not be due to the changes in local temperature of the polymers and they seem to be attributed to the chemical changes of the polymers.

The dependences of luminescence with the ion fluence observed at 360 nm, 450 nm, and 620 nm for poly(2-vinly naphthalene) are shown in Figure 2(a), (b) and (c), respectively. As seen in the figures, the decreasing rates of fluorescences from the two kinds of excimers ((a) and (b)) are similar to the growing rate of the luminescence(c), which indicates a correlation between the decrease of the excimer fluorescence and the growth of the new luminescence. The decrease of excimer fluorescence could be interpreted with the chemical change at the excimer formation sites or with the interruption of the energy migration to the excimer sites by some irradiation products. As the ion pulse radiolysis study(4) showed the quenching of excimer fluorescence of polystyrene, it is likely that some chemical reaction occurred at the excimer sites and this is the reason for the decreased fluorescence yield. On the other hand, the dependence of the new luminescence with the ion fluence shows that the concentration of the site for this luminescence increases with the fluence. Therefore, the new luminescent site is formed in an ion track and the luminescence could come from the site in another ion track. It can be an overlapping effect of ion tracks, where the ion track described here includes the excitation energy transfer, for example, the ranges of δ-rays and the range of energy migration over and through polymer chains. The new luminescent site probably serves as a radiative trap for singlet excitation energy, as well as the excimer sites, in the excitation process of the polymer. The similar behaviors of luminescence concerning to excimers and the new one were also observed for polystyrene and the new luminescence was found by Kouchi et al.(4) for polystyrene. But at present no identification has been made for this new luminescent sites.

Figure 3 shows the dependence of the yield of emitted molecules with the ion fluence for 200 keV He$^+$ irradiated poly(2-vinyl naphthalene). The main components of the emitted products by ion irradiation were hydrogen(M/e=2) and acetylene(M/e=26). Naphthalene(M/e=128) was also released, which showed the very small yield. In the previous work(1), hydrogen and acetylene were also detected from ion irradiated polyethylene and polystyrene. And benzene was detected from ion irradiated polystyrene. As the yield of acetylene was much higher than that of naphthalene, if this product is released from the main chain of poly(2-vinyl naphthalene), naphthalene chromophores will be condensed in this polymer. It is consistent with the graphitization of ion irradiated polymers(2). Furthermore, the entity for the new luminescence could be attributed to one of the candidates for the precursor of the graphitization by ion beam. However, the excitation process and structural changes of polymer under ion irradiation are not clear yet, so that further investigation is necessary especially in the intermediate fluence region.

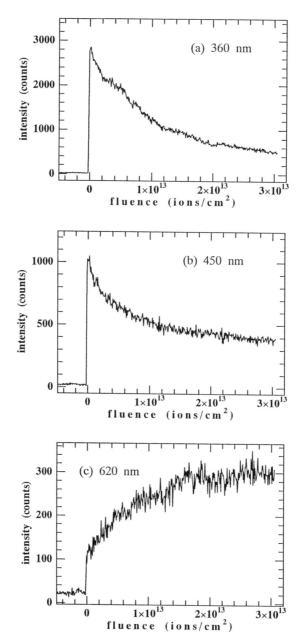

Figure 2 The dependence of luminescence with ion fluence in 200 keV He$^+$ irradiation on poly(2-vinyl naphthalene). (a) 360 nm, (b) 450 nm and (c) 620nm.

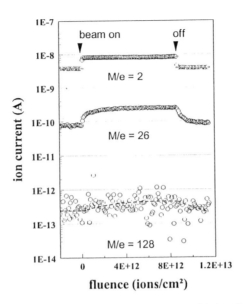

Figure 3 The dependence of yield of emitted molecules with ion fluence in 200 keV He+ irradiation on poly(2-vinyl naphthalene). The broken line for M/e=128 is a guide for eyes.

Literature Cited

(1) Venkatesan, T.; Calcagno, L.; Elman, B.S.; Foti, G. In Ion Beam Modification of Insulators; Beam Modification of Materials, 2; ELSEVIER: Amsterdam, 1987; pp 301-379.

(2) Calcagno, L.; Foti, G. Nucl. Instr. and Meth. **1991**, B59/60, 1153.

(3) Aoki, Y.; Kouchi, N.; Shibata, H.; Tagawa,S.; Tabata, Y.; Imamura, S. Nucl. Instr. and Meth. **1988**, B33, 799.

(4) Kouchi, N.; Tagawa, S.; Kobayashi, H.; Tabata, Y. Radiat. Phys. Chem., **1989**, 34, 453.

(5) Itagaki, H.; Horie, K.; Mita, I.; Washio, M.; Tagawa, S.; Tabata, Y. Radiat. Phys. Chem., **1989**, 34, 597.

(6) Irie, M.; Kamijo, T.; Aikawa, M.; Takemura, T.; Hayashi, K.; Baba, H. J. Phys. Chem., **1977**, 81, 1571.

(7) Locke, R.J.; Lim, E.C. Chem. Phys. Lett., **1989**, 160, 96.

(8) Tagawa, S.; Washio M.; Tabata, Y. Chem. Phys. Lett., **1979**, 68, 276.

(9) Roberts, A.J.; Cureton, C.G.; Phillips, D. Chem. Phys. Lett., **1980**, 72, 554.

(10) Ghiggino, K.P.; Archibald, D.A.; Thistlethwaite, P.J. J. Polym. Sci., **1980**, 18, 673.

RECEIVED September 13, 1994

RESIST SCIENCE AND TECHNOLOGY

Chapter 5

New Directions in the Design of Chemically Amplified Resists

E. Reichmanis, M. E. Galvin, K. E. Uhrich, P. Mirau, and S. A. Heffner

AT&T Bell Laboratories, 600 Mountain Avenue, Murray Hill, NJ 07974

The design of a robust, manufacturable, deep-UV resist requires a fundamental understanding of the interactions between the varied components of such a material at a molecular level. Two factors that have a critical impact on the ability of a given resin to be used in a chemically amplified resist formulation are polymer/additive miscibility and inter- and intra-chain hydrogen bonding interactions. Experiments with trimethylsilyloxystyrene containing materials suggest that efficient chemically amplified resist systems should consist of a polar PAG and polar matrix in order to maximize the interaction between the photogenerated acid and the polymer. Similarly, intra- and intermolecular interactions between polymer chains may be important in determining the solubility characteristics of a matrix. For instance, the substitution pattern on a given chain will be shown to perturb hydrogen bonding interactions and concomitantly, dissolution characteristics. It will be demonstrated how issues such as those described above will affect resist performance and the groundwork will be laid for the design of resist chemistry through understanding and manipulation of polymer structure, molecular properties and synthetic methods.

The progress that has been made in the fabrication of microelectronic devices in general, and the lithographic technology used to generate the high-resolution circuit elements that are characteristic of those devices in particular, can only be classified as remarkable. Even more remarkable is that the technology of choice for fabricating those devices remains photolithography. Thus, conventional photolithography will be able to print features as small as 0.35 μm and will remain the dominant lithographic technology well into this decade (1). However, the ultimate resolution of a printing technique is governed, at the extreme, by the wavelength of light (or radiation) used to form the image (2). It is this physical limitation that is now driving the industry to explore and develop new lithographic technologies. Each of the alternatives, be it 248 or 193 nm photolithography, X-ray or electron-beam lithography, will require not only the development of a manufacturing worthy tool, but the design and development of a manufacturable resist material and process (3,4).

Upon examination of each of the alternative lithographic technologies, it appears unlikely that a material which undergoes only one radiation induced chemical event per absorbed unit radiation dose will provide sufficient sensitivity to ensure adequate throughput. A new class of resists that achieves differential solubility from acid catalyzed chemical reactions was discovered by Ito, Willson and Frechet (5-8) and independently presented by Crivello (9) who utilized the fact that arylonium salts efficiently produce strong acids upon irradiation (10-12). These resists are nominally classified as chemically amplified resists and are compatible with current lithographic process technology. Chemically amplified resists generally exhibit high contrast, good process latitude, excellent thermal stability and good dry-etching resistance. The process sequence for these chemically amplified materials is similar to that for conventional positive resists, although the post-exposure bake (PEB) assumes a different role. For these resists, the exposure and PEB steps play an equally important role in effecting differential solubility between the exposed and unexposed regions of a resist film.

Presented here will be selected design isssues that must be addressed in order to effectively build a manufacturable, production worthy chemically amplified resist material.

DESIGN ISSUES

Lithographic resists must be carefully designed to meet the specific requirements of a given lithographic technology. Although these requirements vary according to the radiation source and device processing sequence, the following resist properties are common to all lithographic technologies: *sensitivity, contrast, resolution, optical density (for UV resists), etching resistance, purity and manufacturability.* These properties can be achieved by careful manipulation of polymer structure, molecular properties and synthetic methods.

The materials issues that must be considered in designing resists with the appropriate properties are given below (13). The polymer resin must

i.   exhibit solubility in solvents that allow the coating of uniform, defect free, thin films, or be amenable to vapor-deposition to achieve the same result,

ii. be sufficiently thermally stable to withstand the temperatures and conditions used with standard processes,

iii. exhibit no flow during pattern transfer of the resist image into the device substrate,

iv. possess a reactive functionality that will facilitate pattern differentiation after irradiation, and;

v. for UV exposure, have absorption characteristics that will permit uniform imaging through the thickness of a resist film.

Using the development of a deep-UV resist as an example, the performance criteria for a production worthy deep-UV resist are given in Table 1 (4,13).

Table 1: Deep-UV resist performance criteria

PARAMETER	CRITERIA
Sensitivity	$<100mJ/cm^2$
Contrast	>5
Resolution	$<0.25\mu$ m
Optical Density	$<0.4/\mu$ m
T_g	$>140°$ C
Etching Resistance	~ novolac materials
% Volatilization Upon PEB	$<15\%$
% Film Loss Upon Development	0
Cost	~ I-line Resists

DESIGN OF MATERIALS CHEMISTRY

The *tert*-butoxycarbonyloxystyrene (TBS) unit has been extensively studied for its application in chemically amplified resist materials. Ito et al. (5-8) were the first to report on the use of this monomer in the design of a poly(*tert*-butoxycarbonyloxystyrene)/onium salt based resist that exhibited deep-UV sensitivities less than 10 mJ/cm^2 (Figure 1). Such materials do however have several drawbacks, namely: i) unacceptable film thickness loss upon PEB (~40%), ii) marginal adhesion due to the hydrophobic nature of the *tert*-butoxycarbonyl appendage, and iii) low flow resistance due to a $T_g < 140°$ C. The T_g of the TBS based resists was easily raised upon incorporation of SO_2 into the polymer backbone which additionally provided sites for radiation induced main chain scission allowing access to single component radiation sensitive chemically amplified resists (14-16) (Figure 2). While poly(*tert*-butoxycarbonyloxystyrene-*co*-sulfone) (PTBSS) based materials exhibited excellent resolution and photospeed, along with high thermal image stability, their adhesion to most device substrates was still marginal due to the highly hydrophobic nature of the matrix and they underwent significant thickness loss upon exposure and PEB which would affect process performance during steps such as dry pattern transfer (17).

Several approaches may be utilized to achieve a decrease in the extent of resist film thickness loss upon exposure and PEB. First, the acid sensitive monomer, i. e., TBS

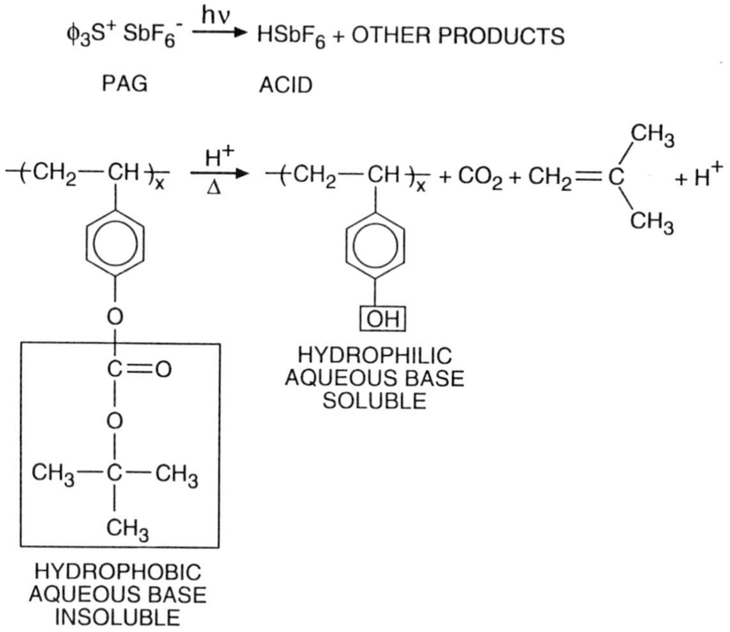

$$\phi_3S^+ \, SbF_6^- \xrightarrow{h\nu} HSbF_6 + \text{OTHER PRODUCTS}$$

PAG   ACID

HYDROPHILIC
AQUEOUS BASE
SOLUBLE

HYDROPHOBIC
AQUEOUS BASE
INSOLUBLE

Figure 1:   Schematic representation of the chemistry associated with poly(*tert*-butoxycarbonyloxystyrene)-onium salt based resists.

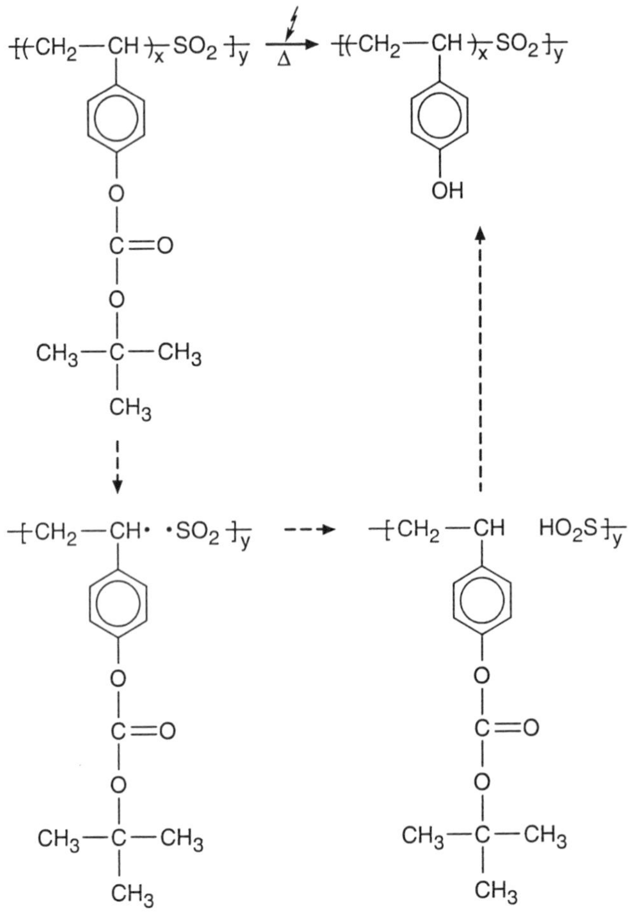

Figure 2:   Schematic representation of the PTBSS radiation induced
               chemistry.

in the above example, may be copolymerized with alternative materials that are unreactive in the presence of acid but allow aqueous base dissolution of the polymer that is generated upon acidolysis. For example, acetoxystyrene has been incorporated into PTBSS to afford poly(acetoxystyrene-co-*tert*-butoxycarbonyloxystyrene-co-sulfone) (PASTBSS), an effective matrix resin for deep-UV resist aplications (18,19). Upon formulation with a photoacid generator (PAG), PASTBSS undergoes acid induced cleavage of the *tert*-butoxycarbonyl protective group after exposure and PEB. Subsequent immersion in an aqueous base developer effects hydrolysis of the acetoxy appendage and concomitant dissolution of the material. Figure 3 presents an outline of the relevant chemistry. Second, the protective group could be changed such that it is either non-volatile and remains in the film, or represents a smaller percentage by weight of the overall mass of the resist. Third, both of the above approaches could be used in the design of a single polymer matrix resin. Each of these alternatives requires careful manipulation of the chemical structure of the polymer to ensure phase compatibility of the resist matrix resin with a given PAG and aqueous base solubility of the acidolysis product. Examples of chemistries that highlight these issues are described below.

Substituent Effects on Miscibility.   One avenue that may be taken to address the question of weight loss in TBS based resists is to dilute the concentration of the *tert*-butoxycarbonyl (BOC) containing monomer (18,20). For example, by copolymerizing TBS with another styrene unit containing an alternate protective group, the amount of BOC groups can be decreased while the overall solubility properties of the resist are maintained. A candidate monomer is trimethylsilyloxystyrene (TMSS) (21-24). The silyl ether group exhibits stability towards highly basic reagents, yet undergoes selective cleavage in even mildly acidic environments. As a trimethylsilyl (TMS) ether is more labile than a BOC group, weaker acids are necessary to generate the phenol. Utilization of a weaker acid may in turn influence the post-exposure delay effects noted with most chemically amplified resists. Upon acid catalyzed hydrolysis, the TMS units react to form hexamethyldisiloxane, a non-volatile by-product (Figure 4). As this unit remains in the film, the percent film thickness loss noted upon exposure and PEB is expected to be reduced.

Various TMSS containing materials were synthesized by free radical polymerization of the silyl ether with TBS and/or sulfur dioxide. As evidenced by examination of Table 2, high molecular weight, low polydispersity materials were readily prepared. The TGA data for the various polymers indicate that incorporation of a silyl ether into the polymer greatly enhances the thermal stability. The onset temperature of polymer decomposition decreases with a decrease in the percent composition of the silyl styrene. Concurrently, the onset temperature of thermal deprotection of the BOC group is highest in the polymers with the highest percentage of silyloxy groups in the polymer. The incorporation of sulfone into the polymers decreases their thermal stability relative to the purely styrenic materials. The onset temperature of decomposition is approximately 160° to 180°C lower in the polymers containing sulfone than for polymers without sulfone. Similarly, the onset temperature of BOC group deprotection is also lowered by about 30-40°C in polymers with sulfone.

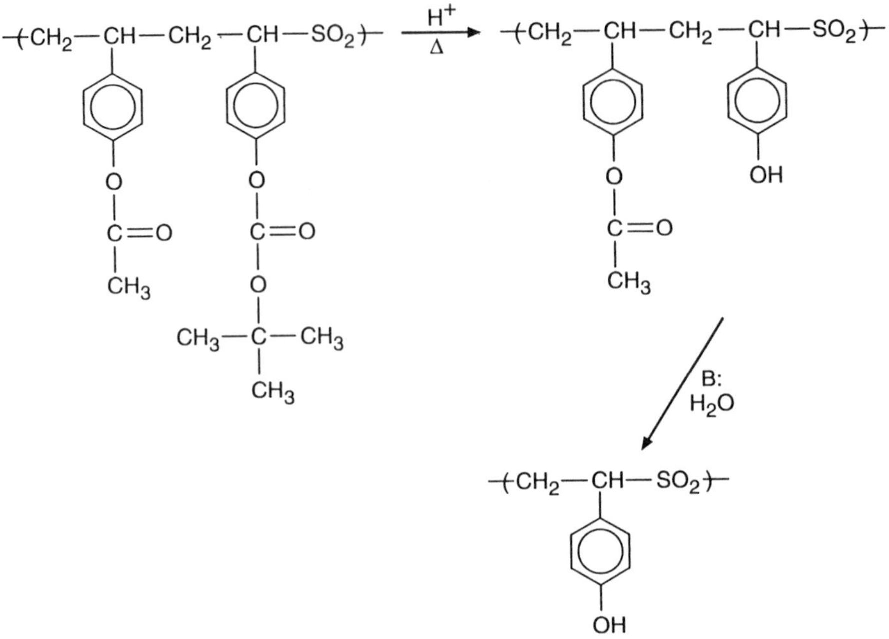

Figure 3:   Schematic representation of the mechanism of PASTBSS resist action.

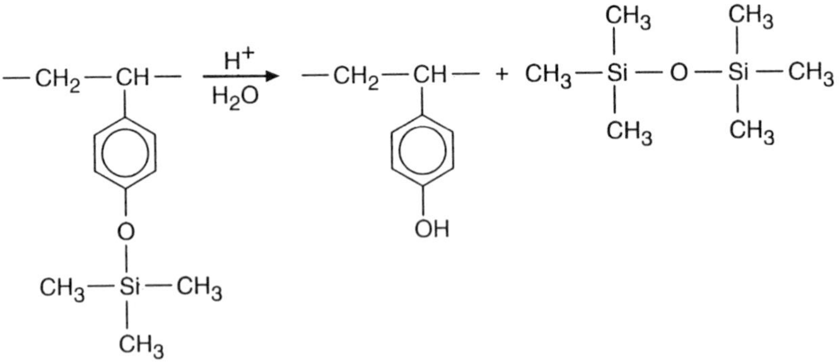

Figure 4:   Schematic representation of the acid catalyzed hydrolysis of trimethylsilyloxystyrenes.

Table 2: TMSS/TBS/SO$_2$ copolymer molecular characteristics

POLYMER	M$_w$	D	T-BOC[a]	T$_{decomp}$[a]	T$_g$[a]
poly(TMSS)	55200	1.2	-	419	78
poly(TMSS$_3$-TBS)	68300	1.3	229	429	85
poly(TMSS-TBS)	64500	1.3	213	396	106
poly(TMSS-TBS$_3$)	75500	1.6	206	399	117
poly(TMSS$_2$-SO$_2$)	40500	1.6	-	254	-
poly(TMSS$_3$-TBS-SO$_2$)	36700	1.7	186	246	164[b]
poly(TMSS-TBS-SO$_2$)	58100	3.3	180	240	178[b]
poly(TMSS-TBS$_3$-SO$_2$)	38600	2.0	164	233	184[b]

[a] Onset temperature, in °C, of the thermal deprotection of the *tert*-butoxycarbonyl group.
[b] T$_g$ of polymer after thermal removal of the *tert*-butoxycarbonyl group.

With respect to the T$_g$, incorporation of trimethylsilyloxy groups tends to lower the T$_g$ due to increased flexibility of the polymers. For the sulfone materials, the T$_g$'s were higher than the onset temperatures for thermal deprotection of the BOC groups.

Upon formulation with a PAG, lithographic evaluation of the series of silylated materials afforded some interesting results. Notably, the sulfone containing polymers consistently afforded enhanced resolution capability with respect to the corresponding non-sulfone resins. For instance, Figure 5 depicts line/space patterns obtained with poly(TMSS-TBS$_3$) and poly(TMSS-TBS$_3$-SO$_2$). In each case, the resists were formulated by dissolving the PAG (15 wt % relative to the polymer) in a solution of the matrix polymer (15 wt/vol %) in ethyl ethoxypropionate. The resists were prebaked at 120°C for 30s, exposed and postexposure baked at 120°C for 60s. The imaged substrates were developed in aqueous TMAH. The styrenic copolymer displays shallow, poorly resolved features, while its sulfone counterpart clearly resolves 0.35μm images. Notably, no imaging could be achieved with the homopolymer, poly(trimethylsilyloxystyrene). These observations led to an examination of the miscibility of PAG materials with a variety of silicon containing matrices (25).

The distribution of representative examples of several families of PAG's in poly(trimethylsilyloxystyrene) was initially investigated using Rutherford back scattering (RBS) to follow the depth distribution of either As or S in the PAG. This technique gives quantitative results with a depth resolution of 200 Å and is especially suited to analysis of heavy elements distributed in a light element matrix. A comparison was made between samples that were baked after spin-coating and samples that were exposed to UV irradiation, followed by a four minute post-exposure bake (PEB). The dose to effect complete dissolution of large area features

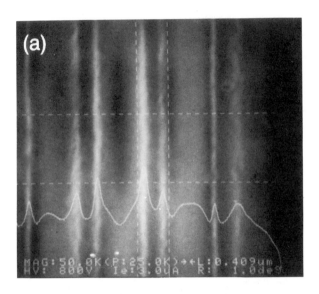

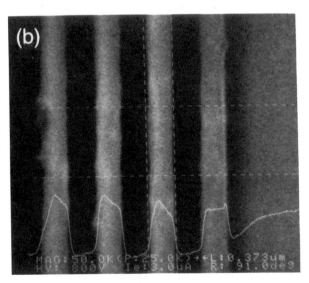

Figure 5:   SEM photomicrographs depicting coded line/space images
              obtained in (a) poly(TMSS-TBS$_3$) with 0.4 μm resolution and
              (b) poly(TMSS-TBS$_3$-SO$_2$) with 0.3 μm resolution.

was first determined on all resist formulations and the exposure dose used for the RBS experiments was approximately twice the clearing dose.

The RBS results are summarized in Table 3.   For the silyloxy homopolymer formulated with triphenylsulfonium hexafluoroarsenate, the arsenic species is strictly confined to a thin, 500 Å thick layer at the polymer/substrate interface, with no arsenic appearing in the top 7600 Å of the film.   This distribution is unchanged after exposure and PEB, suggesting phase incompatibility between the highly polar arsenate salt and the hydrophobic polymer.   As a less polar PAG may be more miscible with the non-polar silyloxystyrene material, the distribution of a nitrobenzyl ester PAG was also examined.   Formulation of poly(trimethylsilyloxystyrene) with 2,6-dinitrobenzyl tosylate affords a film that has an uneven distribution of the PAG. Auger Electron Spectroscopy (AES) provides additional evidence that the PAG materials are immiscible with the silylated polymer.   A polymer film containing the silyloxy homopolymer and the tosylate PAG was baked at 120 °C for 30 seconds to remove solvent.   The AES surface survey (Figure 6) clearly shows that the elements found in the silyl ether homopolymer, namely, carbon, silicon and oxygen, are present. Sulfur, which emanates from the PAG, is noticeably absent.   Upon exposure followed by a PEB at 120 °C for four minutes, carbon and oxygen, as well as sulfur and nitrogen originating from the PAG are detected at the surface.   After exposure then, either the PAG or the photo-generated acid has migrated to the polymer surface effecting cleavage of the silyl ether at the air/polymer interface.   These results are consistent with the RBS data previously discussed.   Further control studies showed that it is the acid that migrates as opposed to the PAG (25).

The data presented above clearly demonstrate the importance of polymer/PAG design in effecting miscibility of even small molecules with polymer matrices.   The combined RBS and Auger data show that the PAG's used in this study are not miscible with the non polar, silylated homopolymer and readily phase segregate affording films where the air/polymer interface is depleted in PAG.   In the case of the nitrobenzyl ester PAG, AES results showed that upon generation of tosic acid, the acid itself was prone to diffuse through and volatilize from the polymer film. This phenomenon of phase immiscibility between polymer materials and selected additives has been reported previously.   In a related example, Hult et al., reported the appearance of phase separation between a styrenic polymer and an onium salt PAG (26). Specifically, the migration of onium salts was followed by ESCA in polystyrene systems, and the air/polymer interface was found to be depleted in salt.

Incorporation of *tert*-butoxycarbonyloxy groups into the silicon bearing polymer alters the interaction between the PAG's and the polymer matrix.   While formulation with an arsenate PAG still leads to phase segregation of the PAG and polymer for a 1:1 copolymer, the less polar tosylate PAG is now more evenly distributed throughout the polymer film and no difference in sulfone concentration was noted between the bottom 1000Å of the film vs. the remaining portion.   Incorporation of 75% of the TBS monomer affords a material that effects miscibility with the arsenate PAG.   Use of a more polar monomer in place of the TBS component, i.e., hydroxystyrene, changes the point at which the PAG becomes miscible with the matrix. For instance, the arsenate onium salt appears fully miscible with a 1:1 copolymer of TMSS and hydroxystyrene.

Table 3:   Compositional characteristics of selected polymer/PAG films as determined by RBS analysis

POLYMER	PAG	EXPOSURE[a]	THICKNESS[b]	%Si[c]	%S[c]	%As[c]
P(TMSS)	arsenate[d]	none	7600	3.3	-	0
			500	3.3	-	0.13
		yes	7600	3.3	-	0
			500	3.3	-	0.13
P(TMSS)	tosylate[e]	none	2000	3.1	0.16	-
			6300	3.1	0.335	-
		yes	6000	0.46	0.29	-
P(TMSS-TBS)	arsenate[d]	none	6600	1.8	-	0.012
			1000	1.8	-	0.062
		yes	6300	1.5	-	0.010
			1000	1.5	-	0.078
P(TMSS-TBS)	tosylate[e]	none	8550	1.6	0.31	-
		yes	6750	0.85	0.33	-
P(TMSS$_3$-TBS)	arsenate[d]	none	6500	2.7	-	0.003
			1500	2.7	-	0.078
		yes	6700	2.65	-	0.0037
			1200	2.65	-	0.079
P(TMSS-TBS$_3$)	arsenate[d]	none	7600	0.91	-	0.024
		yes	7600	0.79	-	0.023
P(TMSS$_{1.5}$-HS[f])	arsenate[d]	none	10,800	2.06	-	0.023
		yes	10,500	1.85	-	0.024
	tosylate[e]	none	12,500	1.26	0.30	-
		yes	13,000	1.3	0.29	-

[a] Exposure at 248 nm; [b] Thickness in Å; [c] Given in atomic %; [d] Denotes triphenylsulfonium hexafluoroarsenate; [e] Denotes 2,6-dinitrobenzyl tosylate; [f] HS denotes hydroxystyrene.

Of the several systems examined here, the most unusual behavior was observed with the most hydrophobic polymer, namely the silylated homopolymer, when formulated with a polar, ionic PAG, i.e., triphenylsulfonium hexafluoroarsenate. Several factors may explain the non-uniform distribution of this small molecule in the polymer matrix. First, the non-polar polymer may tend to exclude the polar onium salt. Second, the hygroscopic arsenate PAG may migrate to the polymer/substrate interface which is most likely to have the highest concentration of water and is therefore a polar surface. Third, the arsenate salt is more soluble in the casting solvent than in the non-polar polymer. During the spin-coating process, solvent evaporates from the air-polymer interface generating a polymer-rich surface. The concentration gradient may cause the dissolved arsenate salt to migrate towards the more polar region at the polymer/substrate interface. Fourth, the low surface tension of silicon may influence the distribution of PAG in the polymer matrix. Overall, the highly polar arsenate salt may have limited solubility in the relatively hydrophobic polymer.

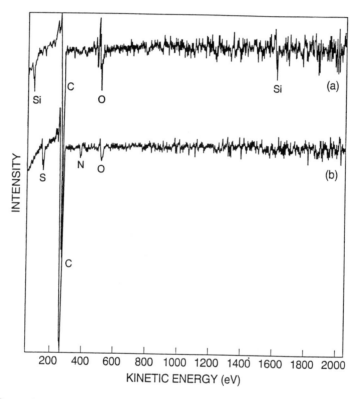

Figure 6:   AES surface survey scan of poly(TMSS) formulated with a nitrobenzyl ester based PAG (a) before exposure, (b) after exposure with PEB.

The nitrobenzyl tosylate PAGs are polar, non-ionic materials which lend themselves to improved miscibility with the polymers examined here. However, as noted above, even though the nitrobenzyl ester based PAGs are miscible, additional concerns arise once the acid is generated. Specifically, the highly polar acid may itself be immiscible. Under such circumstances, the acid may not be as efficient as it would be were the acid in a polar environment and as seen in the cases of the silyloxy homopolymer and the tosylate PAG, the acid may in fact volatilize from the matrix.

The above observations lead to the conclusion that efficient chemically amplified photoresist systems based upon acidolytic reactions should consist of polar PAGs and polar matrices, as it is this combination that maximizes the interaction between the photo-generated acid and the polymer.

The Effects of Intra- and Inter-chain Molecular Interactions. The inherent molecular interactions that are present within a polymer chain are equally important in determining the suitability of a given material for lithographic applications. One example will be presented that is particularly instructive. The desire to reduce the % film thickness lost upon exposure and PEB has led to the need to identify alternate substituted styrene monomers that would be both thermally and acidolytically stable, yet sufficiently polar to allow aqueous base solubility of a deprotected matrix resin. The nature of the individual polymer components plays a decisive role in determining the range of materials that may be used. It has been demonstrated that poly(acetoxystyrene-co-*tert*-butoxycarbonyloxystyrene) (PASTBS) and PASTBSS both undergo acid induced *tert*-butoxycarbonyl cleavage upon exposure and PEB in the presence of PAG materials (18). However, of the resultant hydroxystyrene analogs, only the sulfone containing polymer exhibits aqueous base solubility, a characteristic that is required if a material is to be considered for lithographic applications (18).

Initial studies of these systems revealed that the acetoxy moiety is cleaved during aqueous base development only if the sulfone moiety is present. IR studies of the two polymers after exposure and PEB revealed that hydrogen bond formation between the hydroxy group and the carbonyl of the acetoxy group was occurring in both systems, but was enhanced in the polymer without sulfone (PASTBS) (18). This lead to the proposal that the increased hydrogen bonding in PASTBS was preventing the base from penetrating the film in concentrations high enough to allow for base hydrolysis of the acetoxy group, thereby preventing base solubility. IR, however, does not provide an indication of how many hydroxy groups are participating in hydrogen bond formation. It is also incapable of revealing whether the hydrogen bond formation is intra- or interchain. These questions are better answered by two dimensional nuclear Overhauser effect spectroscopy (2D NOESY).

In 2D NOESY , peaks appear in the NMR spectra at off diagonal positions if the protons of the two groups are within 5 Å of each other. For example, a cross peak would appear between the hydroxystyrene OH ring proton and the acetoxystyrene methyl protons when a hydrogen bond is formed between the hydroxystyrene OH proton and the acetoxystyrene carbonyl. This interaction would bring the methyls within 5 Å of the hydroxystyrene OH proton. If these cross peaks occur in solutions whose polymer concentration is below that required for overlap between two different polymer chains, then the observation of hydrogen bond cross peaks must arise from intra-chain hydrogen bonds. By studying the concentration dependence of hydrogen bond formation, it is therefore possible to evaluate whether the hydrogen bonds arise fron intra- or inter-chain interactions.

Figure 7 plots the cross peak section of the 2D sectra of thermally deprotected PASTBSS and PASTBS copolymers at varying concentrations. Solutions were prepared in DMSO$_{D6}$. NMR spectra were acquired at 500 MHz on a JEOL GX-500 spectrometer. The 90° pulse widths were 20 μs and the sweep widths were set to 7 Hz. The 2D NOESY spectra were obtained with the (90°-t_1,-90°-τ_m-90°-t_2) pulse sequence in the phase sensitive mode. Typically, 256 complex t_1, points and 512 complex t_2 points were acquired with a mixing time of 0-5s and a recycle delay time of 4s. The data were processed with 5 Hz line broadening in each dimension. Note in Figure 7 that the presence of a hydrogen bond cross peak between the acetoxy methyl protons and the hydrogen proton of hydroxystyrene appears at a polymer concentration of 0.2 wt% for PASTBS and at a concentration of 20 wt % for PASTBSS. The appearance of the cross peak at 0.2 wt % for PASTBS indicates that this system is strongly hydrogen bonded. The PASTBSS system exhibits concentration dependence similar to that of only weakly hydrogen bonded systems (27-29). These experiments show that one effect of introducing sulfone into the copolymer is to reduce hydrogen bond formation. This results in more free hydroxy groups which increase base solubility and perhaps assist in base hydrolysis of the acetoxy moiety. In PASTBS the hydroxy groups are consumed in hydrogen bonds and do not permit base penetration into the resist film.

The NMR experiments also provide insight into the nature of the hydrogen bond interaction. Table 4 compares the concentration dependence of the cross peaks for three systems. These include the thermally BOC deprotected PASTBS and PASTBSS, and a physical blend of poly(hydroxystyrene) and poly(acetoxystyrene). In the physical blend, the cross peak occurs at a polymer concentration of 2 wt% and is not apparent at a concentration of 0.2 wt%. This percentage then defines the limit for detection of interchain hydrogen bond formation at 2 wt%. The existence of hydrogen bonds at concentrations of 0.2 wt % PASTBS indicates that some of the hydrogen bonding observed in this system is intrachain. In PASTBSS, hydrogen bonding is not observed until a polymer concentration of 20 wt% is reached, indicating that in the sulfone polymer, the hydrogen bonding is very weak in nature. The incorporation of the sulfone moiety could decrease hydrogen bond formation via two mechanisms. The presence of sulfone in a polymer is known to increase polymer chain stiffness and might, therefore, make if energetically unfavorables for these systems to adopt a geometry that promotes the formation of hydrogen bonds. Alternatively, the sulfone unit might also increase the hydrophilic nature of the polymer enough to stabilize free hydroxy units.

Table 4:   The observation of hydroxyl-methyl cross peaks in selected hydroxystyrene containing copolymers by 2D NMR

POLYMER	0.2WT%[a]	2 WT%[a]	10 WT%[a]	20 WT%[a]	40 WT%[a]
Poly(AS-HS-SO$_2$)[b]	-	-	-	+	+
Poly(AS-HS)[b]	+	+	+	+	+
Poly(AS)/Poly(HS)[b]	-	+	+	+	+

[a] The percentage refers to the weight % of a given polymer in solution. [b] AS refers to acetoxystyrene and HS refers to hydroxystyrene.

These results provide important insight into the design of future generations of materials. Among the key aspects of the hydroxystyrene/ acetoxystyrene/sulfone based materials is their dissolution in an aqueous base developer, and the conversion of acetoxystyrene to hydroxystyrene that increases the base solubility while minimizing weight loss. The penetration of solvent into the resist matrix depends on the local polarity and, therefore, on the state of the hydrogen bond donors and acceptors. Based on the above studies, it would appear that for resins containing polar groups such as hydroxystyrene and acetoxystyrene, it would be desirable that they exhibit only weak or no hydrogen bonding. In the case of PASTBSS, the free hydroxyl and carbonyl groups are available to hydrogen bond to the aqueous solvent and promote the conversion of acetoxystyrene to hydroxystyrene with concomitant dissolution of the polymer. This interpretation is supported by the observation that the hydroxystyrene/ acetoxystyrene copolymer, which lacks the sulfone group and has been shown to exhibit strong hydrogen bonding, is not soluble in aqueous base. Here, the acetoxy unit is converted to hydroxystyrene only with great difficulty.

CONCLUSION

The design of a robust, manufacturable, deep-UV resist requires a fundamental understanding of the interactions between the varied components of such a material at a molecular level. Two factors that have been demonstrated to have a critical impact on the ability of a given resin to be used in a chemically amplified resist formulation are polymer/additive miscibility and inter- and intrachain hydrogen bonding interactions. Based on experiments with trimethylsilyloxystyrene polymers, it appears that efficient chemically amplified photoresist systems based upon acidolytic reactions should consist of polar PAG's and polar matrices in order to maximize the interaction between the photogenerated acid and the polymer. The miscibility of all chemical components of the resist, ranging from starting resist components through photoproducts and acidolysis products is important in determining the lithographic behavior of a given system. Similarly, inherent molecular interactions present is such polar materials, i. e., hydrogen bonding interactions present both within and between polymer chains, are important to consider in the design of a material. As noted above, the presence or absence of internal hydrogen bonding may significantly affect the solubility of a given material in aqueous base. In the example provided, the

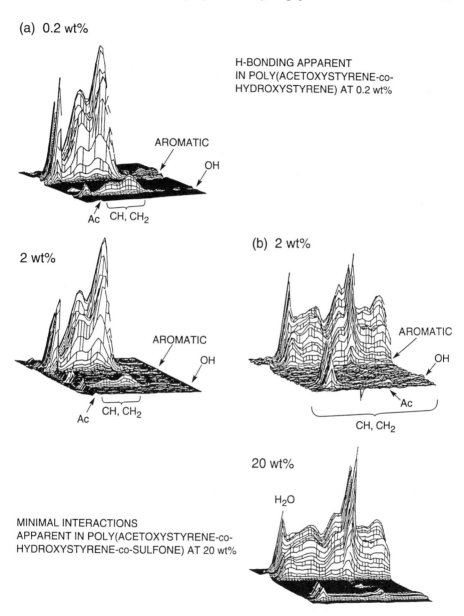

Figure 7: 2D NOESY spectra of (a) poly(acetoxystyrene-co-hydroxystyrene) at solution concentrations of 0.2μs and 2 wt% and (b) poly(acetoxystyrene-co-hydroxystyrene-co-sulfone) and solution concentrations of 2 and 20 wt%. This section of the 2D spectrum shows the cross peaks between the hydroxyl and aromatic protons on the top with the aliphatic and methyl protons on the bottom.

incorporation of sulfur dioxide into an acetoxystyrene/hydroxystyrene resin perturbs the inter- and intramolecular hydrogen bonds present in the system allowing the basic media to penetrate the matrix, facilitating dissolution. The information gathered concerning the various molecular level interactions present within lithographic materials candidate components can be further used in the design of resist materials chemistry.

LITERATURE CITED

1. McCoy, J. H., Lee, W., Varnell, G. L., *Solid State Technology*, 1989, **32(3)**, 87.
2. Thompson, L. F., Willson, C. G., Bowden, M. J., "Introduction to Microlithography", *ACS Symposium Series* **219**, ACS, Washington, D. C., 1983, pp. 2-85.
3. Reichmanis,E., Novembre, A. E., *Annual Review of Materials Science*, 1993, **23**, 11-44.
4. Reichmanis, E., Thompson, L. F., *AT&T Technical Journal,* Nov/Dec 1990, pp 32-45.
5. Ito, H., Willson, C. G., In "Polymers in Electronics", ACS Symposium Series **242**, Davidson, T., Ed., ACS Washington, D. C. 1984, pp 11-23.
6. Frechet, J. M. J., Eichler, E., Ito H., Willson, C. G., *Polymer*, 1980, **24**, 995.
7. Ito, H., Willson, C. G., Frechet, J. M.J., Farrall, M. J. Eichler, E., *Macromolecules*, 1983, **16**, 1510.
8. Ito, H., Willson, C. G., *Polym. Eng. Sci.*, 1983, **23**, 1012.
9. Crivello, J. V., In "Polymers in Electronics", ACS Symposium Series 242, Davidson, T., Ed., ACS, Washington, D. C., 1984, pp 3-10.
10. Crivello, J. V., Lee, J. L. Coulon, D. A., *Macromol. Chem. Makromol. Symp.*, 1988, **1314**, 145.
11. Crivello, J. V., Lam, J. H. W., *Macromolecules*, 1977, **10**, 1307.
12. Crivello, J. V., Lam, J. H. W., *J. Polym. Sci., Polym. Chem. Ed.*, 1979, **17**, 977.
13. Reichmanis, E., Thompson, L. F., In "Materials Chemistry: An Emerging Subdiscipline ", *ACS Advances in Chemistry Series*, Interrante, L. V., Ed., ACS Washington, D. C., in press.
14. Tarascon, R. G., Reichmanis, E., Houlihan, F. M., Shugard, A., *Polym. Eng. Sci.*, 1989, **29(13)**, 850.
15. Kanga, R. S., Kometani, J. M., Reichmanis, E., Hanson, J. E., Nalamasu, O., Thompson, L. F., Heffner, S. A., Tai, W. W., Trevor, P., *Chem. Mater.*, 1991, **3**, 662.
16. Novembre, A. E., Tai, W. W., Kometani, J. M., Hanson, J. E., Nalamasu, O., Taylor, G. N., Reichmanis, E., Thompson, L. F., *Chem. Mater.*, 1992, **4**, 278.
17. Nalamasu, O., Reichmanis, E., Cheng, M., Pol, V., Kometani, J. M., Houlihan, F. M., Neenan, T. X., Bohrer, M. P., Mixon, D. A., Thompson, L. F., *Proc. SPIE*, 1991, **1466**, 13-25.
18. Kometani, J. M., Galvin, M. E., Heffner, S. A., Houlihan, F. M., Nalamasu, O., Chin, E., Reichmanis, E., *Macromolecules*, 1993, **26**, 2165.
19. Nalamasu, O., Kometani, J. M., Cheng, M., Timko, A. G., Reichmanis, E., *J. Vac. Sci. Technol.*, 1992, **10(6)**, 2536.

20. Uhrich, K. E., Reichmanis, E., Heffner, S. A., Kometani, J. M., Nalamasu, P., *Chem. Mater.*, submitted.
21. Yamaoka, T., Nishiki.M., Koseki, K., Koshiba, M., *Polym. Eng. Sci.*, 1989, **29**, 856.
22. Bonfils, F., Giral, L., Montginoul, C., Sagnes, R., Schue, F., *Angew. Makromol. Chem.*, 1992, **198**,123.
23. Bonfils, F., Giral, L., Montginoul, C., Sagnes, R., Schue, F., *Makromol. Chem.*, 1992, **193**, 143.
24. Bonfils, F., Giral, L., Montginoul, C., Sagnes, R., Schue, F., Vinet, F., *Makromol. Chem.*, 1992, **193**, 1289.
25. Uhrich, K. E., Reichmanis, E., Baiocchi, F. A., *Chem. Mater.*, submitted.
26. Hult, A., MacDonald, S. A., Willson, C. G., *Macromolecules*, 1985, **18**, 1804.
27. Kogler, G., Mirau, P., *Macromolecules*, 1991, **24**,.
28. Crowther, M., Levy, G., *Macromolecules*, 1990, **23**, 2924.
29 Mirau, P., Bovey, F. A., *Macromolecules*, 1990, **23**.

RECEIVED September 13, 1994

Chapter 6

Dual-Tone and Aqueous Base Developable Negative Resists Based on Acid-Catalyzed Dehydration

Hiroshi Ito and Yasunari Maekawa[1]

Research Division, IBM Almaden Research Center, 650 Harry Road, San Jose, CA 95120-6099

A dual tone resist based on a polarity change from a polar to a nonpolar state has been successfully designed. Poly[4-(1-hydroxy-1-phenylethyl)styrene] undergoes exceptionally clean and efficient acid-catalyzed intramolecular dehydration to form a 1,1-diphenylethylene structure in solution or in the solid state, which does not react any further in the presence of acid even upon heating. Thus, the two-component resist containing 1.5 wt% of triphenylsulfonium trifluoromethanesulfonate (triflate) offers at 0.2 mJ/cm^2 of 254 nm radiation positive images when nonpolar xylene is used as the developer as well as negative images with a polar alcohol developer.

Furthermore, aqueous base developable two-component negative resists have been formulated, employing acid-catalyzed dehydration as the imaging mechanisms. One system consists of a copolymer of 4-hydroxystyrene and 3-methyl-2-(4-vinyl-phenyl)-2,3-butanediol and triphenylsulfonium triflate, which is based on pinacol rearrangement of the polar diol unit to a less polar, dissolution inhibiting ketone unit. The other negative resist functions on the basis of crosslinking via acid-catalyzed self-condensation and O- and C-alkylations of phenolic groups of a copolymer of 4-hydroxystyrene and 4-(1-hydroxyethyl)styrene through acid-catalyzed intermolecular dehydration.

[1]Current address: Hitachi Research Laboratory, Hitachi, Ltd., Hitachi, Ibaraki 319-12, Japan

The chemical amplification concept (*1*) based on the use of photochemical acid generators has led to the birth of an entire family of advanced resist systems. Photochemically-induced acid-catalyzed deprotection reactions result in a change of a polarity of a polymer from a nonpolar to a polar state, allowing dual tone imaging depending on the polarity of the developer solvent or positive imaging with aqueous base. A classical example is the tBOC resist based on conversion of poly(4-*t*-butoxycarbonyloxystyrene) (PBOCST) to poly(4-hydroxystyrene) (PHOST) (*2*). The clean irreversible nature of the acidolysis reaction is the key to the high resist contrast. Furthermore, the polarity change is equivalent to alteration of the reactivity with the reactive polar functionality unmasked by the deprotection reaction, which provides a basis for high contrast silylation of the exposed areas for dry development with oxygen reactive ion etching (*3*). Thus, the polarity change from a nonpolar to a polar state induced by acid-catalyzed deprotection has become the primary imaging mechanism for modern advanced resist systems (*4*).

We have been also interested in incorporating a reverse polarity change mechanism in the chemically amplified resist design, which involves a change of the polarity from a polar to a nonpolar state (*5*). One such successful example is based on pinacol rearrangement, which is an acid-catalyzed dehydration of pendant polar *vic*-diol to less polar ketone or aldehyde (Scheme I), allowing negative-tone imaging with use of a polar alcohol as the developer (*6*).

Another interesting chemical amplification resist that has been successfully formulated on the basis of the polarity change from a polar to a nonpolar state utilizes acid-catalyzed intramolecular dehydration of *tertiary* alcohol to lipophilic olefin (*7*) (Scheme II). In contrast, treatment of a polymeric *secondary* alcohol with acid resulted in crosslinking *via* ether linkage through intermolecular dehydration without involving a polarity change (*7*).

However, positive imaging of the above resist systems based on the reverse polarity change was not possible due to concomitant minor crosslinking. In this paper is described a dual tone resist that provides negative and positive images upon selection of a developer solvent owing to a clean polarity change that is achieved through dehydration without a crosslinking component.

Development with aqueous base is almost mandated in semiconductor manufacturing. In an attempt to design aqueous base developable negative resists, pinacol rearrangement of small *vic*-diol in a novolac matrix resin was successfully utilized to generate dissolution inhibiting ketone or aldehyde through acidolysis (*6,8*). However, while the novolac resins are too opaque for deep UV exposure, PHOST performed only meagerly as a matrix resin for the three-component resists. In this paper are reported approaches to incorporation of the acid-catalyzed dehydration mechanisms in deep-UV-transparent PHOST systems for aqueous base development, which involve copolymerization of HOST with a styrene bearing pendant *vic*-diol or *secondary* alcohol.

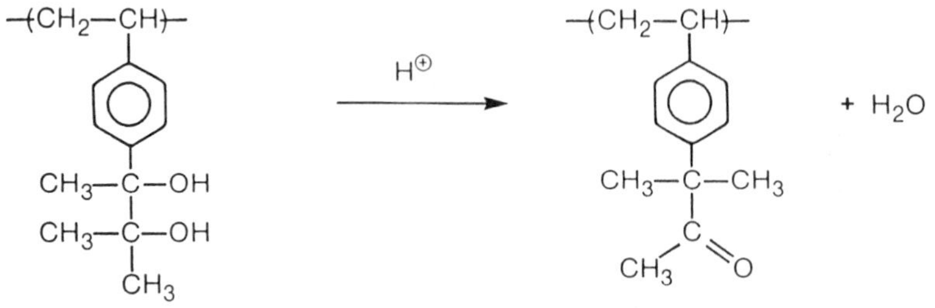

Scheme I. Pinacol rearrangement of polymeric *vic*-diol.

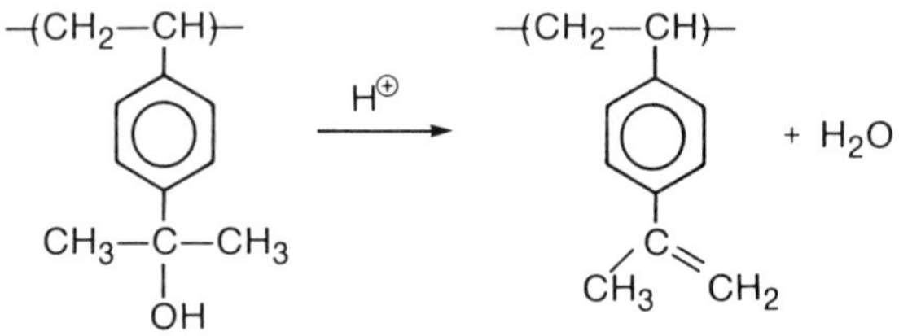

Scheme II. Acid-catalyzed intramolecular dehydration.

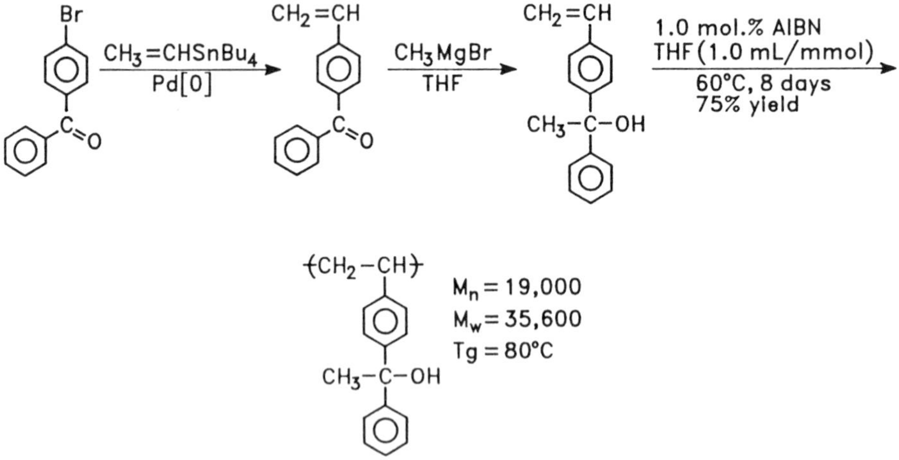

Scheme III. Synthesis of poly[4-(1-phenyl-1-hydroxy)styrene].

Experimental

Materials. 4-(1-Phenyl-1-hydroxyethyl)styrene was synthesized by reacting a methyl Grignard reagent with 4-benzoylstyrene, which had been in turn prepared by Stille's Pd-catlyzed tin coupling reaction (9) on 4-bromobenzophenone (Scheme III). When acetophenone was treated with 4-vinylphenylmagnesium chloride, the desired monomer could not be separated from an aldol condensation byproduct.

Synthetic procedures for 3-methyl-2-(4-vinylphenyl)-2,3-butanediol and 4-(1-hydroxyethyl)styrene have been reported previously (6,7). 4-Acetoxystyrene (ACOST) was commercially obtained from Hoechst Celanese Co.

Polymerization of the methyl diphenyl carbinol monomer was carried out with 1 mol% of 2,2'-azobis(isobutyronitrile) (AIBN) in tetrahydrofuran (THF, 4.5 mL/g monomer) at 60 °C for 8 days to give a polymer with $M_n = 19,000$ and $M_w = 35,600$ in 75 % yield. The polymer was purified by column chromatography followed by precipitation in water/methanol (95/5).

The *secondary* alcohol monomer was copolymerized with ACOST (feed ratio = 20/80) using AIBN (1 mol%) in THF and the copolymer was cleanly and quantitatively converted to a *sec*-alcohol/HOST (15/85) copolymer with $M_n = 13,500$ and $M_w = 23,500$ by hydrolysis with ammonium hydroxide in methanol.

The *vic*-diol monomer was copolymerized with ACOST (feed ratio = 20/80) using AIBN (1 mol%) in THF at 60 °C. The ACOST copolymer was cleanly deacylated with ammonium hydroxide in methanol to afford a pinacol/HOST copolymer (17/83) with $M_n = 25,900$ and $M_w = 42,400$.

Triphenylsulfonium trifluoromethanesulfonate (triflate) employed as the deep UV acid generator in our formulations was synthesized according to the literature (10). Cyclohexanone was used as our casting solvent. The aqueous base developers used in the imaging experiments were tetramethylammonium hydroxide solutions (MF319 or MF321).

Lithographic Evaluation and Imaging. Resist solutions were formulated by dissolving the polymers (at ~16 wt%) and triphenylsulfonium triflate (1.5 or 5.0 wt% of the total solid) in cyclohexanone and filtered down to 0.2 μm.

Resist films were spin-cast onto Si wafers for imaging experiments, NaCl plates for IR studies, and quartz discs for UV measurements, and then baked at 100 °C for 2 min. The deep UV exposure system employed in these studies was an Optical Associate, Inc. apparatus. The exposed resist films were postbaked at 100 °C for 2 min and developed or subjected to IR or UV analyses.

Measurements. Molecular weight determination was made by gel permeation chromatography (GPC) using a Waters Model 150-C chromatograph equipped with 4 ultrastyragel columns at 40 °C in THF. Thermal analyses were performed on a Du Pont 910 at 10 °C/min for differential scanning ca-

lorimetry (DSC) and on a Perkin Elmer TGS-2 at a heating rate of 5 °C/min
for thermogravimetric analysis (TGA) under nitrogen atmosphere. IR spectra
of the resist films were obtained with an IBM IR/32 FT spectrometer using
1-mm thick NaCl discs as substrates. UV spectra were recorded on a Hew-
lett-Packard Model 8450A UV/VIS spectrometer using thin films cast on
quartz plates.   NMR spectra were obtained on an IBM Instrument
NR-250/AF spectrometer. Film thickness was measured on a Tencor al-
pha-step 200.

Results and Discussion

Dual Tone Resist Based on Polarity Change. Although 4-(2-hydroxy-2-prop-
yl)styrene and 4-(1-hydroxyethyl)styrene can be readily prepared by reacting
4-vinylphenylmagnesium chloride with acetone and acetaldehyde, respectively
(7), a similar reaction of the styrenic Grignard reagent with acetophenone
produced an aldol condensation byproduct which could not be separated from
the desired monomer, 4-(1-hydroxy-1-phenylethyl)styrene. Therefore, 4-ben-
zoylstyrene was synthesized first by the Stille's Pd-catalyzed tin coupling reaction (9) on 4-bromobenzophenone and then reacted with methylmagnesium
chloride (Scheme III). As the radical polymerization of the *tertiary* and *secondary* alcohol monomers resulted in gelation when the monomer concentration or conversion was high (7), the methyl diphenyl carbinol monomer was
polymerized at a low monomer concentration of 1.0 mL THF/mmol monomer
to 75 % conversion. Furthermore, according to our previous experience (7),
the polymerization mixture was subjected to column chromatography first to
remove the unreacted monomer and then the polymer was precipitated in
water/methanol (95/5).   The polymer thus obtained had $M_n = 19,000$ and
$M_w = 35,600$ ($M_w/M_n = 1.87$) according to GPC using polystyrene as a standard, the structure of which was confirmed by ^1H and ^{13}C NMR and IR
spectroscopies. The polymer did not show any weight loss below 200 °C according to TGA.
    The polymer and 1.5 wt% of triphenylsulfonium triflate were dissolved
in cyclohexanone. The spin-cast resist film baked at 100 °C for 2 min has an
optical density (OD) of 0.42/μm at 248 nm and becomes extremely opaque
upon postbake (100 °C, 2 min) after exposure to 2 mJ/cm^2 of 254 nm radiation (Figure 1). The high OD of the exposed/postbaked film indicates formation of a conjugated structure. IR spectra of the resist film presented in
Figure 2 demonstrate facile intramolecular dehydration. The resist film prebaked at 100 °C for 2 min shows a large OH absorption at ~3500 cm^{-1}, which
completely disappears at 2 mJ/cm^2 upon postbake at 100 °C (2 min). A new
peak at ca. 1620 cm^{-1} is due to conjugated olefinic double bonds. Thus, the
spectroscopic studies clearly suggest that the pendant methyl diphenyl carbinol structure is cleanly converted to a pendant diphenylethylene structure in
the solid state with photochemically generated triflic acid.

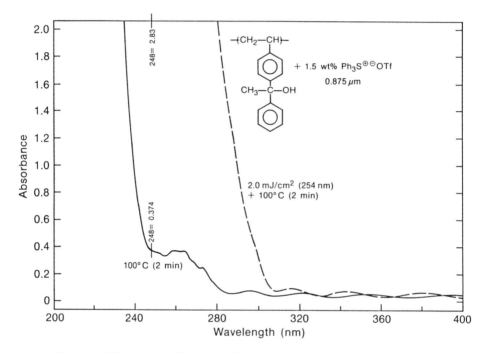

Figure 1. UV spectra of methyl diphenyl carbinol resist before and after deep UV exposure/postbake.

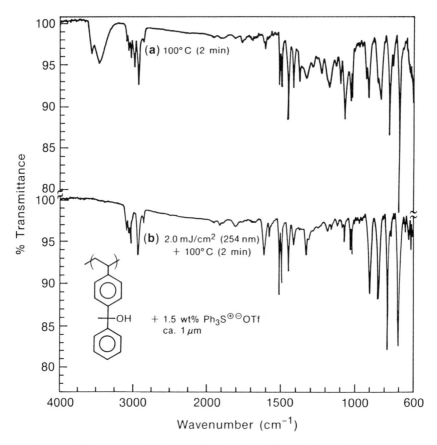

Figure 2. IR spectra of methyl diphenyl carbinol resist before (a) and after (b) deep UV exposure /postbake.

In order to further confirm the reaction pathway and to examine the possibility of side reactions, NMR spectroscopic studies on a model reaction of methyl diphenyl carbinol with triflic acid (TfOH) in $CDCl_3$ were carried out. Both ^1H and ^{13}C NMR spectra clearly indicated that the carbinol was rapidly converted to 1,1-diphenylethylene upon addition of a drop of triflic acid at room temperature and that the olefinic product did not undergo any further reaction upon standing at room temperature for 24 hrs (Figure 3). In Figure 3 the methyl (32 ppm) and quaternary carbon (77 ppm) resonances disappear and new peaks due to olefinic carbons appear at 114 and 142 ppm upon addition of triflic acid. The spectrum of the diphenylethylene product (Figure 3b) remains unchanged for at least 24 hrs. In the case of dimethyl phenyl carbinol, similar NMR studies indicated that in addition to α-methylstyrene linear and cyclic (indane) dimers were produced as shown in Scheme IV, which contributed to concomitant minor crosslinking and prevented the *tertiary* alcohol resist in Scheme II from being imaged in a positive mode. In contrast, the methyl diphenyl carbenium ion does not seem to add to the 1,1-diphenylethylene product presumably due to the steric hindrance (Scheme V). When methyl diphenyl carbinol was heated with a small amount of triflic acid at 60 °C for 30 min, 1,1-diphenylethylene was the sole product. Thus, the chemical amplification resist consisting of poly[4-(1-phenyl-1-hydroxyethyl)styrene] and triphenylsulfonium triflate is expected to provide both negative and positive images due to the clean polarity change without involving crosslinking.

The resist film containing 1.5 wt% of triphenylsulfonium triflate exhibits its maximum shrinkage of ca. 8 % at 0.2 mJ/cm^2 and at this dose the exposed area becomes completely insoluble in a 1:1 mixture of isopropanol (IPA) and ethanol (EtOH) when pre- and postbaked at 100 °C for 2 min while the unexposed film is cleanly soluble in the alcohol developer, providing negative images with a very high contrast (γ) of 12 (Figure 4). Furthermore, when the maximum shrinkage is attained at 0.2 mJ/cm^2, the exposed regions become completely soluble in *p*-xylene, while the unexposed areas are totally insoluble in the developer, resulting in positive tone imaging with a high γ of 7 (Figure 5). The small shrinkage (8 %) associated with dehydration may be advantageous over the large thinning that is observed upon postbake in the deprotection resist systems (the tBOC resist could lose as much as 45 % of its thickness upon postbake due to liberation of carbon dioxide and isobutene). In Figure 6 are presented positive (a) and negative (b) images contact-printed in the methyl diphenyl carbinol resist containing 1.5 wt% of triphenylsulfonium triflate at 0.50 and 0.41 mJ/cm^2 of 254 nm radiation using *p*-xylene and a 1:1 mixture of IPA-EtOH, respectively. Thus, this is the first and only example of dual tone imaging materials based on the polarity change from a polar to a nonpolar state and is perhaps the most sensitive resist currently known.

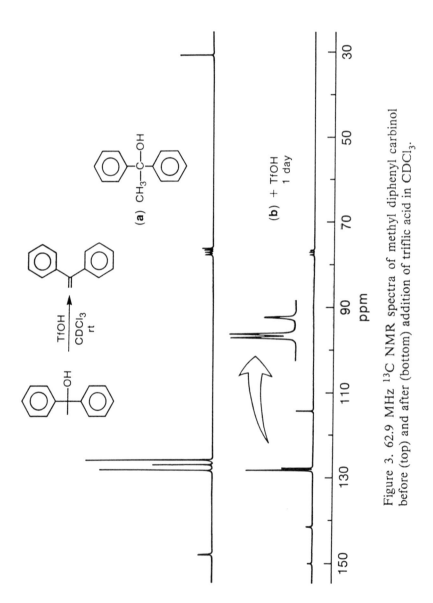

Figure 3. 62.9 MHz ^{13}C NMR spectra of methyl diphenyl carbinol before (top) and after (bottom) addition of triflic acid in CDCl$_3$.

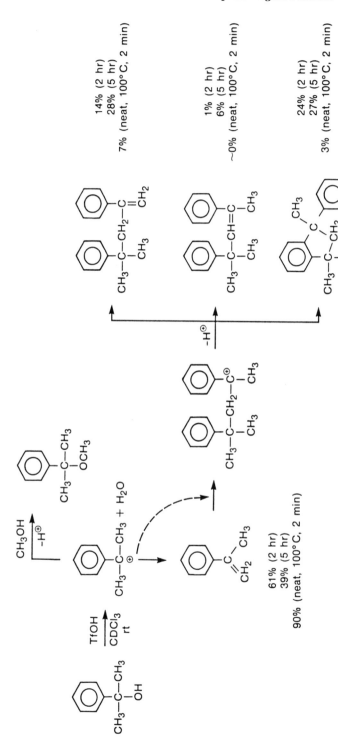

Scheme IV. Pathways in acid-catalyzed dehydration of dimethyl phenyl carbinol.

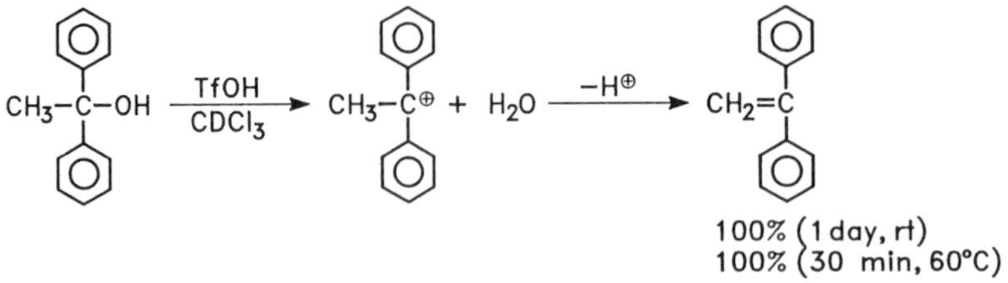

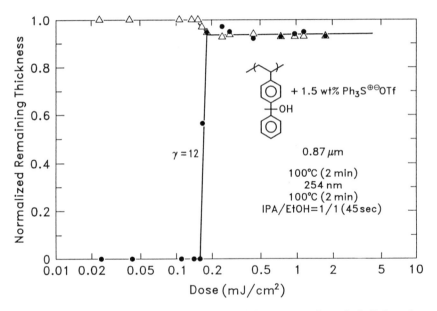

Scheme V. Acid-catalyzed intramolecular dehydration of methyl diphenyl carbinol.

Figure 4. Negative-tone deep UV sensitivity curve of methyl diphenyl carbinol resist containing 1.5 wt% $Ph_3S^+{}^-OTf$; film thickness measured after postbake ($\triangle$) and after development ($\bullet$).

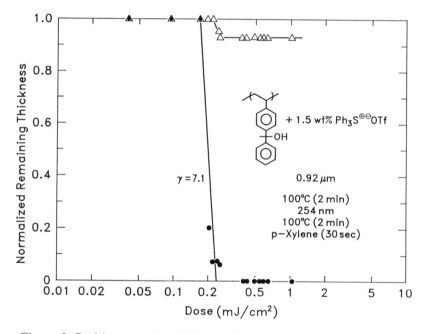

Figure 5. Positive-tone deep UV sensitivity curve of methyl diphenyl carbinol resist containing 1.5 wt% Ph_3S^+OTf; thickness measured after postbake ($\triangle$) and after development ($\bullet$).

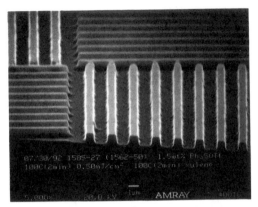

(a) positive, 0.50 mJ/cm^2, p-Xylene (30 sec)

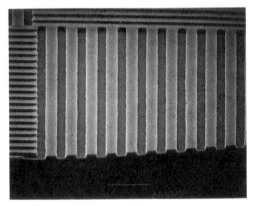

(b) negative, 0.41 mJ/cm^2, IPA/EtOH = 1/1 (45 sec)

Figure 6. Scanning electron micrographs of positive (a) and negative (b) images contact-printed in methyl diphenyl carbinol resist.

Aqueous Base Developable Negative Resists. In order to make use of the deep-UV-transparent and base-soluble PHOST structure, we decided to take a two-component approach rather than a three-component (blend) approach in the design of aqueous base developable systems. To prepare our resist resins, a protected HOST was radically copolymerized with a monomer carrying a pendant functionality that undergoes acid-catalyzed dehydration, followed by deprotection to unmask the phenolic OH functionality.

Polarity Change through Pinacol Rearrangement. As mentioned earlier, pinacol rearrangement of small *vic*-diol in a novolac resin to generate dissolution inhibiting ketone or aldehyde has been successfully employed to design aqueous base developable negative resists (*6,8*). However, PHOST, which is highly transparent at 248 nm and is therefore more attractive as a deep UV matrix resin, performed only marginally in the three-component design (*6*). Furthermore, 3-methyl-2-phenyl-2,3-butanediol, the model compound of the polymeric pinacol in Scheme I, was too volatile although it functioned as a dissolution promoter of PHOST (*6*). Incorporation of such a structure in the polymer chain by copolymerization could overcome the volatility problem.

3-Methyl-2-(4-vinylphenyl)-2,3-butanediol was copolymerized with ACOST (feed ratio = 20/80) using AIBN in THF at 60 °C. The ACOST copolymer with $M_n = 30,200$ and $M_w = 51,500$ was cleanly hydrolyzed with ammonium hydroxide in methanol without complication to afford the desired pinacol/HOST copolymer (17/83, $M_n = 25,900$ and $M_w = 42,400$) (Scheme VI). Thermal deprotection at 170 °C of a BOCST copolymer resulted in a slightly crosslinked HOST copolymer.

The pinacol/HOST copolymer and 5.0 wt% of triphenylsulfonium triflate were dissolved in cyclohexanone. The spin-cast films were prebaked at 100 °C for 2 min (OD = 0.366/μm at 248 nm), exposed to 254 nm radiation, postbaked at 100 °C for 2 min, and developed with MF319 for 45-60 sec. The deep UV sensitivity curve presented in Figure 7 indicates that the resist film does not exhibit any thinning upon postbake ($\triangle$) but becomes totally insoluble in MF319 for at least 60 sec at ca. 1 mJ/cm^2 with γ of 4.1. A scanning electron micrograph of negative images contact-printed in the copolymer resist at 4.5 mJ/cm^2 using MF319 as the developer is presented in Figure 8.

In order to elucidate the imaging mechanism and to examine whether or not the phenolic functionality participates in the reaction, we carried out model reactions of a 1:1 mixture of 3-methyl-2-phenyl-2,3-butanediol and 4-isopropylphenol with triflic acid in CDCl$_3$. The ^{13}C NMR spectra in Figure 9 clearly indicate that the *vic*-diol is cleanly and quantitatively converted to 3-methyl-3-phenyl-2-butanone as evident from the disappearance of the resonances at 75.7 and 78.7 ppm due to the quaternary carbons of the diol and appearance of a quaternary and carbonyl carbon resonances at 52 and 232 ppm. The resonances due to 4-isopropylphenol remained unchanged. The ketone product was isolated by column chromatography from a similar reaction mixture and spectroscopically identified. It is interesting to note that

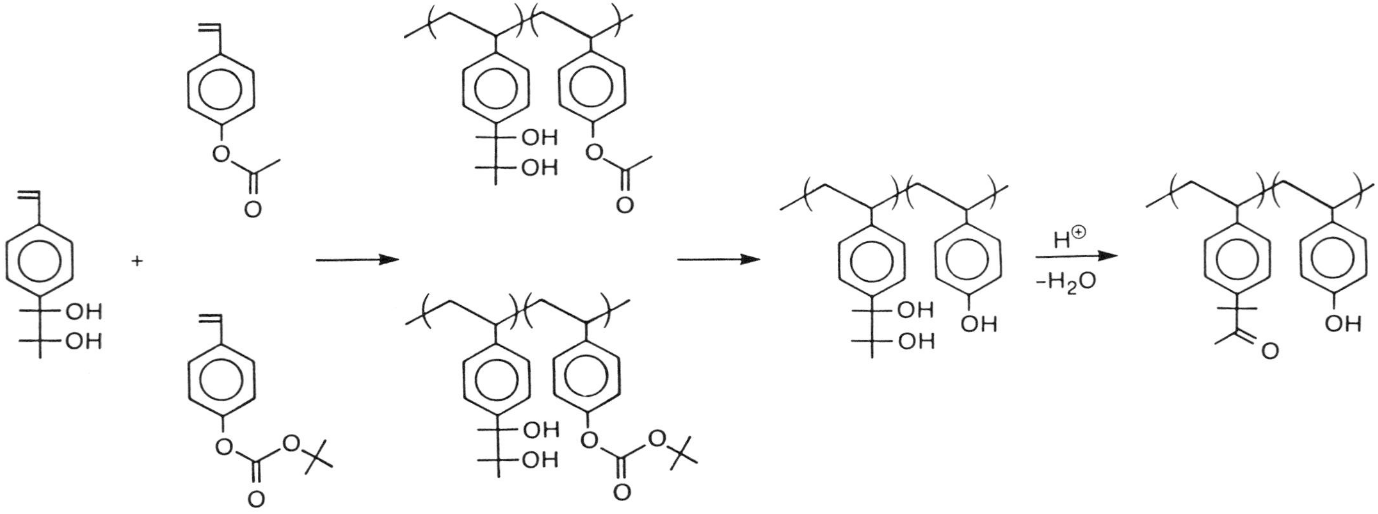

Scheme VI. Preparation of HOST copolymer bearing *vic*-diol and its pinacol rearrangement.

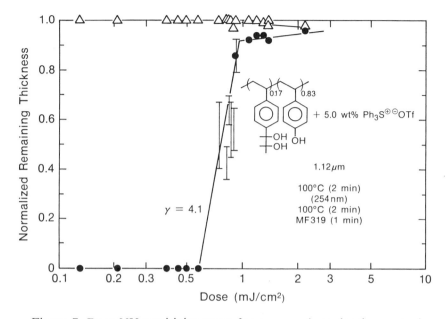

Figure 7. Deep UV sensitivity curve for aqueous base development of HOST copolymer resist bearing *vic*-diol; film thickness measured after postbake ($\triangle$) and after development ($\bullet$).

Figure 8. Scanning electron micrograph of negative images contact-printed at 4.5mJ/cm^2 in pinacol/HOST copolymer resist containing 5.0 wt% Ph$_3$S$^+$-OTf.

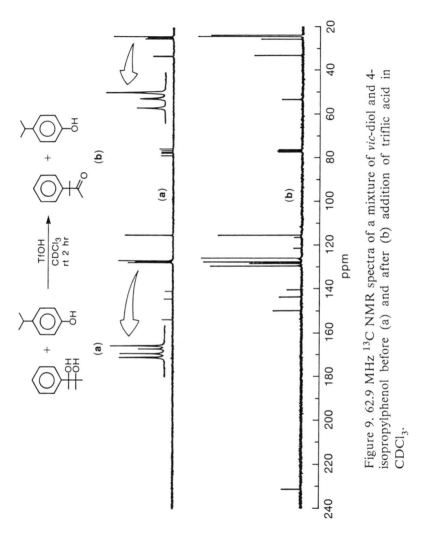

Figure 9. 62.9 MHz ^{13}C NMR spectra of a mixture of *vic*-diol and 4-isopropylphenol before (a) and after (b) addition of triflic acid in CDCl$_3$.

the ketone mixed with 4-isopropylphenol exhibits its carbonyl resonance at a lower field by as much as 20 ppm than the isolated ketone in $CDCl_3$ (212 ppm), indicating a strong hydrogen bonding interaction between the ketone and the phenol in $CDCl_3$. Thus, it is clear that the primary mechanism of the negative imaging of the pinacol/HOST copolymer resist is generation of a dissolution inhibiting ketone *via* polarity change by pinacol rearrangement without involving the phenolic functionality.

Intermolecular Dehydration (Condensation). 4-(1-Hydroxyethyl)styrene, a styrene derivative bearing a *secondary* alcohol structure, was copolymerized with ACOST (feed ratio = 20/80) and the ACOST copolymer ($M_n = 16,400$, $M_w = 30,900$, and $M_w/M_n = 1.88$) was cleanly and quantitatively converted to a *sec*-alcohol/HOST (15/85) copolymer with $M_n = 13,500$, $M_w = 23,500$, and $M_w/M_n = 1.74$ by hydrolysis with ammonium hydroxide in methanol (Scheme VII). In the case of 4-(2-hydroxy-2-propyl)styrene bearing pendant *tertiary* alcohol, a BOCST copolymer could not be deprotected without crosslinking either by thermolysis at 180 °C, by refluxing in glacial acetic acid, or with hydrazine (Scheme VII). Its copolymer with ACOST ($M_n = 30,100$, $M_w = 43,000$, and $M_w/M_n = 1.43$) was cleanly hydrolyzed with ammonium hydroxide in methanol to afford a soluble HOST copolymer (a tiny amount of an insoluble fraction) with $M_n = 27,700$, $M_w = 40,900$, and $M_w/M_n = 1.48$. However, the copolymer solution containing 5.0 wt% of triphenylsulfonium triflate in cyclohexanone could not be filtered.

The *sec*-alcohol/HOST copolymer and 5.0 wt% of triphenylsulfonium triflate were dissolved in cyclohexanone. The spin-cast films were prebaked at 100 °C for 2 min (OD = 0.439/μm at 248 nm), exposed to 254 nm radiation, postbaked at 100 °C for 2 min, and developed with MF319 for 90 sec or with MF321 for 105 sec. Figure 10 exhibits deep UV sensitivity curves for the *sec*-alcohol system. The film thickness was measured after postbake (△) and after development with MF319 for 90 sec (●). The resist film does not exhibit any thinning upon postbake as is the case with the homopolymer resist (7) but achieves retention of its full thickness at ca. 0.9 mJ/cm^2 with γ of 7.2. In Figure 11 is presented a scanning electron micrograph of 1-μm line/space negative images contact-printed at 1.1 mJ/cm^2 in the *sec*-alcohol copolymer resist by development with MF321 for 105 sec.

In the case of the *sec*-alcohol homopolymer resist, crosslinking *via* intermolecular dehydration (condensation) to form a di(α-methylbenzyl) ether structure (an enantiomeric mixture) is the primary imaging mechanism (7). Model reactions were carried out by treating a 1:1 mixture of α-methylbenzyl alcohol and 4-isopropylphenol with a drop of triflic acid in $CDCl_3$ or $CHCl_3$. The NMR spectra of the reaction mixture was quite complex, with significant amounts of α-methylbenzyl alcohol and 4-isopropylphenol remaining unreacted, but allowed us to identify three products. The self-condensation product was minor and the major product after 6 hrs at room temperature was α-methylbenzyl 4-isopropylphenyl ether. A C-alkylation product (a) and

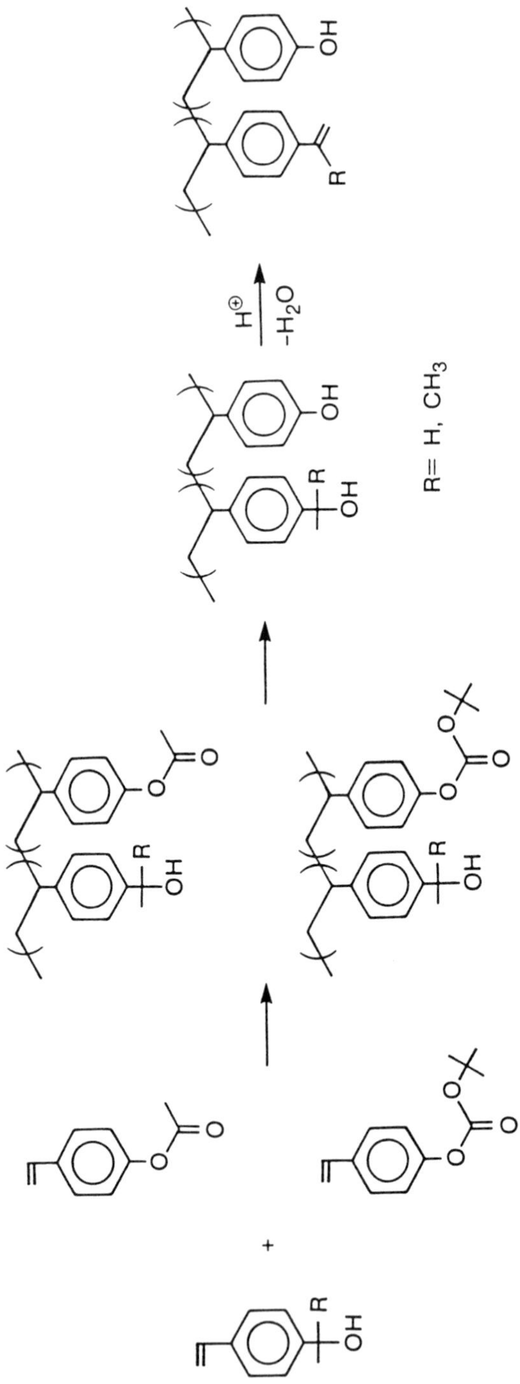

Scheme VII. Preparation of HOST copolymer bearing pendant *tertiary* and *secondary* alcohol.

R= H, CH₃

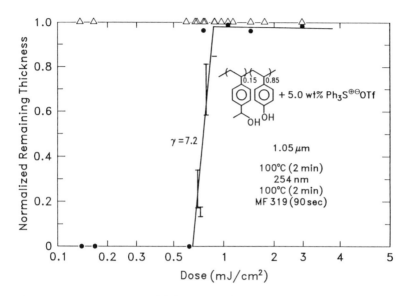

Figure 10. Deep UV sensitivity curve for aqueous base development of HOST copolymer resist bearing pendant *sec*-alcohol; film thickness measured after postbake ($\triangle$) and after development ($\bullet$).

Figure 11. Scanning electron micrograph of negative images contact-printed at 1.1 mJ/cm^2 in *sec*-alcohol/HOST copolymer resist containing 5.0 wt% Ph$_3$S$^+$OTf.

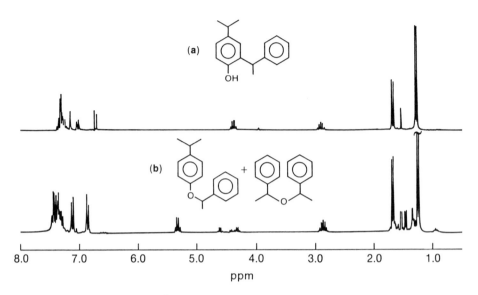

Figure 12. 250 MHz ^{1}H NMR spectra in CDCl$_3$ of C-alkylation product (a) and O-alkylation products (b) isolated from a mixture of α-methylbenzyl alcohol and 4-isopropylphenol treated with triflic acid by column chromatography.

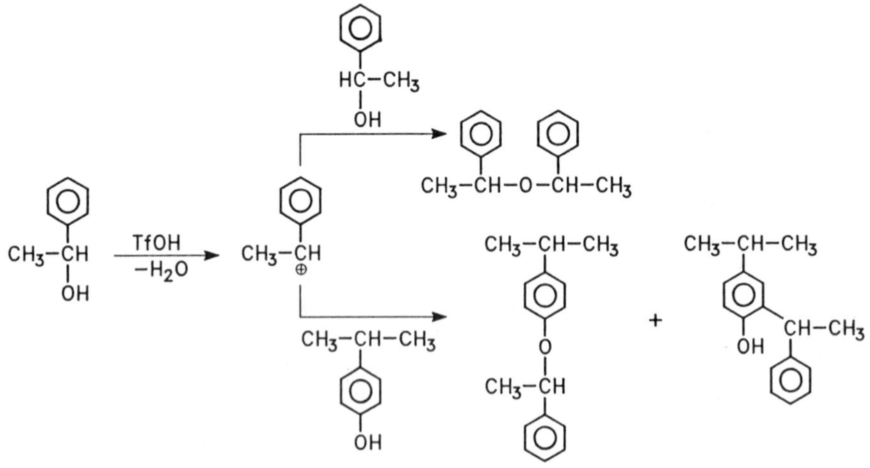

Scheme VIII. Acid-catalyzed intermolecular dehydration of α-methylbenzyl alcohol and 4-isopropylphenol.

a mixture of *O*-alkylation products (ethers) (b) were isolated by subjecting the reaction mixtures to column chromatography and their ^{1}H NMR spectra are presented in Figure 12. The ratio of the self-condensation to the cross-condensation was 1/3 in the ether formation (Figure 12b). When the mixture of α-methylbenzyl alcohol and 4-isopropylphenol was treated with triflic acid at 60 °C, the *C*-alkylation was predominant, with a very small amount of a styrene structure and no phenyl ether produced. Thus, as summarized in Scheme VIII, the carbocation produced by slow dehydration of α-methylbenzyl alcohol reacts with *sec*-alcohol to form di(α-methylbenzyl) ether or undergoes *C*- and *O*-alkylation on to the phenolic group, which all contribute to crosslinking if the *sec*-alcohol and phenol groups are pendant from a polymer chain. Since the model reaction suggests that the carbocation prefers the phenolic group over the alcohol functionality and because the concentration of the phenol group is much higher than the alcohol concentration in the copolymer resist (85/15), the cross-condensation is the most likely mechanism for the negative imaging.

Summary

1. Acid-catalyzed intramolecular dehydration of a pendant methyl diphenyl carbinol structure has provided the first and only dual tone resist that can be developed in a positive mode with a nonpolar solvent and in a negative mode with a polar solvent. The dual tone imaging is based on a polarity change from a polar alcohol structure to a nonpolar olefin structure without crosslinking.
2. A HOST copolymer bearing a pendant *vic*-diol functionality provides an aqueous base developable negative resist due to acid-catalyzed pinacol rearrangement of the diol group to a less polar dissolution inhibiting ketone group. This polarity change involves intramolecular dehydration.
3. Intermolecular dehydration (condensation) of a HOST copolymer bearing a pendant *sec*-alcohol functionality results in self- and cross-*O*-alkylations and *C*-alkylation of the phenol group, allowing negative imaging with aqueous base due to crosslinking.

Acknowledgments

The authors thank G. May for his NMR maintenance and H. Truong for her GPC and thermal analyses.

Literature Cited

1. Ito, H.; Willson, C. G. *Polym. Eng. Sci.* **1983**, *23*, 1012.
2. Ito, H.; Willson, C. G. In *Polymers in Electronics*; Davidson, T., Ed.; Symposium Series 241; American Chemical Society: Washington, D. C., 1984, pp 11-23.

3. MacDonald, S. A.; Schlosser, H.; Ito, H.; Clecak, N. J.; Willson, C. G. *Chem. Mater.* **1991**, *3*, 435.
4. Ito, H. In *Radiation Curing in Polymer Science and Technology*; Fouassier, J. P.; Rabek, J. F., Eds.; Elsevier: London, 1993, Vol. 4, Chapter 11, pp 237-359.
5. Ito, H. In *Irradiation of Polymeric Materials*; Reichmanis, E.; Frank, C. W., O'Donnell, J. H., Eds.; Symposium Series 527; American Chemical Society: Washington, D. C., 1993, pp 197-223.
6. Ito, H.; Sooriyakumaran, R.; Mash, E. A. *J. Photopolym. Sci. Technol.* **1991**, *4*, 319.
7. Ito, H.; Maekawa, Y.; Sooriyakumaran, R.; Mash, E. A. In *Polymers for Microelectronics*; Thompson, L. F.; Willson, C. G.; Tagawa, S., Eds.; Symposium Series 537; American Chemical Society: Washington, D. C., 1994, pp 64-87.
8. Uchino, S.; Iwayanagi, T.; Ueno, T.; Hayashi, N. *Proc. SPIE* **1991**, *1466*, 429.
9. McKean, D. M.; Parrinelo, G.; Renaldo, A. F.; Stille, J. K. *J. Org. Chem.* **1987**, *52*, 422.
10. Miller, R. D.; Renaldo, A. F.; Ito, H. *J. Org. Chem.* **1988**, *53*, 5571.

RECEIVED June 28, 1994

Chapter 7

Importance of Donor–Acceptor Reactions for the Photogeneration of Acid in Chemically Amplified Resists

Nigel P. Hacker

Research Division, IBM Almaden Research Center, 650 Harry Road, San Jose, CA 95120–6099

The photophysical interactions between the polymer and photoinitiator in resist systems can play an important role in the photogeneration of acid in chemically amplified resists. If there is no photophysical interaction acid is generated by direct photodecomposition, but if the polymer can act as an excited state electron donor an electron transfer can occur. Fluorescence spectroscopy and photoproduct analyses are used as mechanistic process for electron transfer reactions.

The invention of new photoresists using photochemically generated acid to catalyze reactions in polymer films has traditionally been a vertical process. The process involves the synthesis of a new polymer that can undergo an acid catalyzed chemical reaction. A requirement of the acid catalyzed reaction is a change in polarity of the polymer, e.g. a hydrophobic to hydrophilic reaction, that renders the polymer more soluble in the development solvent. Once the polymer dissolution properties are optimized to give good contrast, i.e. is good dissolution inhibitor before, and gives enhanced dissolution after the acid catalyzed process, the cationic photoinitiator is added. The role of the cationic photoinitiator is to make the resist system photosensitive. The chemistry of the cationic photoinitiator was considered to be innocuous because the photoinitiator concentration is low, c.a. 1-2 wt %. For example the changes in dissolution properties of a formulation that may be caused by a chemical change to an additive at 1 wt % loading in a high contrast

0097–6156/94/0579–0093$08.00/0

resist, are expected to be minimal. Although the role of the photoinitiator for changing the wall profiles of sub-micron features of resists has been observed, the mechanism is not well understood. It has always been appreciated, however, that the UV absorption of the photoinitiator should be optimized such that the initiator absorbs all of the incident light. Also under these conditions the polymer should have zero absorbance. In most deep UV resists the polymer has aromatic pendent groups that absorb some of the incident light and it is generally assumed that light absorption by the polymer is wasted and thus is detrimental to resist performance. In this paper the photochemical and photophysical properties of the polymer and photoinitiator will be described and the consequences of these properties on the photoinitiation reaction will be discussed. The goal is to better understand photoinitiation processes in chemically amplified resists and clarify the roles played by both polymer and photoinitiator. This knowledge, coupled with learning physical properties, e.g. dissolution properties, solubility and photophysics, of the individual components should lead to a horizontal approach for the development of new photoresists.

Photochemistry

The solution photochemistry of triarylsulfonium salts has been studied in detail. Direct photolysis of triphenylsulfonium salts (TPS) gives 2-, 3- and 4-phenylthiobiphenyls (PTB), diphenylsulfide, and acid. (1 - 4) The PTB's are formed by in-cage fragmentation-recombination reactions, whereas diphenylsulfide formation is a cage-escape process. The general trend that the yield of PTB isomers increased relative to diphenylsulfide in more viscous solvents, presented strong evidence for the cage-versus-escape reactivity from direct photolysis of TPS. Accompanying the change in relative yields of PTB's is a decrease in quantum yield in more viscous solvents. This is because a process that regenerates sulfonium salt, fragmentation followed by recombination on sulfur, becomes predominant. Figure 1 plots the decrease in relative quantum yield and the increase in ratio of PTB's : diphenylsulfide versus increasing viscosity. Also the viscosity effect is observed in polymers where it was found that the ratio of PTB's : diphenylsulfide were higher in films of poly(methyl methacrylate) and poly(vinyl alcohol) than in non-viscous solutions. (3)

Triphenylsulfonium salts have a triplet energy of 74 kcal mole^{-1} and undergo triplet sensitized photolysis to give diphenylsulfide, benzene and acid by a cage-escape reaction. (5) In contrast to direct photolysis, which occurs

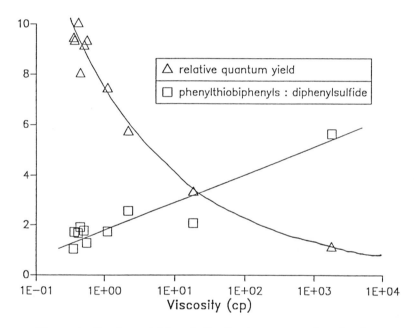

Figure 1. Product yield and distribution versus viscosity from photolysis of triphenylsulfonium salts at $\lambda = 254$ nm.

by initial heterolysis of the carbon-sulfur bond, the triplet sensitized reaction proceeds by homolytic cleavage to give the phenyl radical - diphenylsulfinyl radical cation, triplet radical pair which react with solvent to give escape products faster than undergoing spin inversion to the singlet radical pair for recombination. Triphenylsulfonium salts are good electron acceptors (E_{red} = -1.2 V) and react with excited state electron donors by an electron transfer reaction. For example the singlet excited state for anthracene reacts with TPS to give anthracene radical cation and triphenylsulfur radical, this pair of intermediates fragments to give the triad of anthracene radical cation, phenyl radical and diphenylsulfide. (6) Acid is produced from the latter triad by an in-cage recombination reaction to give 1-, 2- and 9-phenylanthracenes or by reaction with solvent to give cage-escape products. Triplet energy transfer from the excited state of anthracene will not occur because the energy of the triplet excited state of anthracene, 42 kcal mole^{-1}, is too low for energy transfer. Electron transfer from the singlet excited state of anthracene is exothermic by 23 kcal mole^{-1}. Further evidence for electron transfer from the singlet excited state of anthracene comes from fluorescence spectroscopy. Figure 2 shows the effect on the fluorescence spectrum of anthracene by adding triphenylsulfonium salt. The wavelengths of the anthracene fluorescence peaks are not shifted but do decrease in intensity upon addition of sulfonium salt. A Stern-Volmer plot of this fluorescence quenching reveals that the reaction occurs at close to the diffusion controlled rate in acetonitrile solution. In contrast to the triplet energy sensitized reaction, the electron transfer reaction gives diphenylsulfide as a by-product from both in-cage recombination and cage-escape reactions.

Evidence for the intermediates described from the above product studies for the direct, triplet sensitized and photo-induced electron transfer reactions of TPS salts has been also presented using photo-CIDNP and nanosecond flash photolysis techniques. (7)

Interaction between Polymer and Photoinitiator

The photochemistry of TPS in poly[4-[(tert-butoxycarbonyl)oxy] styrene] (poly-TBOC) was studied to better understand the photochemistry of chemically-amplified resists. (8, 9) From detailed product studies in solution and in films, it was determined that the polymer was sensitizing the photodecomposition of TPS because considerably less PTB relative to diphenylsulfide was produced in the presence of poly-TBOC. Table 1 shows the effects the polymers can have on the yields of PTB's and diphenylsulfide

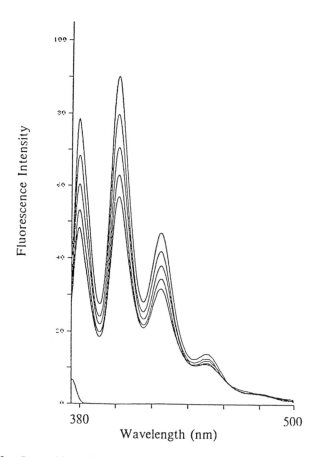

Figure 2. Quenching of anthracene fluorescence by triphenylsulfonium salt.

Table 1: Photoproduct Distribution from Irradiation of Triphenylsulfonium Salts (Concentration x 10^5 M)

		Ph_2S	Ph-PhSPh	Cage/Escape
	$\lambda = 254$ nm			
1	1.0%, TBOC, Film	0.95	1.95	2.04
2	1.0%, PMMA, Film	1.92	6.75	3.51
3	1.0%, TBOC, CH_3CN	13.24	10.42	0.79
4	10.0%, TBOC, CH_3CN	110.9	82.8	0.75
5	0.01M, CH_3CN	121.5	137.3	1.13
6	0.01M, CH_3CN + 0.1M Anisole	221.4	75.0	0.34
	$\lambda = 300$ nm			
7	1.0%, TBOC, Film	0.70	0.55	0.78
8	1.0%, TBOC, CH_3CN	1.28	trace	----
9	10.0%, TBOC, CH_3CN	8.96	0.37	0.04
10	0.01M, CH_3CN	3.70	1.29	0.35
11	0.01M, CH_3CN + 0.1M Anisole	14.43	0.79	0.06

in solution and the solid state. Entries 1 and 2 show that relatively more PTB's are formed after photolysis of TPS salts at 254 nm in poly(methylmethacrylate) (PMMA) than in poly-TBOC. The change in PTB formation may be due to a sensitization process which occurs in poly-TBOC and not PMMA. Alternatively if the microviscosity of PMMA is larger than poly-TBOC, then a similar trend is expected for the sulfide photoproducts. To eliminate the microviscosity effect, TPS / poly-TBOC formulations were dissolved in acetonitrile and irradiated. Entries 3 and 4 show that the relative yield of PTB's markedly decrease to less than from direct photolysis of TPS in the absence of polymer (entry 5). These results suggest that a sensitization reaction is occurring. To further probe the sensitization reaction, TPS formulations were irradiated at 300 nm where the polymer absorbs more of the incident light than the photoinitiator. The presence of poly-TBOC significantly lowers the relative yield of PTB's (entries 7-10) in both films and solution. For example the ratio of PTB : diphenylsulfide (entry 9) is 0.04 in the presence of poly-TBOC whereas without the polymer the ratio is 0.35 under similar conditions (entry 10). Experiments using anisole (entries 6 and 11) seem to mimic the poly-TBOC influence on TPS photochemistry in solution, suggesting anisole is a good candidate as a model monomeric compound for poly-TBOC.

If anisole is used as a model, energy transfer from the triplet excited state of anisole (E_T = 80.8 kcal mole^{-1}) to TPS (E_T = 74 kcal mole^{-1}) is viable from energetic arguments. Similarly an electron transfer reaction from either the singlet or triplet excited state of anisole to TPS (- 44 and -22 kcal mole^{-1} respectively) is also thermodynamically favorable. To determine the nature of the sensitization reaction the fluorescence spectroscopy of the polymers was studied. Figures 3 and 4 show the fluorescence spectra of a number of substituted poly(styrenes) in solution. All of these polymers emit in the 280 - 450 nm region. In particular poly(4-hydroxystyrene) (poly-HOST), poly(4-methoxystyrene) (poly-MOST), and poly-TBOC all fluoresce in the 300 - 350 nm region in both solution and as films. Addition of TPS to solutions of 4-oxystyrene polymers does not shift the emission peaks but results in a decrease in the emission intensities. In solution the fluorescence from these polymers is quenched by TPS to give linear Stern-Volmer plots (Figure 5). This quenching is at close to diffusion controlled rate based on the lifetimes of model monomers and an experimentally obtained value for poly-TBOC (Table 2).

Table 2: Quenching constants for TPS in 4-oxystyrene polymers

Polymer	k_qT (M^{-1})	
	Gradient[a]	Estimate
poly-MOST	166	166[b]
poly-HOST	80	40 - 148[c]
poly-TBOC	90	50[d]

a. from Stern-Volmer plots ($\Phi_0 / \Phi = I_0 / I = 1 + k_qT[Q]$ (for acetonitrile K_q = 2 x 10^{10} L s^{-1}). b. for anisole, a model monomer, T = 8.3 ns. c. for phenol, a model monomer, T = 2.1 - 7.4 ns. d. for poly-TBOC, T = 2.5 ns.

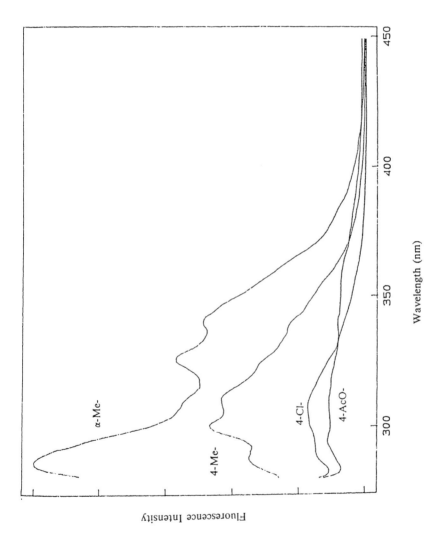

Figure 3. Fluorescence spectra of α-methyl-, 4-methyl-, 4-chloro- and 4-acetoxy- styrene polymers.

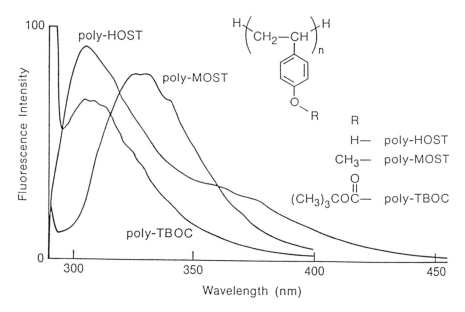

Figure 4. Fluorescence spectra of poly(4-oxystyrene) derivatives.

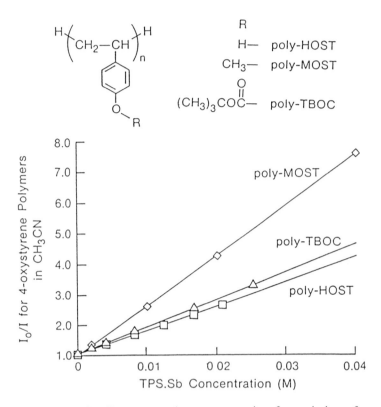

Figure 5. Plot of I_0/I versus molar concentration for emission of poly(4-oxystyrene) derivatives in acetonitrile solution in the presence of triphenylsulfonium salt.

$$P + Ph_3S^+X^- \xrightarrow{h\nu} [P]^* + Ph_3S^+X^- + [Ph_3S^+X^-]^* + P$$

$$[Ph_3S^+X^-]^* \longrightarrow PhPhSPh + Ph_2S + HX$$

$$[P]^* + Ph_3S^+X^- \longrightarrow P^{+\cdot} + Ph_3S^\cdot + X^-$$

$$P^{+\cdot} + Ph_3S^\cdot + X^- \longrightarrow P^{+\cdot} + Ph^\cdot + Ph_2S + X^-$$

$$P^{+\cdot} + Ph^\cdot + X^- \longrightarrow P\text{-}Ph + HX$$

where P = poly[4-[(tert-butoxycarbonyl)oxy]styrene]

Figure 6. Dual photoinitiation mechanism for triphenylsulfonium salts in poly-TBOC resist.

The results from fluorescence spectroscopy indicate that the sensitization reaction is a photoinduced electron transfer reaction from the singlet excited state of the polymer to TPS salts. However the photoproduct studies show that PTB's, products expected from direct photolysis of TPS salts, are formed, albeit in lower than expected yields. The combined photoproduct and fluorescence spectroscopy studies suggest that acid formed by both direct photolysis and photoinduced electron transfer reaction in the poly-TBOC/TPS resist and a Dual Photoinitiation Mechanism (DPM) is proposed for acid formation (Figure 6). At 1 - 10 wt % TPS loadings both the polymer and the photoinitiator absorb the incident light. The light absorbed by TPS generates acid by the direct photolysis mechanism whereas the light absorbed by the polymer also produces acid by an electron transfer reaction from singlet excited state of poly-TBOC to TPS. The latter reaction occurs by mechanism similar to the reaction described for anthracene sensitization of TPS salts.

Excited State Polymer and Ground State Photoinitiator Reactions

The ability of the polymer to generate acid by a photoinduced electron transfer reaction with the cationic photoinitiator is not exclusive to TPS derivatives. Figure 7 shows the UV absorption spectra of non-ionic photo-acid generators. The concentrations of each initiator are adjusted for maximum absorbance. The pyrogallol and succinimidoyl derivatives have extinction coefficients of 10^2 M^{-1} at 250 nm, about an order of magnitude less than substituted polystyrenes ($\varepsilon = 10^3$ M^{-1} cm^{-1}). As the photoacid generator is only 1-2 wt % of the formulation in chemically-amplified systems, it can be concluded that the polymer absorbs 99 % of the incident light in resists formulated with these initiators. The Stern-Volmer plots for quenching of poly(4-hydroxystyrene) fluorescence by non-ionic photoinitiators are shown in Figure 8. The pyrogallol sulfonate derivative exhibits similar quenching behavior to TPS salts, i.e. quenching is close to the diffusion controlled rate. While the succinimidoyl derivative is 2-3 times slower than TPS salts, substituted derivatives of these imides can exhibit more efficient quenching. The relatively weak absorbances of these photoinitiators in the deep UV and their ability to efficiently quench the polymer fluorescence suggest that the polymer absorbs the incident photon and sensitizes the decomposition of the initiator. Kasai has reported that excitation of the D-line of sodium atoms in argon matrices with imidoyl triflates results in a dissociative electron transfer reaction. (*10*) A similar reaction occurs in resists. The singlet excited state of the polymer donates an electron to the photoinitiator which dissociates and gen-

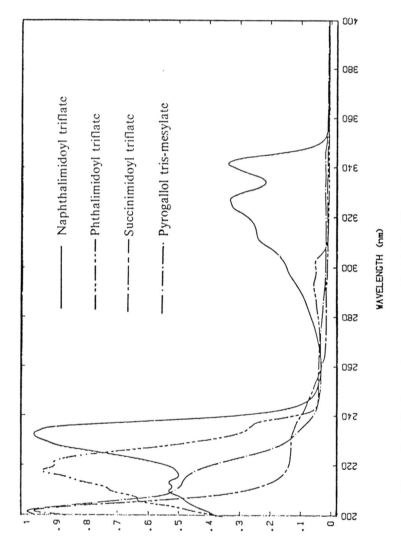

Figure 7. UV absorption spectra of imidoyl triflates and pyrogallol tris-mesylate.

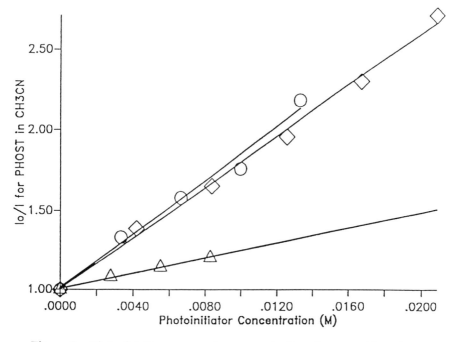

Figure 8. Plot of I_0/I versus molar concentration for quenching the fluorescence from poly(4-hydroxystyrene) in acetonitrile solution in the presence of cationic photoinitiators.

erates acid. There is no direct photolysis component, which is observed with TPS salts, because of the weak absorbances of the non-ionic photoinitiators at 250 nm. This rationalizes the observation of quantum yields > 10 for pyrogallol sulfonate derivatives resist formulations by Ueno who concluded that a sensitization reaction must occur. (*11*)

Ground State Polymer and Excited State Photoinitiator Reactions

The UV spectrum of the naphthalimidoyl triflate shows a long wavelength absorbance at 300 - 360 nm (Figure 7). The fluorescence spectrum at room temperature and phosphorescence spectrum at 77 K are shown in Figure 9. The observation of fluorescence from a photoinitiator at room temperature is unusual because most photoactive compounds e.g. TPS salts photodecompose rather than fluoresce. However if phenol is added to acetonitrile solutions of naphthalimide triflate, the fluorescence intensity decreases (Figure 10). If the quenching is diffusion controlled, a fluorescence lifetime of 0.75 ns is estimated from the Stern-Volmer plot. It is known that naphthalimide triflate is good electron acceptor and undergoes dissociative electron transfer reactions in the presence of an electron donor. (*10*) If phenol is considered as a model compound for poly-HOST, it is proposed that the excited state of naphthalimide photoinitiator accepts an electron from the ground state polymer and dissociates to generate acid. Wallraff has reported that naphthalimide triflate is a poor photoacid generator in poly(alkyl acrylate) resists relative to onium salts. (*12*) Acrylates are poor electron donors and thus will not sensitize the photodecomposition of photoinitiators. Onium salts decompose by the direct photolysis mechanism in acrylate polymers and do not require the presence of an electron donor to generate acid.

Conclusion

From these studies it can be concluded that light absorption by the polymer can result in acid formation from the photoinitiator in chemically amplified resists. Onium salts are capable of generating acid by direct absorption of light and also by electron transfer sensitization from the polymer excited state. Pyrogallol and succinimidoyl sulfonate derivatives are weakly absorbing and generate acid by a photoinduced electron transfer reaction from the singlet excited state of the polymer. In contrast while naphthalimidoyl sulfonates do absorb the incident light, photogeneration of acid occurs by an electron transfer of the polymer ground state to singlet excited state of the photoinitiator. From these studies it is concluded that the photochemistry

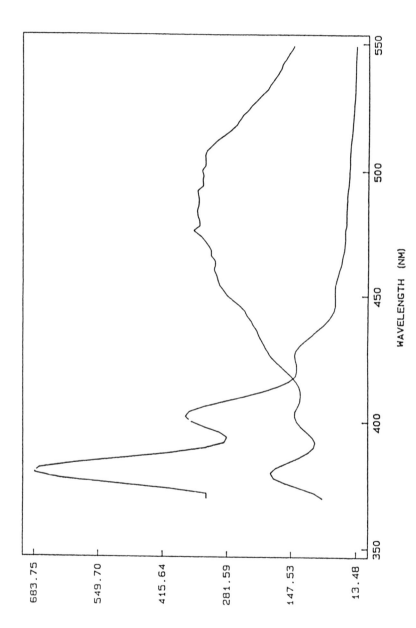

Figure 9. Fluorescence and phosphorescence spectra of naphthalimidoyl triflate in ethanol/methanol.

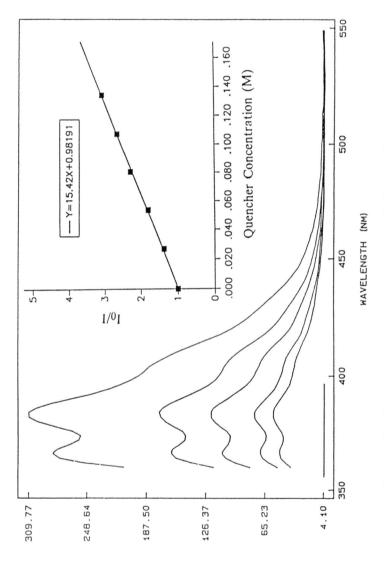

Figure 10. Fluorescence quenching of naphthalimidoyl triflate by phenol in acetonitrile (inset: Stern-Volmer plot).

and photophysics of both the polymer and cationic photoinitiator need to be optimized for maximum performance from a new photoresist design.

References

1. N. P. Hacker, in *New Aspects of Radiation Curing in Polymer Science and Technology, Volume 2*, J. P. Fouassier and J. F. Rabek, Eds. Elsevier Science, 1993, 473; N. P. Hacker, in *Photopolymers, Mechanisms, Development and Applications* A. D. Trifunac and V. V. Krongauz, Eds. Chapman and Hall, in press.
2. Dektar, J. L.; Hacker, N. P. *J. Chem. Soc., Chem. Commun.* **1987**, 1591.
3. Hacker, N. P.; Dektar, J. L. *Polym Prepr.* **1988**, *29*, 524.
4. Dektar, J. L.; Hacker, N. P. *J. Am. Chem. Soc.* **1990**, *112*, 6004.
5. Dektar, J. L.; Hacker, N. P. *J. Org. Chem.* **1988**, *53*, 1833.
6. Dektar, J. L. ; Hacker, N. P. *J. Photochem. Photobiol., A. Chem.*, **1989**, *46*, 233.
7. Welsh, K. M.; Dektar, J. L.; Garcia-Garibaya, M. A.; Hacker, N. P.; Turro, N. J. *J. Org. Chem.*, **1992**, *57*, 4179.
8. Hacker, N. P.; Welsh, K. M.; *Macromolecules* **1991**, *24*, 2137.
9. Hacker, N. P.; Welsh, K. M. *Structure-Property Relations in Polymers: Spectroscopy and Performance*, ACS Advances in Chemistry Series No. 236, Urban, M. W.; Claver, C. D. Eds.; American Chemical Society, Washington D. C. 1993, 557.
10. Kasai, P. H. *J. Am. Chem. Soc.* **1992**, *114*, 2875.
11. Schlegel, L.; Ueno, T.; Shiraishi, H; Hayashi, N.; Iwayanagi, T. *Chem. Mater.*, **1990**, *2*, 299.
12. Wallraff, G.; Allen, R.; Hinsberg, W.; Larson, C.; Johnson, R. DiPietro, R.; Breyta, G.; Hacker, N.; Kunz, R. R. *J. Vac. Sci. Technol. B*, **1993**, *11*, 2783.

RECEIVED September 13, 1994

Chapter 8

Acid Generation in Chemically Amplified Resist Films

Takeo Watanabe[1], Yoshio Yamashita[1], Takahiro Kozawa[2], Yoichi Yoshida[3], and Seiichi Tagawa[3]

[1]SORTEC Corporation, 16–1 Wadai, Tsukuba-shi, Ibaraki 300–42, Japan
[2]Nuclear Engineering Research Laboratory, Faculty of Engineering, University of Tokyo, 2–22 Sirakata-Sirane, Tokai-mura, Ibaraki 319–11, Japan
[3]Institute of Scientific and Industrial Research, Osaka University, 8–1 Mihogaoka, Ibaraki, Osaka 567, Japan

A new interpretation of radiation-induced acid generation processes in polymer films is reported. For the investigation of the acid generation in chemically amplified resist films we employed a model system. In order to analyze the acid generation process, we utilized the visible absorption characteristics from a conventional spectrophotometer and a nanosecond pulse radiolysis system. The acid generation mechanisms of triphenylsulfonium triflate in m-cresol novolac and p-cresol novolac systems are discussed on the basis of absorption spectra. The acid-catalyzed reaction during post-exposure bake is also discussed in terms of the absorption spectra of the m-cresol novolac system.

Chemically amplified resists (1) for X-ray and electron beam lithography can provide the high performance required for fabrication of LSI's of quarter-micron feature size and below. Many papers have reported the effects in chemically amplified resists such as the processing of prebake, development, post-exposure bake (PEB), and delay time between exposure and PEB (1-8). As their sensitivity and resolution strongly depend on acid generation upon exposure and subsequent acid diffusion during PEB, it is important to understand the behavior of acid in the resist films.

The acid photogeneration mechanisms of onium salts have been investigated (9,10). However, they have not been fully verified yet. Acid generation by photoinduced and radiation-induced reactions was studied in solution with several acid indicators. Quantum yields for acid generation have been determined from photolysis of onium salts in acetonitrile solution (11). For the analysis of acid generated by photoinduced and radiation-induced reactions in chemically amplified resists, an IBM group (12,13) employed the merocyanine dye technique (14). For the analysis of the radiation-induced mechanisms in chemically amplified resists, a University of Tokyo group used pyrene as a model compound for crosslinkers and dissolution inhibitors with the experiment on nanosecond pulse radiolysis (15,16).

The present paper describes a new interpretation of acid generation process by radiation-induced reaction in polymer films in the absence of acid indicators. In order to investigate the acid generation process, we utilized the visible absorption characteristics from a conventional spectrophotometer and a nanosecond pulse radiolysis system.

0097–6156/94/0579–0110$08.00/0

Experimental

Two kinds of model systems which were examined consisted of m-cresol or p-cresol novolac as the base resin. M-cresol novolac and p-cresol novolac, which were provided by Sumitomo Durez Co., have a weight average molecular weight (Mw) of 2,000 and that of 700, respectively. Each model system consisted of triphenylsulfonium triflate ($\phi_3SCF_3SO_3$) as the acid generator (PAG) and methyl 3-methoxypropionate (MMP) as the solvent. Triphenylsulfonium triflate was purchased from Midori Kagaku Co.. Methyl 3-methoxypropionate was purchased from Tokyo Ohka Co.. The content of the novolac resin was 40 wt.% of MMP, and the content of $\phi_3SCF_3SO_3$ was 5 wt.% of the novolac resin. Two-μm-thick films were coated on 3-inch-diameter quartz wafers. Prebake and PEB were carried out on a hot plate for 120 sec at temperatures of 120 °C and 110 °C, respectively. A SORTEC synchrotron radiation (SR) ring was used as the exposure source (17). Energy and critical wavelength are 1 GeV and 1.55 nm, respectively. The beamline has an oscillating Pt mirror and a 40-μm-thick Be window. The peak wavelength of SR irradiation through the mask membrane was 0.7 nm and the wavelength range was 0.2-1.5 nm. The exposures were carried out in helium at 1 atm.

For the films with and without PAG, absorption spectra before exposure, after exposure, and after PEB were recorded on a spectrophotometer, Hitachi model U-3410. It took about 1 min to set a wafer in spectrophotometer after exposure. Data collection time was about 5 min per absorption spectrum.

In order to investigate the irradiation effect in the time range of nanoseconds, an experiment on nanosecond pulse radiolysis was carried out. The nanosecond pulse radiolysis system for optical absorption spectroscopy is composed of a 28 MeV linac as an irradiation source, a Xe lamp as analyzing light, a monochromator, a Si photodiode, a Ge photodiode, and a transient digitizer. The width of an electron pulse was 2 nsec. The solid samples were exposed by an electron pulse with a current of 2 nC. The details of the nanosecond pulse radiolysis system were reported previously (18).

Results and discussion

For the m-cresol novolac system, absorption spectra of the films were measured before exposure, after exposure, and after PEB, in the time range of 1 min by the conventional spectrophotometer. For films without PAG, the spectra before exposure, after exposure with a dose of 540 mJ/cm^2, and after PEB are shown in Figure 1. The weak absorption peaks at wavelengths less than 400 nm may be products of decomposition of the m-cresol novolac by SR irradiation. The weak absorption became weaker after PEB.

For films with PAG, the spectra before exposure, after exposure with a dose of 540 mJ/cm^2, and after PEB are shown in Figure 2. There are no characteristic absorption peaks before exposure. After exposure there is a strong absorption peak at 542 nm. The exposure dose dependence of the absorbance difference at 542 nm before exposure and after exposure is shown in Figure 3. We can fit the curve with the equation

$$\Delta E = 4.29 \times 10^{-1} (1 - \exp(-7.77 \times 10^{-4} X))$$

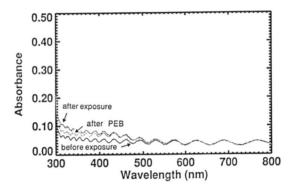

Figure 1: Absorption spectra before exposure, after exposure with a dose of 540 mJ/cm^2, and after PEB, for m-cresol novolac without PAG.

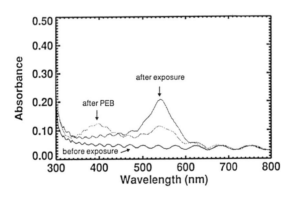

Figure 2: Absorption spectra before exposure, after exposure with a dose of 540 mJ/cm^2, and after PEB, for m-cresol novolac with PAG.

for 98 % confidence level (C.L.), where ΔE and X (mJ/cm^2) are absorbance difference and exposure dose, respectively. The equation indicates that a first-order reaction occurred by exposure.

Two 1.5-mm-thick films of m-cresol novolac without and with 30 wt.% PAG were prepared for the experiment on nanosecond pulse radiolysis. Absorption-time-dependent behaviors at 540 nm for the film with and without PAG are shown in Figure 4. Absorption at 540 nm for the film without PAG is weak and includes a large amount of noise, while absorption at 540 nm for the film with PAG is strong. Absorption-time-dependent behaviors at 540 nm and 700 nm for the film with PAG are shown in Figure 5. The decay at 540 nm is very slow, while the decay at 700 nm is very fast. The absorption at 700 nm may be due to the existence of cationic species of m-cresol novolac. For m-cresol novolac with PAG, the strong absorption near 542 nm may be due to the existence of a protonated intermediate. The protonated intermediate may be a product from the m-cresol novolac and acid (proton).

The spectra after exposure and after PEB, as shown in Figure 2, indicate that after PEB the protonated intermediate decomposed and then an absorption peak at 394 nm appeared. The exposure dose dependence of the absorbance difference at 542 nm after exposure and after PEB is shown in Figure 6. We can fit the curve with the equation

$$\Delta E = -2.83 \times 10^{-1} (1 - \exp(-6.49 \times 10^{-4} X))$$

for 99 % C.L.. The equation indicates that the decomposition by PEB of the protonated intermediate is a first-order reaction. The exposure dose dependence of the absorbance difference at 394 nm after exposure and after PEB is shown in Figure 7. We can fit the curve with the equation

$$\Delta E = 7.04 \times 10^{-2} (1 - \exp(-1.85 \times 10^{-3} X))$$

for 99 % C.L.. The equation indicates that the decomposition reaction by PEB of the product species is a first-order reaction. The protonated intermediate seemed to be decomposed by PEB and the absorption peak at 542 nm became weaker, while the absorption peak at 394 nm became stronger. The absorption spectrum of m-cresol novolac with 5 wt.% p-toluenesulfonic acid is shown in Figure 8. PEB was carried out on a hot plate for 120 sec at a temperature of 120 °C. An absorption peak near 394 nm appeared. Before PEB absorption peaks were not observed. Therefore during PEB, it could be assumed that protonated adducts of m-cresol novolac was formed by the attachment of a proton. Though it was previously considered that the acid is generated during exposure, comparing the absorption spectrum after PEB for m-cresol novolac which contains PAG with that after PEB for m-cresol novolac which contains p-toluenesulfonic acid, it can be assumed that the acid may be generated not only during exposure but also during PEB in the m-cresol novolac with PAG.

For the p-cresol novolac system, absorption spectra of the films were measured before exposure and after exposure, in the time range of 1 min by the conventional spectrophotometer. For the film without PAG, the spectra before exposure, and after exposure with a dose of 2160 mJ/cm^2 are shown in Figure 9. There are no strong absorption peaks. For the film with PAG, the spectra before exposure and after exposure with a dose of 2160 mJ/cm^2 are shown in Figure 10. There are no strong absorption peaks.

Two 2-mm-thick films of p-cresol novolac without and with 15 wt.% PAG were prepared for the experiment on nanosecond pulse radiolysis. Transient absorption spectra are shown in Figure 11. For the film without PAG, there are no strong absorption peaks in the time range of 450 nsec. The film with PAG was exposed at a

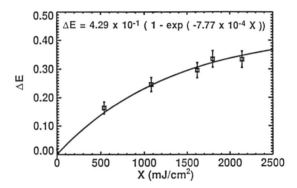

Figure 3: Exposure dose dependence of absorbance difference at 542 nm before and after exposure for m-cresol novolac with PAG.

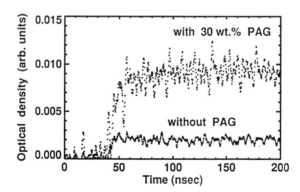

Figure 4: Absorption-time-dependent behaviors at 540 nm for m-cresol novolac with and without PAG by nanosecond pulse radiolysis.

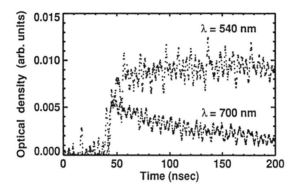

Figure 5: Absorption-time-dependent behaviors at 540 nm and at 700 nm for m-cresol novolac with PAG by nanosecond pulse radiolysis.

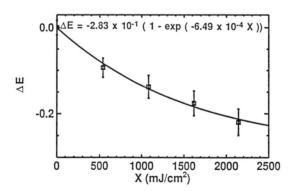

Figure 6: Exposure dose dependence of absorbance difference at 542 nm after exposure and after PEB for m-cresol novolac with PAG.

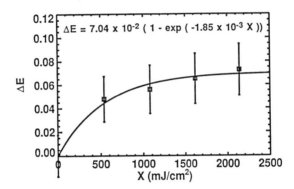

Figure 7: Exposure dose dependence of absorbance difference at 394 nm after exposure and after PEB for m-cresol novolac with PAG.

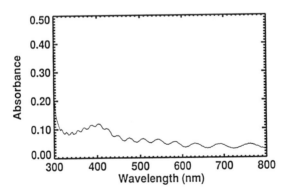

Figure 8: Absorption spectrum of m-cresol novolac with p-toluenesulfonic acid.

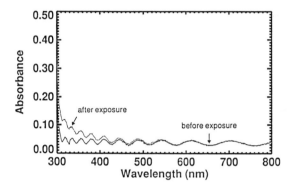

Figure 9: Absorption spectra before exposure and after exposure with a dose of 2160 mJ/cm^2 for p-cresol novolac without PAG.

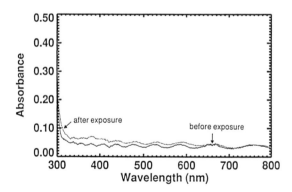

Figure 10: Absorption spectra before exposure and after exposure with a dose of 2160 mJ/cm^2 for p-cresol novolac with PAG.

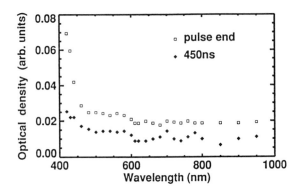

Figure 11: Transient absorption spectra for p-cresol novolac without PAG by nanosecond pulse radiolysis.

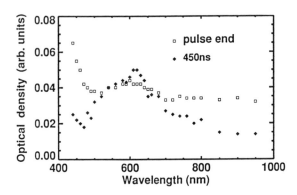

Figure 12: Transient absorption spectra for p-cresol novolac with PAG by nanosecond pulse radiolysis.

temperature of 85 °C. Transient absorption spectra are shown in Figure 12. For the film with PAG, there is a strong absorption peak at 620 nm in the time range of 450 nsec. It is considered that the absorption peak may be due to a protonated intermediate in the time range of 450 nsec after irradiation by 2 nsec electron pulses. A comparison of the experimental results for m-cresol novolac which contains PAG with those for p-cresol novolac which contains PAG indicates that the time range of acid generation may be related to the base resin system.

Conclusions

For the m-cresol novolac with PAG, the absorption peak at 542 nm may be due to the existence of a protonated intermediate. The relationship between absorbance difference ΔE at 542 nm and exposure dose X (mJ/cm^2) can be represented by

$$\Delta E = 4.29 \times 10^{-1} (1 - \exp(-7.77 \times 10^{-4}X))$$

for 98 %C.L..

A comparison of the experimental results for m-cresol novolac which contains PAG with those for p-cresol novolac which contains PAG indicates that the time range of acid generation may be related to the base resin system.

For m-cresol novolac with PAG, comparing the experimental results after exposure with those after PEB, the protonated intermediate seemed to be decomposed by PEB and the absorption peaks at 542 nm became weaker, while the absorption peaks at 394 nm became stronger. Comparing the absorption spectrum after PEB for m-cresol novolac which contains PAG with that after PEB for m-cresol novolac which contains p-toluenesulfonic acid, it can be assumed that the acid may be generated not only during exposure but also during PEB in the m-cresol novolac with PAG.

Literature Cited

(1) Ito, H.; Willson, C. G. *Polym. Eng. Sci.* **1983**, *23*, 1012.
(2) Dammel, R.; Døssel, K.-F.; Lingnau, J.; Theis, J.; Huber, H.; Oertel, H.; Trube, J. *Microelectr. Eng.* **1989**, *9*, 575.
(3) Nalamasu, O.; Kometani, J.; Cheng, M.; Timko, A. G.; Reichmanis, E.; Slater, S.; Blakeney, A. *J. Vac. Sci. Technol.* **1992**, *B10*, 2536.
(4) Nakamura, J.; Ban, H.; Deguchi, K.; Tanaka, A. *Jpn. J. Appl. Phys.* **1991**, *30*, 2619.
(5) Schlegel, L.; Ueno, T.; Hayashi, N.; Iwayanagi, T. *J. Vac. Sci. Technol.* **1991**, *B9*, 278.
(6) Pawlowski, G; Dammel, R.; Przybilla, K.; Röschert, H.; Spiess, W. *J. Photopolym. Sci. Technol.* **1991**, *4*, No. 3, 389.
(7) Nakajima, Y.; Inoguchi, Y.; Padmanaban, M.; Kinoshita, Y.; Dammel, R.; Meier, W.; Pawlowski, G. *Proc. 1991 Int. MicroProcess Conf. (Jpn. J. Appl. Phys., Tokyo, 1991)* **1991**, *JJAP Series 5*, 162.
(8) Ban, H.; Nakamura, J.; Deguchi, K.; Tanaka, A. *J. Vac. Sci. Technol.* **1991**, *B9*, 3387.
(9) Dektar, J. L.; Hacker, N. P. *J. Org. Chem.* **1990**, *55*, 639.
(10) Dektar, J. L.; Hacker, N. P. *J. Am. Chem. Soc.* **1990**, *112*, 6004.
(11) Pappas, S. P.; Pappas, B. C.; Gatechair, L. R.; Schnabel, W. J. *J. Polym. Sci. Polym. Chem. Ed.* **1984**, *22*, 69.
(12) McKean, D. R.; Schaedeli, U. P.; MacDonald, S. A. *J. Polym. Sci.: Part A: Polym. Chem.* **1989**, *27.*, 3927.

(13) McKean, D. R; Allen, R. D.; Kasai, P. H.; Schaedeli, U. P.; MacDonald, S. A. *Proc. SPIE* **1992**, *1672*, 94.
(14) Gaines Jr, G. L., *Anal. Chem.* **1976**, *48*, 450.
(15) Kozawa, T.; Yoshida, Y.; Uesaka, M.; Tagawa, S. *Jpn. J. Appl. Phys.* **1992**, *31*, 1574.
(16) Kozawa, T.; Yoshida, Y.; Uesaka, M.; Tagawa, S. *Jpn. J. Appl. Phys.* **1992**, *31*, 4301.
(17) Kodaira, M.; Awaji, N.; Kishimoto, T.; Usami, H.; Watanabe M. *Jpn. J. Appl. Phys.* **1991**, *30*, 3043.
(18) Yoshida, Y.; Ueda, T.; Kobayashi, T.; Tagawa, S. *J. Photopolym. Sci. Technol.* **1991**, *4*, 171.

RECEIVED September 13, 1994

Chapter 9

Radiation-Induced Reactions of Onium Salts in Novolak

Takahiro Kozawa[1], M. Uesaka[1], Takeo Watanabe[2], Yoshio Yamashita[2],
H. Shibata[3], Yoichi Yoshida[4], and Seiichi Tagawa[4]

[1]Nuclear Engineering Research Laboratory, Faculty of Engineering,
University of Tokyo, 2–22 Sirakata-Sirane, Tokai-mura, Naka-gun,
Ibaraki 319–11, Japan
[2]SORTEC Corporation, 16–1 Wadai, Tsukuba-shi, Ibaraki 300–42, Japan
[3]Research Center for Nuclear Science and Technology, University
of Tokyo, 2–22 Sirakata-Sirane, Tokai-mura, Ibaraki 319–11, Japan
[4]Institute of Scientific and Industrial Research, Osaka University, 8–1
Mihogaoka, Ibaraki, Osaka 567, Japan

Radiation-induced reactions of onium salts in m-cresol, which is a
model compound of phenolic resins, and in a novolak resin, have been
studied by means of pico- and nanosecond pulse radiolyses. The
absorptions due to an oxylradical and cationic species of m-cresol were
observed in the m-cresol solution by electron beam exposure. This
result suggests that the proton adducts of m-cresol are formed in m-
cresol by ion-molecular reactions between m-cresol and its radical
cations. The electron scavenging effect of onium salts delays the
recombination of cationic intermediates with electrons and prolongs the
lifetime of the cationic intermediates. Similar proton transfer occurs in
novolak. Furthermore, the absorptions arising from novolak and proton
(acid) were observed in novolak containing onium salts.

The chemically amplified resist is one of the promising materials for submicron
patterning by electron beam (EB) and X-ray lithography. The chemically amplified
resists based on acid catalytic chain reaction mechanisms show high sensitivity and
high contrast under certain conditions (1). However, chemically amplified resists have
serious problems, in particular, the instability of sensitivity. Many papers have reported
on impurity effects, prebake effects, post exposure bake effects, and delay time effects,
among others (2-7). Many researches and development in this field have been
performed for the improvement of qualities, that is, enhancement and stabilization of
their sensitivity (8-10). Process simulations have also been attempted in order to apply
the chemically amplified resists to production lines. For these works, it is important
and necessary to understand the radiation-induced reaction mechanisms such as the
processes of acid generation. Though the reaction mechanisms of the chemically
amplified resists have been investigated recently (11, 12), the details on these have not
been clarified. In EB and X-ray resists, ionization and excitation of base resins and acid
generators contribute to acid generation, and the contribution of the ionization of base
resins is the most important, because radiation energy is mostly absorbed by the base
resins. In this paper, radiation-induced reactions in novolak films have been studied
using pico- and nanosecond pulse radiolysis techniques to investigate short-lived

0097–6156/94/0579–0121$08.00/0

reactive species. Primary processes of acid generation due to the ionization of base resins are specifically discussed.

Experimental

The irradiation source is the 28 MeV linac at the University of Tokyo. The widths of electron pulses are 10 ps and 2 ns and the absorption doses are 10 and 50 Gy(J/kg), respectively. Details of the pico- and nanosecond pulse radiolysis system for optical absorption spectroscopy were described elsewhere (11,13). The absorption data are obtained by subtracting the absorption value before irradiation from that after irradiation.

Triphenylsulfonium triflate ($\phi_3SCF_3SO_3$, Midori Chemical) and hexafluoroantimonate (ϕ_3SSbF_6, Midori Chemical) were used as acid generators. Tetrahydrofuran (Merck, Uvasol), dichloromethane (Merck, Uvasol) and m-cresol (Wako, S.G.) were used as solvents. m-Cresol was used as a model compound of phenolic resins. Triethylamine (Wako, S.G.) was used as a cation scavenger. The liquid samples deaerated by argon bubbling in quartz cells were irradiated by 10 ps and 2 ns electron pulses at room temperature. The path length of quartz cells is 20 mm. p-Cresol-novolak (Mw=700), m-cresol-novolak (Mw=2000), and poly(methyl methacrylate) (PMMA) were used as base resins. p-Cresol-novolak and m-cresol-novolak were melted at 80°C and 100°C, respectively, and molded into 2-mm-thick films. Ten mm cubes of PMMA were obtained by polymerizing methyl methacrylate in quartz cells. The solid samples were irradiated by 2 ns electron pulses at room temperature.

Results and Discussion

Radiation Chemistry in Liquid m-Cresol.

Radiation-Induced Reactions of Neat m-Cresol. Figure 1 shows the transient optical absorption spectra obtained in the pulse radiolysis of liquid m-cresol, which is a model compound of novolak. Two peaks are observed in the visible (around 400 nm) and infrared (above 500 nm) regions. The time-dependent behavior of intermediate monitored at 806 nm is shown in Figure 2. It can be presumed that the absorption below 360 nm is due to the benzene rings of intermediates. The formation time of the infrared band is very fast (less than 50 ps), however, this band disappears in the presence of triethylamine which is a cation scavenger. Therefore, the infrared band is due to cationic species of m-cresol (e.g. $\phi(CH_3)OH^+$ and $\phi(CH_3)OH_2^+$). On the other hand, the peak at 400 nm is identified as an oxylradical of m-cresol ($\phi(CH_3)O\cdot$) (14). These species contribute to the following reactions:

$$\phi(CH_3)OH \rightsquigarrow \phi(CH_3)OH^+ + e^-, \tag{1}$$

$$\phi(CH_3)OH^+ + \phi(CH_3)OH \longrightarrow \phi(CH_3)O\cdot + \phi(CH_3)OH_2^+. \tag{2}$$

The protonated adducts of the solvent are formed by ion-molecular reactions of the solvent with its radical cations in m-cresol.

Radiation-Induced Reactions of Onium Salts in Liquid m-Cresol. Onium salts scavenge solvated electrons in tetrahydrofuran very efficiently.

$$Onium + e^-_{sol} \longrightarrow Onium^- \tag{3}$$

The reaction of onium salts with solvated electrons in tetrahydrofuran is a diffusion-controlled reaction (11). In tetrahydrofuran, no strong absorption due to onium salts (triphenylsulfonium triflate and hexafluoroantimonate) and their radiolytic products is

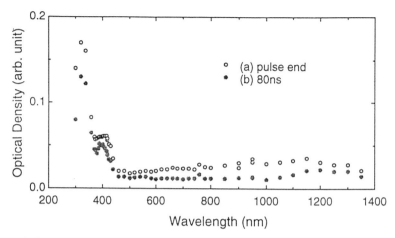

Figure 1. Transient absorption spectra obtained in the pulse radiolysis of liquid m-cresol (a) immediately and (b) 80 ns after electron pulses.

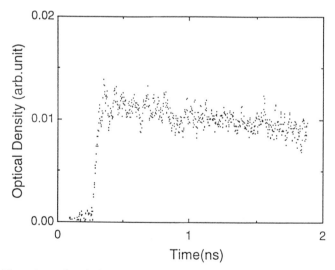

Figure 2. Time-dependent behavior of an intermediate obtained in the pulse radiolysis of m-cresol solution monitored at 806 nm.

observed in the wavelength range from 400 to 1000 nm. The transient absorption spectra obtained for a 50 mM triphenylsulfonium triflate solution in m-cresol are shown in Figure 3. No identifiable absorption due to the onium salts and their radiolytic products can be found in Figure 3. However, it is found that the yields of the cationic species of m-cresol are increased by the addition of the onium salt. This is due to the electron scavenging effect of the onium salts (reaction 3). The cationic intermediates such as a radical cation and proton adduct of m-cresol recombine with electrons and anionic species of m-cresol.

$$\text{Cationic species + anionic species} \longrightarrow \text{recombination} \tag{4}$$

The recombination of radical cations of m-cresol with electrons and other anionic species is delayed because the onium salts scavenge anionic species. Consequently, the yields and lifetimes of cationic species increase.

Radiation Chemistry of Onium Salts in Solid Polymer.

Reaction of Onium Salts with Anionic Species in PMMA. The reactivity of onium salts with electrons in a solid matrix of PMMA was investigated. Electrons and radical cations of PMMA are produced by EB irradiation beam.

$$\text{PMMA} \xrightarrow{\hspace{1cm}} \text{PMMA}^{+} + e^{-} \tag{5}$$

Radical anions of PMMA are generated by the reactions of PMMA with electrons produced from reaction 5.

$$\text{PMMA} + e^{-} \longrightarrow \text{PMMA}^{-} \tag{6}$$

These radical anions exhibit a strong absorption at around 400 nm.
    No identifiable absorption due to the onium salts and their radiolytic products could be found in the transient absorption spectra obtained for PMMA containing 5 wt.% triphenylsulfonium hexafluoroantimonate. Figure 4 shows the time-dependent behavior of radical anions of PMMA obtained in the pulse radiolysis of PMMA monitored at 420 nm in the absence and presence of the onium salts. It is found that the yield of radical anions of PMMA is decreased by the addition of the onium salts. This is because the onium salts scavenge electrons generated by ionization. The yield of radical anions of PMMA is decreased due to competition for electrons between PMMA and the onium salts. The lifetime of radical anions of PMMA, however, changes minimally after and before the addition of onium salts because both the onium salts and radical anion of PMMA can hardly move in the PMMA matrix at room temperature.

Radiation-Induced Reactions of p-Cresol-Novolak. Figure 5 shows the transient absorption spectra obtained in the pulse radiolysis of a 100 mM p-cresol-novolak solution in dichloromethane. The spectra are similar to those of liquid m-cresol as shown in Figure 1. On the other hand, in the transient absorption spectra obtained for the 100 mM p-cresol-novolak solution in tetrahydrofuran, absorption due to the anionic species of p-cresol-novolak could not be found in the time range from 10 to 200 ns and wavelength range from 350 to 1000 nm. Therefore, the broad absorption in the infrared region is due to the cationic species of p-cresol-novolak. The transient absorption spectra obtained in the pulse radiolysis of solid p-cresol-novolak are shown in Figure 6. The spectra below 425 nm are uncertain because the datum at 400 nm is noisy. On the analogy of the experimental results of m-cresol, the similar ion-molecular reactions may occur in p-cresol-novolak (M).

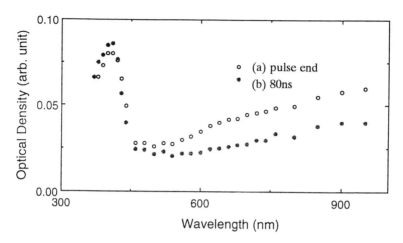

Figure 3. Transient absorption spectra obtained in the pulse radiolysis of 100 mM triphenylsulfonium triflate solution in m-cresol (a) immediately and (b) 80 ns after electron pulses.

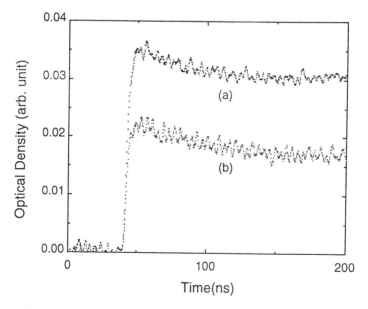

Figure 4. Time-dependent behavior of radical anions of PMMA obtained in the pulse radiolysis of PMMA monitored at 420 nm (a) with no additive and (b) with 5 wt.% triphenylsulfonium hexafluoroantimonate.

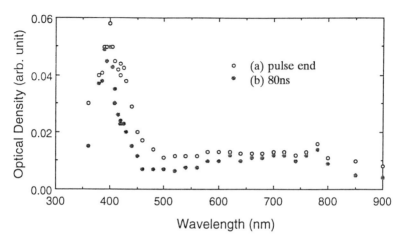

Figure 5. Transient absorption spectra obtained in the pulse radiolysis of 100 mM p-cresol-novolak solution in dichloromethane (a) immediately and (b) 80 ns after electron pulses.

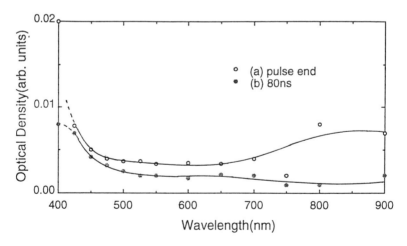

Figure 6. Transient absorption spectra obtained in the pulse radiolysis of p-cresol novolak (a) immediately and (b) 80 ns after electron pulses.

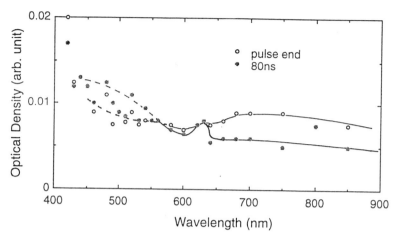

Figure 7. Transient absorption spectra obtained in the pulse radiolysis of p-cresol-novolak containing 15 wt.% triphenylsulfonium triflate (a) immediately and (b) 80 ns after electron pulses.

$$M \xrightarrow{\text{\Large\char`\~\char`\~}} M^+ + e^- \tag{7}$$

$$M + M^+ \longrightarrow MH^+ + M (-H) \tag{8}$$

Proton transfer also may occur intramolecularly.

Radiation-Induced Reactions of Onium Salts in Cresol-Novolak. Figure 7 shows the transient absorption spectra obtained in the pulse radiolysis of solid p-cresol-novolak containing 15 wt.% triphenylsulfonium triflate. The broken lines indicate that the spectra below 560 nm are uncertain because the datum at each wavelength is noisy. The time-dependent behaviors of cationic species of novolak monitored at 700 nm in the absence and presence of onium salts are shown in Figure 8. It is found that the yields of cationic species of novolak are increased by the addition of onium salts. This is due to the electron scavenging effect of onium salts described above. Furthermore, the absorption peak of the intermediate is observed around 620 nm at 80 ns after 2 ns electron pulses as shown in Figure 7. The time-dependent behaviors of the intermediates observed at 620 and 700 nm are shown in Figure 9. The absorption at 620 nm shows a very slow increase. On the other hand, in m-cresol-novolak, the absorption of intermediate is observed at 540 nm. The intermediate observed at 540 nm has a long lifetime over the time scale of minute region. The intermediate may thus come from novolak and proton (acid). The intermediates observed at 540 and 620 nm may be precursors of an acid.

Conclusion

The absorptions due to an oxylradical and cationic species of m-cresol were observed in a m-cresol solution by EB irradiation. This result suggests that the protonated adducts of m-cresol are formed in m-cresol by the ion-molecular reactions of m-cresol with its radical cations. Onium salts scavenge electrons generated by ionization. The electron scavenging effect of onium salts delays the recombination of cationic intermediates with electrons and prolong the lifetime of cationic intermediates. Because

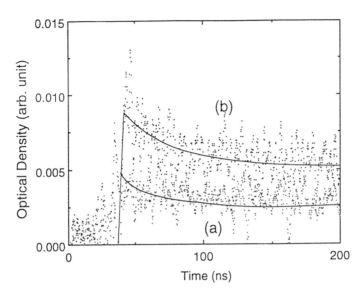

Figure 8. Time-dependent behaviors of cationic species of novolak obtained in the pulse radiolysis of novolak monitored at 700 nm (a) with no additive and (b) with 15 wt.% triphenylsulfonium triflate.

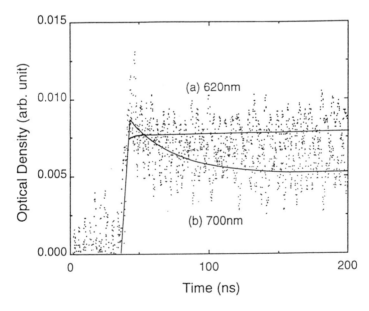

Figure 9. The time-dependent behaviors of intermediates obtained in the pulse radiolysis of p-cresol-novolak containing 15 wt.% triphenylsulfonium-triflate monitored at (a) 620 nm and (b) 700 nm.

of the electron scavenging effects of onium salts, the yields of cationic species were increased. Similar ion-molecular reactions occur in novolak. Furthermore, the absorption arising from novolak and proton (acid) were observed in novolak containing onium salts. The intermediates observed at 540 and 620 nm may be precursors of an acid.

Acknowledgments

The authors would like to express their appreciation to Mr. T. Ueda and Mr. T. Kobayashi of the University of Tokyo for their operation of the linac and to Mr. S. Seki of the University of Tokyo for his assistance in experiments.

Literature Cited

1. Ito, H.; Willson, C. G. *Polym.Eng.Sci.* **1983**, *23*, 1012.
2. Ito, H. *Jpn.J.Appl.Phys.* **1992**, *31*, 4273.
3. Reichmanis, E.; Houlihan, F.M.; Nalamasu, O.; Neenan, T. X. *Chem. Mater.* **1991**, *3*, 394.
4. Dammel, R.; Døssel, K. -F.; Lingnau, J.; Theis, J.; Huber, H.; Oertel, H.; Trube, J. *Microelectr. Eng.* **1989**, *9*, 575.
5. Schlegel, L.; Ueno, T.; Shiraishi, H.; Hayashi, N.; Iwayanagi, T. *Microelectr. Eng.* **1991**, *14*, 227.
6. Hayashi, N.; Tadano, K.; Tanaka, T.; Shiraishi, H.; Ueno, T.; Iwayanagi, T. *Proc. 1990 Int. MicroProcess Conf. (Jpn. J. Appl. Phys., Tokyo, 1990) JJAP Series 4*, p.124.
7. Nakamura, J.; Ban, H.; Tanaka, A. *Jpn. J. Appl. Phys.* **1992**, *31*, 4294.
8. Novembre, A. E.; Tai, W. W.; Kometani, J. M.; Hanson, J. E.; Nalamasu, O.; Taylor, G. N.; Reichmanis, E.; Thompson, L. F.; Tomes, D. N. *J. Vac. Sci. Technol.* **1991**, *B9*, 3338.
9. Shiraishi, H.; Hayashi, N.; Ueno, T.; Sakamizu, T.; Murai, F. *J. Vac. Sci. Technol.* **1991**, *B9*, 3343.
10. Ban, H.; Nakamura, J.; Deguchi, K.; Tanaka, A. *J. Vac. Sci. Technol.* **1991**, *B9*, 3387.
11. Kozawa, T.; Yoshida, Y.; Uesaka, M.; Tagawa, S. *Jpn.J.Appl.Phys.* **1992**, *31*, 4301.
12. Kozawa, T.; Yoshida, Y.; Uesaka, M.; Tagawa, S. *Jpn.J.Appl.Phys.* **1993**, *32*, 6049.
13. Yoshida, Y.; Ueda, T.; Kobayashi, T.; Tagawa, S. *J.Photopolym.Sci.Technol.* **1991**, *4*, 171.
14. Land, E. J.; Ebert, M. *Trans. Faraday Soc.* **1967**, *63*, 1181.

RECEIVED September 13, 1994

Chapter 10

Polymeric Sulfonium Salts as Acid Generators for Excimer Laser Lithography

Katsumi Maeda, Kaichiro Nakano, and Etsuo Hasegawa

Functional Devices Research Laboratories, NEC Corporation, Kanagawa 216, Japan

New polymers having sulfonium salt units in the main chain have been synthesized. Their photoactivity for excimer lasers (KrF, ArF) and their thermal stabilities have been elucidated. Phenyl-type polymeric sulfonium salts exhibit good thermal stability (dec. 274-304 °C). The photo-acid generating efficiency of the phenyl-type polymeric sulfonium salts for the ArF excimer laser (193.4 nm) increases with an increase in x. A phenyl-type polymeric sulfonium salt with x=0.478 exhibits 2.5 times the photoactivity for triphenylsulfonium trifluoromethanesulfonate (TPS) and lower absorption property than TPS. These results suggest that inner-filter effects for phenylene sulfide groups and electron donating properties for neighboring phenylthio groups control the photoactivity for polymers.

Excimer laser (KrF, ArF) lithography (*1-5*) is a key technology for ULSI production with the minimum featuresize below 0.25 μm. However, resists with high photosensitivity must be developed to counteract the high excimer laser cost. Chemically amplified resist systems (*6,7*), proposed by Ito et al., are well-known to be very useful. A typical chemically amplified resist consists of a base polymer, photo-acid generator (PAG), and dissolution inhibitor (or crosslinker). It exhibits high sensitivity through a catalytic reaction by photochemically generated acids. This paper reports syntheses, thermal stabilities, and photoactivities of new polymers having sulfonium salt units as a photo-acid generating group sensitive to deep UV lights (i.e., KrF and ArF excimer lasers).

Experimental Section

Materials.
General Procedure for Preparing Phenyl-type Polymeric Sulfonium Salts (PS-1, 2, and 3).
Preparation of the phenyl-type polymeric sulfonium salts was achieved by a modification of the triarylsulfonium salt syntheses (*8*). A mixture of poly(3,5-dimethyl-1,4-phenylene sulfide) (PDMPS, number-average degree of polymerization: 9) (*9*), diphenyliodonium trifluoromethanesulfonate (Ph$_2$I$^+$-OTf, DPIT), and copper benzoate (0.03 mol% to DPIT) was suspended in chlorobenzene and heated at 125-130 °C under a nitrogen atmosphere for 3 hours. After cooling to room tempera-

0097–6156/94/0579–0130$08.00/0

ture, the mixture was poured into diethyl ether to precipitate the polymeric sulfonium salt. The product was purified by two reprecipitations. By changing the molar ratio of the reagents, polymers with various sulfonium salt unit concentrations (x) were synthesized. The sulfonium salt unit concentrations were determined by ^{1}H-NMR spectroscopy.

PS-1(x=0.180): The PDMPS to DPIT molar ratio was 1:0.5. Dec. 290 °C; ^{1}H-NMR(acetone-d_6)δ2.10-2.53(m, Me), 6.05-8.11(m, aromatic);IR(KBr pellet)3450, 3050, 2970, 2920, 1570, 1445, 1255, 1220, 1150, 1030 cm^{-1}; UV(CH$_3$CN)ε_{248}=39,800, ε_{193}=12,7300

PS-2(x=0.376): The PDMPS to DPIT molar ratio was 1:1. Dec. 304 °C; ^{1}H-NMR(acetone-d_6)δ 2.09-2.51(m, Me), 6.05-8.15(m, aromatic); IR(KBr pellet)3500, 3050, 2970, 2920, 1570, 1440, 1265, 1220, 1150, 1030 cm^{-1}; UV(CH$_3$CN)ε_{248}=27,400, ε_{193}=71,100

PS-3(x=0.478): The PDMPS to DPIT molar ratio was 1:1.5. Dec. 274°C; ^{1}H-NMR(acetone-d_6)δ 2.19-2.51(m, Me), 6.45-8.10(m, aromatic); IR(KBr pellet)3400, 3050, 2920, 1570, 1440, 1265, 1225, 1160, 1030 cm^{-1}; UV(CH$_3$CN)ε_{248}=16,100, ε_{193}=42,000

General Procedure for Preparing Methyl-type Polymeric Sulfonium Salts (PS-4, 5, and 6). Methyl-type polymeric sulfonium salts were prepared by the procedure used in alkyldiarylsulfonium salt syntheses (*10*). PDMPS (number-average degree of polymerization: 9), methyl iodide, and silver trifluoromethanesulfonate were dissolved in methylene chloride, and the mixture was stirred for 3 hours at room temperature. The insoluble silver salts were filtered off, and the filtrate was poured into diethyl ether to precipitate the polymeric sulfonium salt. The product was purified by two reprecipitations. By changing the molar ratio of the reagents, polymers with various sulfonium salt unit concentrations were synthesized.

PS-4(x=0.088): The PDMPS to methyl iodide molar ratio was 1:0.5. Dec. 131 °C; ^{1}H-NMR(acetone-d_6)δ 2.14-2.47(m, Me), 3.99(s, Me-S$^+$), 6.14-7.74(m, aromatic); IR(KBr pellet)3450, 2920, 1570, 1450, 1262, 1220, 1158, 1030 cm^{-1}; UV(CH$_3$CN)ε_{248}=59,500, ε_{193}=153,800

PS-5(x=0.144): The PDMPS to methyl iodide molar ratio was 1:1. Dec. 133 °C; ^{1}H-NMR(acetone-d_6)δ 2.11-2.58(m, Me), 3.95(s, Me-S$^+$), 3.99(s, Me-S$^+$), 6.16-7.77(m, aromatic); IR(KBr pellet)3450, 2920, 1570, 1450, 1265, 1220, 1158, 1030 cm^{-1}; UV(CH$_3$CN)ε_{248}=36,100, ε_{193}=89,300

PS-6(x=0.678): The PDMPS to methyl iodide molar ratio was 1:2. Dec. 134 °C; ^{1}H-NMR(acetone-d_6)δ 2.09-2.48(m, Me), 3.94(s, Me-S$^+$), 3.99(s, Me-S$^+$), 6.17-7.77(m, aromatic); IR(KBr pellet)3475, 2920, 1570, 1445, 1250, 1225, 1160, 1030 cm^{-1}; UV(CH$_3$CN)ε_{248}=10,700, ε_{193}=26,400

(2-Methylphenyl)diphenylsulfonium trifluoromethanesulfonate (TAS-1). TAS-1, 2, and 3 were prepared according to the procedure published in pertinent literature (*8*). Yield 63%; mp 135-137 °C; ^{1}H-NMR(CDCl$_3$)δ 2.53(s, 3H), 6.82-7.83(m, 14H); IR(KBr pellet)3050, 1475, 1440, 1275, 1260, 1150, 1030 cm^{-1}; m/z(%)277(100), 186(21), 91(16); UV(CH$_3$CN)ε_{248}=15,400, ε_{193}=68,500

(2,6-Dimethylphenyl)diphenylsulfonium trifluoromethanesulfonate (TAS-2): Yield 46%; mp 137-139 °C; ^{1}H-NMR(CDCl$_3$)δ 2.33(s, 6H), 7.10-7.77(m, 13H); IR(KBr pellet)3060, 1475, 1460, 1443, 1275, 1260, 1150, 1030 cm^{-1}; m/z(%)291(100), 214(26), 186(71); UV(CH$_3$CN)ε_{248}=13,000, ε_{193}=72,100

Bis(2-methylphenyl)phenylsulfonium trifluoromethanesulfonate (TAS-3): Yield 52%; mp 134-135 °C; ^{1}H-NMR(CDCl$_3$)δ 2.52(s, 6H), 6.97-7.05(m, 2H), 7.07-7.67(m, 11H); IR(KBr pellet)3050,1470, 1460, 1440, 1275, 1260, 1145, 1030 cm^{-1}; m/z(%)291(100), 200(40), 91(47); UV(CH$_3$CN)ε_{248}=15,900, ε_{193}=70,800

Measurements. ^{1}H-NMR spectra, infrared spectra, and ultraviolet or visible absorption spectra were measured on a Burcker AMX-400(400 MHz) spectrometer, a Shimadzu IR-470 spectrometer, and a Shimadzu UV-365 spectrometer, respectively.

Decomposition temperatures were measured on a Mack Science Thermal Analysis System 001 (TG-DTA2000). Exposure of KrF and ArF excimer laser (248 and 193.4 nm) was carried out with the NEC MEX excimer laser apparatus and the Lumonics HE-460-SM-A excimer laser apparatus, respectively.

Measurements of acid generation efficiencies. A solution of a polymeric sulfonium salt (3 mL, 1 mmol/L of the sulfonium salt unit) in acetonitrile was irradiated in a quartz cell (path-length: 1cm) by an excimer laser (KrF, ArF). The photolyzed solution was added to a 2 mL acetonitrile solution of tetrabromophenol blue sodium salt (*11*) (1.6 mmol/L), followed by dilution with acetonitrile to 20 mL. Then, the absorbance at 619 nm of the resulting solution was measured. The acid generation efficiencies were determined from the absorbance decrease at 619 nm. Transmittance values of acetonitrile (path-length: 1cm) at 248 and 193 nm were 100% and 58%, respectively. When acetonitrile was irradiated by an ArF excimer laser (dose: 40 mJ/cm^2), no acid was detected.

Results and discussion

Molecular design for polymer sulfonium salts. Acid generation of triphenylsulfonium salt has been reported to occur via hetelolytic cleavage of a carbon-sulfur bond along with some homolytic cleavage (*12-15*). In the former case, a phenyl cation and diphenyl sulfide are produced. Then, an electrophilic substitution reaction of the phenyl cation onto diphenyl sulfide occurs, generating an acid. This mechanism suggests that, for improvement in the photo-acid generating efficiency, the carbon-sulfur bond cleavage and the electrophilic substitution reaction must be facile. From these viewpoints, we designed a new PAG, based on the following concept. First, for high photosensitivity, an electron donating phenylthio group was introduced into the aromatic ring. An electron donor is expected to activate the reactivity for the electrophilic substitution reaction and stabilize the phenyl cation formed by the carbon-sulfur bond cleavage. Second, for good solubility, alkyl substituents were introduced. Third, especially for ArF excimer lithography, PAG must be highly photoactive but also highly transparent. Therefore, polymeric sulfonium salts were employed to reduce light absorption per sulfonium salt unit (Scheme 1).

Syntheses and thermal stabilities of polymeric sulfonium salts. Scheme 2 shows the synthesis route for polymeric sulfonium salts (PS). Linear, soluble, and white poly(3,5-dimethyl-1,4-phenylene sulfide (PDMPS) (*9*) was used as a starting material. The phenyl-type PS (**PS-1, 2,** and **3**) were prepared by reacting PDMPS with DPIT in chlorobenzene at 125-130 °C. The methyl-type PS (**PS-4, 5,** and **6**) were prepared with methyl iodide in methylene chloride at room temperature. By changing the molar ratio of the reagents, polymers with various sulfonium salt unit concentrations were synthesized (x=0.180-0.478 for the phenyl-type PS, and X=0.088-0.678 for the methyl-type PS). They have good solubility in polar solvents such as tetrahydrofuran, N-methylpyrrolidone, and acetone. Figure 1 shows UV absorption spectra for **PS-3** and triphenylsulfonium trifluoromethanesulfonate (TPS) (*9*). **PS-3** has lower absorption intensity at 193nm than TPS. That is, the absorption intensity of **PS-3** is reduced to 3/5 of that of TPS. Thus, **PS-3** has an advantage in absorption property for an ArF excimer laser lithography.

A photo-acid generator must be thermally stable in the lithography process. Therefore, in order to examine the thermal stability, polymeric sulfonium salts were analyzed by thermogravimetry. In the case of the methyl-type PS, the decomposition occurs in two steps, while in the phenyl-type PS, it occurs in one step (Figure 2). The phenyl-type PS is more thermally stable than the methyl-type PS. Then, replacing the methyl group with the phenyl group raises the decomposition temperature by 160 °C (Table I). The sulfonium salt concentration scarcely affects the first and the

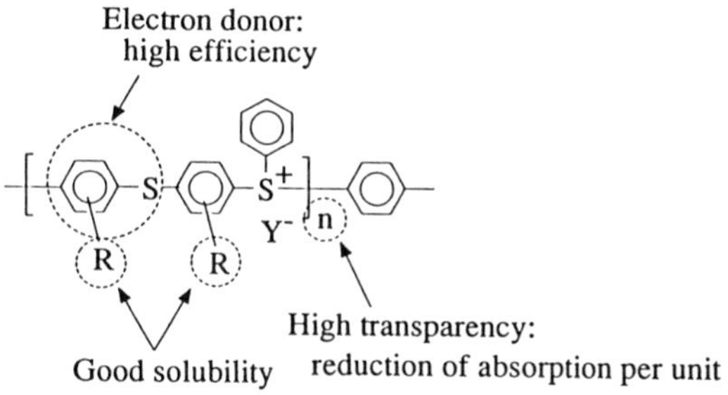

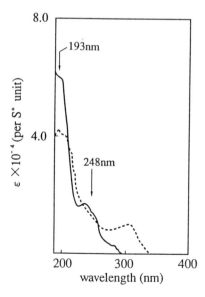

Electron donor:
high efficiency
Y⁻
n
R
R
High transparency:
reduction of absorption per unit
Good solubility

Scheme 1. Molecular design concept.

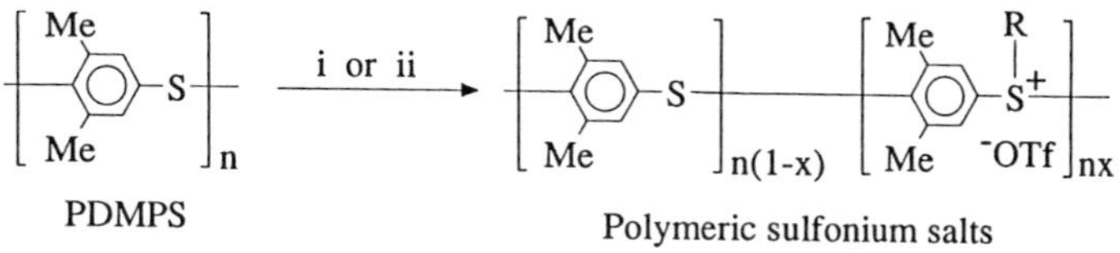

Me
S
Me
n
PDMPS
i or ii
n(1-x)
R
⁻OTf
nx
Polymeric sulfonium salts

Scheme 2. Synthesis routes for polymeric sulfonium salts (PS). *Reagents*: i, $Ph_2I^{+-}OTf/(PhCOO)_2Cu$; ii, MeI/AgOTf

8.0
4.0
193nm
248nm
ε ×10⁻⁴ (per S⁺ unit)
200
300
400
wavelength (nm)

Figure 1. UV absorption spectra of PS-3 and TPS in acetonitrile; (——): TPS, (- - -): PS-3.

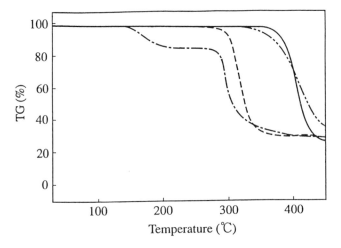

Figure 2. Thermogravimetric curves for various sulfonium salts; (——):
TPS, (- - -): PS-1, (– - –): PS-4, (– -- –): PDMPS.

Table I. Thermogravimetric analysis of PS

Compounds	R	x	Dec.(1)/°C	TG(%)[a]	Dec.(2)/°C
PS- 1	Ph	0.180	290	72	-
PS- 2		0.376	304	75	-
PS- 3		0.478	274	81	-
PS- 4	Me	0.088	131	14	270
PS- 5		0.144	133	8	275
PS- 6		0.678	134	11	270
TPS		-	370	72	-

Conditions; 10°C/min, N_2 -flow

a) Weight loss in the first decomposition

Table II. Photo-acid Generating Efficiency of PS

Compounds	R	x	H^+ (KrF)[a]	H^+ (ArF)[a]
PS- 1	Ph	0.180	0.13	0.09
PS- 2		0.376	0.13	0.56
PS- 3		0.478	0.46	2.51
PS- 4	Me	0.088	0.02	0.07
PS- 5		0.144	0.05	0.08
PS- 6		0.678	0.06	0.08
TPS		-	1.00	1.00

a) Relative values (at $40 mJ/cm^2$)

second decomposition temperatures and weight changes in the both polymers (Table I). These results indicate that the thermal stability of the phenyl-type PS, although inferior to that of TPS, is sufficient for use in a typical lithography process.

Photo-acid generating activity in polymeric sulfonium salts. To evaluate the photo-acid generating efficiencies of PS, a PS in acetonitrile (1 mmol/L of sulfonium salt unit) solution was irradiated in a quartz cell with an excimer laser (KrF, ArF). Generated acids were spectrophotometrically determined by the dye method (*11*). In the case of the methyl-type PS, the photo-acid generating efficiencies for KrF and ArF excimer lasers were very low (0.02-0.08 times that of TPS) and were scarcely influenced by the sulfonium salt unit concentration (Table II). These results reveal that diphenylmethyl sulfonium salt unit is less effective as an acid generating group. The photo-acid generating efficiencies of the phenyl-type PS are higher than those of the methyl-types and increase with increasing sulfonium salt unit concentration (Table II). This tendency appears more marked in ArF excimer laser exposure (Figure 3). Especially, the **PS-3** with x=0.478 shows 2.5 times the efficiency of TPS (Table II). In order to investigate the factor governing the photoactivity of the polymeric sulfonium salts, we synthesized TPS derivatives (**TAS-1, 2, and 3**) substituted with methyl groups as model compounds (Scheme 3) and measured their photo-acid generating efficiencies for use in an ArF excimer laser. Their efficiencies are reduced to 1/2 to 3/4 that of TPS (Table III). These results reveal that the methyl-substituents have a negative effect on the acid generation.

Next, we tried to clarify the relationship between the efficiencies and sulfonium salt unit concentrations or absorption coefficients. Figure 4 shows the relationship between sulfonium salt unit concentrations, the photo-acid generating efficiencies, and absorption coefficients in the phenyl-type PS. The absorption intensity of the PS decreases with an increase in x. This tendency seems to be due to different phenylene sulfide group contents per sulfonium salt unit. The acid generation efficiencies exhibit the reverse relationship. From these observations, the results are interpreted as follows. When x is small (case A, scheme 4), the phenylene sulfide groups having a majority of dye residues absorb the greater part of the excimer laser beams. On the contrary, in a PS compound with larger x values (case B, scheme 4), absorption by the phenylene sulfide residues decreases, resulting in more efficient absorption by the sulfonium salt units. The electron donating phenylthio groups assist in improving the photoactivity, although the effect is expected to disappear when applying much higher x values (x>0.5) (Scheme 4). Thus, these results suggest that two main factors, inner-filter effects for phenylene sulfide groups and electron donating

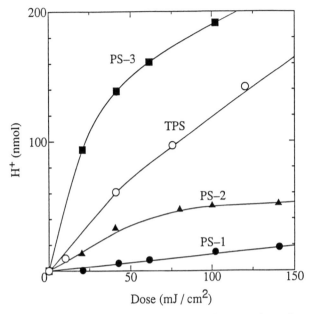

Figure 3. Acid generation dependence of phenyl-type polymeric sulfonium salts on ArF excimer laser dose.

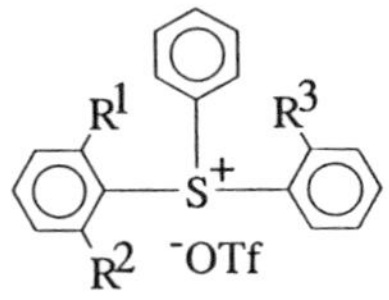

Scheme 3. Triphenylsulfonium trifluoromethanesulfonate (TPS) derivatives substituted by methyl groups.

Table III. Acid Generating Efficiency for TPS Derivatives

Compounds	R^1	R^2	R^3	H^+ (ArF)[a]
TPS	H	H	H	1.00
TAS-1	H	H	H	0.74
TAS-2	Me	Me	H	0.49
TAS-3	H	H	Me	0.61

a) Relative values (at 40mJ/cm^2)

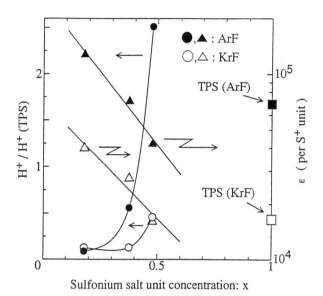

Figure 4. Dependence of acid generating and absorption properties of PS on sulfonium salt unit concentration.

(A) In small x value

(B) In large x value

Scheme 4. Polymeric sulfonium salt structures with different x values.

properties of neighboring phenylthio groups, control the acid generation from polymeric sulfonium salts.

Conclusion. This paper has reported polymers having sulfonium salt units as a photosensitive group in a main chain. Phenyl-type polymeric sulfonium salts exhibit good thermal stability, high photosensitivity, and lower absorption property, and suited for use in ArF excimer laser lithography.

Literature Cited

(1)M. Sasago, K. Yamashita, M. Endo, T. Koizumi, T. Matsuo, K. Matsuoka, A. Katsuyama, S. Kobayashi, and N. Nomura, *J. Photopolym. Sci. Technol.* **1993**, *6*(4), 443.

(2)R. R. Kunz, M. A. Hartney, M. W. Horn, C. L. Keast, M. Rotschild, and D. C. Shaver, *J. Photopolym. Sci. Technol.* **1993**, *6*(4), 473.

(3)Y. Kaimoto, K. Nazaki, S. Takechi, and N. Abe, *Proc. SPIE(Adv. Resist Technol. Process.)* **1992**, *1672*, 66.

(4)N. Nomura, H. Nakagawa, Y. Tani, K. Araki, T. Sato, and M. Sasago, *Microelectronic Engineering* **1990**, *11*, 183.

(5)T. Hattori, L. Schlegel, A. Imai, N. Hayashi, and T. Ueno, *Proc. SPIE(Adv. Resist Technol. Process.)* **1993**, *1925*, 146.

(6)H. Ito, C. G. Willson, and J. M. Frechet, *Digest of Technical paper of 1982 Symposium on VLSI Technology.* **1982**, 86.

(7)T. Iwayanagi, T. Ueno, S. Nonogaki, H. Ito, and C. G. Willson, *Electronic and Photonic Applications of Polymers*; M. J. Bowden and S. R. Turner eds.; ACS Advances in Chemistry Series 218; American Chemical Society: 1988, ch3.

(8)J. V. Crivello and J. H. W. Law, *J. Org. Chem.* **1978**, *43*, 3055.

(9)K. Yamamoto, M. Jikei, J. Katho, H. Nishide, and E. Tsuchida, *Macromolecules* **1992**, *25*, 2698.

(10)F. D. Saeva and B. P. Morgan, *J. Am. Chem. Soc.* **1984**, *106*, 4121.

(11)G. Pawloski, R. Dammel, C. R. Lindley, H. J. Merren, H. Roschert, and J. Lingnau *Proc. SPIE(Adv. Resist Technol. Process.)* **1990**, *1262*, 16.

(12)J. L. Dektar and N. P. Hacker, *J. Am. Chem. Soc.* **1990**, *112*, 6004.

(13)R. S. Davidson and J. W. Goodwin, *Eur. Polym. J.* **1982**, *18*, 487.

(14)J. L. Dektar and N. P. Hacker, *J. Chem. Soc., Chem. Commun.* **1987**, 1951.

(15)K. M. Welsh, J. L. Dektar, M. A. Garcia-Garibaya, N. P. Hacker, and N. J. Turro, *J. Org. Chem.* **1992**, *57*, 4179.

RECEIVED September 13, 1994

Chapter 11

Application of Triaryl Phosphate to Photosensitive Materials
Photoreaction Mechanism of Triaryl Phosphate

I. Naito[1], A. Kinoshita[1], Y. Okamoto[2], and S. Takamuku[2]

[1]Department of Photography, Kyushu Sangyo University, Matsukadai, Higashi-ku, Fukuoka 813, Japan
[2]ISIR, Osaka University, Mihogaoka, Ibaraki, Osaka 567, Japan

Five kinds of methoxy-substituted diaryl phosphate (DAP) and triaryl phosphates (TAP) were studied for application to positive-acting photosensitive resist materials. Photoirradiation of these compounds with 254 nm light gave a biaryl and phosphoric acid (DAP) or its monoaryl ester (TAP) equivalently with good quantum yields. During the photoirradiation, a strong emission was detected. The emission spectrum has two emission maxima at 310 nm (monomer fluorescence) and around 350 nm (intramolecular excimer fluorescence). Photolysis of bis(4-methoxyphenyl) phosphate were carried out in methanol in the presence of oxygen as a quencher. Because a quenching rate (Stern-Volmer constant; $kq\,\tau$) of the biaryl formation (214 M^{-1}) agrees with that of the excimer fluorescence (248 M^{-1}), the reaction is believed to proceed through the intramolecular excimer. Phosphate-sensitized thin films of poly(4-trimethyl-siloxystyrene) (PSSt) were made for the photosensitivity study. The most sensitive formulation was a 3-μm-thick film of PSSt containing 5.0 wt/wt% tris(4-methoxyphenyl) phosphate that required a dose of 12 mJ cm^{-2} at 254 nm.

New positive-acting photosensitive resist materials consisting of a photo-acid generator and an acid-sensitive polymer have attracted our interest.[1-3] In reactions of these materials, the photoinitiator initially generates an acid in the thin film by irradiation, as shown in eq. 1. The polymer in the irradiated film undergoes an acid-catalyzed deprotection reaction by thermal treatment. A relief image can be formed in the nonirradiated polymer that remains after

NOTE: This is Part II in a series.

0097–6156/94/0579–0139$08.00/0

development, as shown in eq. 2.

$$\text{Photoinitiator} \xrightarrow{\;h\nu\;} \text{Acid} \qquad (1)$$

$$\text{Base-insoluble Polymer} \xrightarrow{\;\overset{+}{H},\,\Delta\;} \text{Base-soluble Polymer} \qquad (2)$$

Recently, we reported the photoreaction of TAP to yield a bisaryl compound and phosphoric acid monoaryl ester, equivalently, with good quantum yield.[4,5]   An emission spectrum of TAP exhibited two maxima around 310 and 350 nm, which were assigned to a monomer fluorecence and an intramolecular exicimer fluorescence, respectively. The photochemical reaction is proposed to proceed through the intramolecular excimer. Because phosphoric acid monoaryl ester is a strong acid, TAPs were studied for an application to positive-acting photosensitive resist materials.[5]

We synthesized methoxy-substituted bisaryl phosphates (DAPs) and TAPs to study the reaction mechanism and to prepare a highly sensitive film [(R_1O)$_2$ PO (OR_2); $R_1 = R_2 =$ 4-methoxyphenyl (TMP); $R_1 =$ 4-methoxyphenyl, $R_2 = H$ (DMP); $R_1 =$ 4-methoxyphenyl, $R_2 =$ ethyl (DMEP); $R_1 = R_2 = 3,4,5$-trimethoxyphenyl (TTMP); $R_1 = 3,4,5$-trimethoxyphenyl, $R_2 = H$ (DTMP)]. Photoirradiation of DAPs yields phosphoric acid, a strong and non-volatile acid. Poly(4-trimethylsiloxystyrene) (PSSt) was used as matrices for the sensitivity experiments.

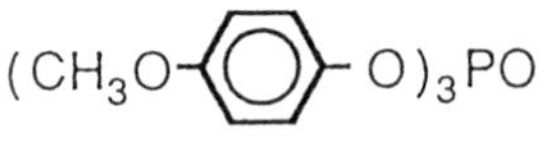

$$(CH_3O\!-\!\langle\bigcirc\rangle\!-\!O)_3PO$$

(TMP)

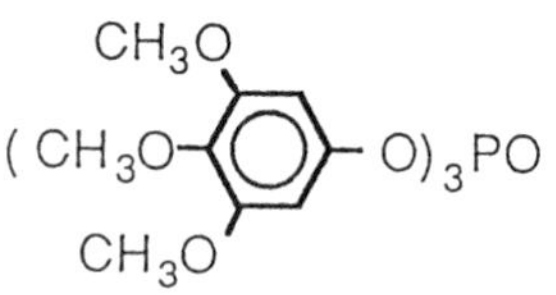

$$(\,CH_3O\!-\!\underset{CH_3O}{\overset{CH_3O}{\langle\bigcirc\rangle}}\!-\!O)_3PO$$

(TTMP)

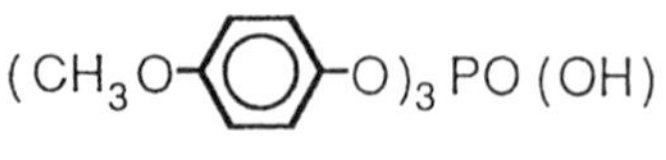

$$(CH_3O\!-\!\langle\bigcirc\rangle\!-\!O)_3\,PO\,(OH)$$

(DMP)

$$(CH_3O\!-\!\langle\bigcirc\rangle\!-\!O)_2PO\,(OC_2H_5)$$

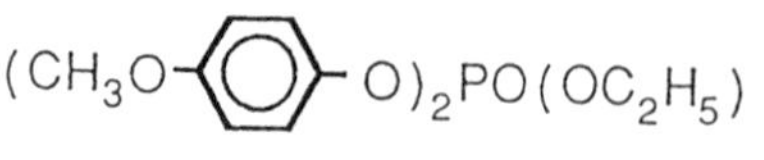

(DMEP)

(DMEP)

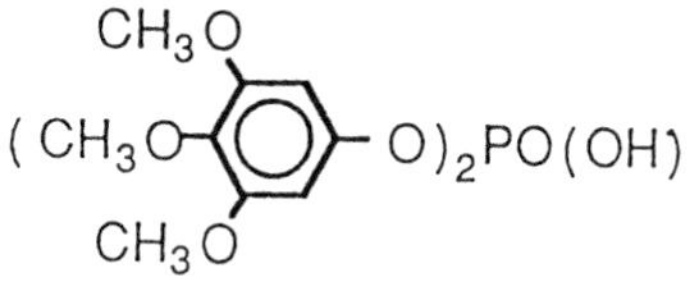

$$(CH_3O-\underset{CH_3O}{\overset{CH_3O}{\bigcirc}}-O)_2PO(OH)$$

(DTMP)

$$(-CH_2-CH-)_m-(-CH_2-CH-)_n$$

OH (23 mol%)

$$CH_3-\underset{CH_3}{\overset{O}{Si}}-CH_3$$ (77 mol%)

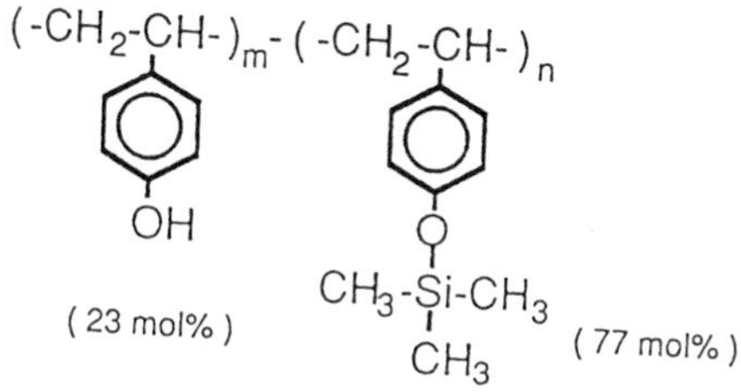

(PSSt)

Experimental

Materials   Syntheses of TMP and DMEP were described previously.[5]

TTMP was obtained by the reaction of phosphorus trichloride with 3,4,5-trimethoxyphenol, by the same method used for TMP.[5]   The product was identified by elemental analysis and by IR and NMR spectroscopies.   The yield was 45.2 %.

4-Methoxyphenol (12.41 g) and phosphorus trichloride (6.88 g) were refluxed for 20 hours in dioxane (200 mL). After dropwise addition of water (0.90 g) in dioxane (20 mL), the solution was refluxed for 2 hours. The solution was dried over anhydrous sodium sulfate, and then, dioxane was removed by evaporation.   The product was purified using a neutral alumina column with 1,2-dichloroethane as the eluent, further purified by two recrystallizations from 1,2-dichloroethane (2.74 g, yield; 37.6 %), and identified by IR and NMR spectroscopies, and elemental analysis.

DTMP was obtained by reacting phosphorus trichloride with 3,4,5-trimethoxyphenol, by the same method used for DMP (yield: 21.4 %).

4,4'-Dimethoxybiphenyl was purified by two recrystallizations of the commercial product from methanol.

Commercial phenol was used after two distillations.

Spectroscopic grade methanol, dioxane, and 1,2-dichloroethane were used without further purification.

PSSt was obtained by treatung poly(4-hydroxystyrene) (Maruzen Petrochemical Co., Ltd, Mn = 5 x 10³) with 1,1,1,3,3,3-hexamethyldisilazane.[6] A highly substituted polymer was obtained by repeating the reaction.   The conversion of the reaction was determined to be 77 mol% by NMR spectroscopy.

Other materials were used without further purification.

Photolysis of Bis(4-methoxyphenyl) Phosphate   DMP (198.94 mg) was irradiated in methanol (500 mL) with a N_2 purge for three hours with a 20 W low pressure Hg lamp. Filters were not used since DMP does not absorb light with a wavelength longer than 300 nm. After evaporation of the solvent, the products were dissolved in ether and washed several times with a 2 % sodium hydroxide aqueous solution. After drying the solution over anhydrous sodium sulfate, the ether was evaporated. Sublimation yielded 66.34 mg of pure product, which was identified as 4,4'-dimethoxybiphenyl by IR and NMR spectroscopies (Yield: 64.8 %).

DMP (ca. 200 mg) was exposed in methanol (500 mL) with a N_2 purge for three hours to 254 nm radiation. After evaporating the solvent, the remaining products were reacted with diazomethane in ether as previously reported.[5] The product was identified to be trimethyl phosphate by gas chromatography (column: silicon grease OV-17) by comparison with a standard material.

Emission Spectrum   Emission spectra of TAPs and DAPs were measured in degassed or nitrogen-saturated methanol by 254 nm light excitation ($d_{254nm} = 0.150 \pm 0.005$). The spectra of 4-methoxyphenol and 3,4,5-trimethoxyphenol were also measured. The quantum yield of the emission (ϕ_F) was determined by means of integrated emission intensities and ϕ_F of phenol was previously reported ($\phi_F = 0.066$).[7]

Determination of Quantum Yield   Determinations of reaction quantum yields were carried out using the 20 W low pressure Hg lamp and a Schott glass filter UG-5. The cycloreversion reaction of r-1,t-2,t-3,c-4-tetraphenylcyclobutane (to yield trans-stilbene; $\phi = 0.67$ in 1-butyl chloride [8]) was used as the actinometer. The reactions were run to 10 % conversion. The yield of bisaryl was determined by using the gas chromatography on an OV-1 column. The acids were titrated with 0.01 N aqueous sodium hydroxide using a phenolphthalein indicator.

Quenching   The reactions of DMP were carried out in bubbling a N_2-O_2 mixture and the effect of oxygen on the bisaryl formation was measured. A quenching of the emissions were also studied by using oxygen as a quencher. Emission intensities at 312 and 370 nm were monitored after bubbling a N_2-O_2 gas mixture.

<u>Single Photon Counting</u>   The decay profiles of the emissions were monitored by the aid of a Horiba fluorescence lifetime measuring apparatus NAES-700F. A H_2 flash lamp was used with a glass filter and solution filters (excitation: pulse width: ca. 1.5 ns, λ_{ex} = ca. 280 nm, λ_{em} < 340 nm). TAPs and DAPs were measured in methanol after N_2 or N_2-O_2 mixed gas bubbling. The optical densities at 280 nm were adjusted to 0.60 ($\pm$0.01). Monitored traces were analyzed by using a program for analysis of triple exponential decay processes.

<u>Sensitivity</u>   Commercial pre-sensitized plates for printing (Fuji FPS type) were used as grained aluminium plates immediately after exposure and development. A solution of polymer (0.50 g in 20 mL of dioxane) and TAP or DAP (50 mg) was spin-coated on a 1.00×10^{-2} m^2 square grained aluminium or quartz plate. After drying under vacuum, the plate coated with 3.0 ($\pm$ 0.02) g m^{-2} of the materials was used. The plates were pre-baked at 100 ℃ for 3 min, and then irradiated with 254 nm light.   The light intensity was monitored with a Spectronics Joul-meter (DRC-100X) with a radiometer sensor (DIX-254). The exposed plates were heated at 100 ℃ for 5 min, and then dipped in an aqueous solution of tetrabutylammonium hydroxide (1.0 % for 5 min). The sensitivity of the film was determined as the minimum energy for satisfactory development of the irradiated polymer.

<u>Results and Discussion.</u>
<u>Photolysis of Aryl Phosphate</u>   Figure 1 shows the absorption spectra of TAPs and DAPs in methanol. Each compound has absorption maxima around 220 and 270 nm. Since the absorption tail ends around 300 nm, the photoreaction of TAPs was carried out by 254 nm irradiation in methanol.

Photolysis of TMP and DMP yielded 4,4'-dimethoxybiphenyl as a main product [ϕ = 0.17 (TMP), 0.12 (DMET)]. [4)]   The photoirradiation of DMP was carried out with 254 nm light for four hours in methanol. The reaction produced acid and 4,4'-dimethoxybiphenyl (A) in good yields (isolated yield: 33.3 %). The quantum yield of the reaction was determined to be 0.05.   Addition of diazomethane-ether solution to the reaction products gave trimethyl phosphate and bis(4-methoxyphenyl) methyl phosphate. These products are derived from main reaction products must be A and phosphoric acid (B), as shown in eq. 3.

$$(Aryl\text{-}O)_2PO(OH) \xrightarrow[\text{in MeOH}]{h\nu} Aryl\text{-}Aryl + H_3PO_4 \quad (3)$$

$$(A) \qquad\qquad (B)$$

<u>Emission Spectrum</u>   During the photoirradiation of DMP, a strong emission was detected. The emission spectra of DMP, DTMP, and TTMP were measured in degassed methanol by 254 nm excitation (d_{254nm} = 0.150 $\pm$ 0.004). The spectra exhibit two emission maxima at 310 nm and around 350 nm (DMP: 370 nm), as shown in Figure 2. The spectra are not affected by dilution of the solution. Accordingly, the maximum at the shorter wavelength is assigned to the monomer fluorescence and the longer wavelength maximum to the intra-molecular excimer as previously assigned in the TMP and DMEP emission spectra.[5]

The fluorescence quantum yield (ϕ_F) of phenol was reported to be 6.6 x 10^{-2}.[7] The ϕ_F values of TTMP, DMP, and DTMP were estimated by

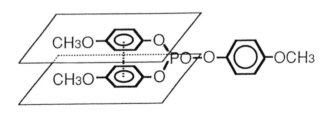

Intramolecular Excimer

comparison with the ϕ_F value of phenol and the integrated emission intensities ($\int I_F d\nu$), listed in Table I [ϕ_{Fex}: ϕ_F of the excimer, ϕ_{Fmono} : that of the monomer]. The spectra of the monomer and the excimer fluorescences have approximately the same shape as those of the phenol derivative and TMP, respectively.

<u>Quantum Yield of Acid Formation</u>   TAPs produced phosphoric acid monoaryl ester upon photoirradiation with 254 nm light and the photoreaction of DAP yielded phosphoric acid. The yield of the acid was determined titremetrically

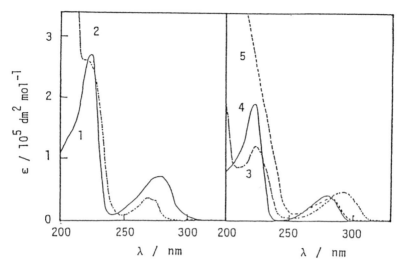

Figure 1. Absorption spectra of TMP (1), DMP (2), DMEP (3), TTMP (4), and DTMP (5) in methanol.

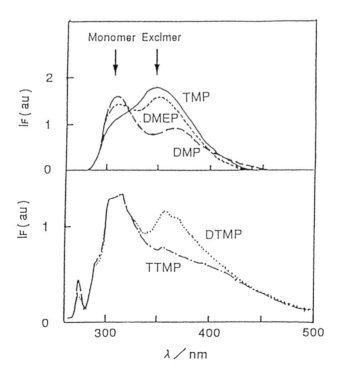

Figure 2. Emission spectra of TMP (1), DMP (2), DMEP (3), TTMP (4), and DTMP (5) in methanol. $\lambda_{ex.}$ = 254 nm, d = 0.150 ±0.005.

Table I. Results of the emissions and the lifetimes of the excited states in methanol

Compound	Monomer Fluorescence		Excimer Fluorescence		Single Photon Counting	
	λ_{max} (nm)	ϕ_{Fmono}	λ_{max} (nm)	ϕ_{Fex}	τ_s (ns)	τ_{ex} (ns)
TMP	305	0.01	350	0.12	4.2	11.4
DMP	312	0.05	370	0.02	3.8	9.8
DMEP	307	< 0.05	350	> 0.02	4.3	12.2
TTMP	312	5×10^{-4}	ca. 360	—	5.9	30.9
DTMP	315	4×10^{-4}	355	3×10^{-4}	5.3	23.0

with 0.01 N aqueous sodium hydroxide solution. The quantum yields for the acid formation (ϕ_{acid}) are listed in Table II. The ϕ_{acid} value of DMP is about twice the value of the bisaryl formation. Phosphoric acid reacts with two molar equivalents of sodium hydroxide to yield Na_2HPO_4. It is concluded that the reactions of the each compound proceeded unimolecularly.

Quenching   The photoreactions of DMP were carried out with N_2-O_2 mixed gas bubbling. When the rates of the quantum yields of the biaryl formation in the absence and presence of oxygen (ϕ_0 / ϕ; ϕ_0: in the absence, ϕ: in the presence of O_2) were plotted against oxygen concentration ($[O_2]_{saturated} = 1.02 \times 10^{-2}$ M in methanol at 25 °C [9]), a linear relationship was obtained, as shown in Figure 3a. The quenching rate ($kq \tau$) was determined to be 214 M^{-1}. The quenching of the DMP emissions was also carried out. The fluorescence

Table II. Quantum yields of the biaryl formation and
the acid formation, and the sensitivity

Compounds	$\phi_{bisaryl}$	ϕ_{acid}	E (mJ m^{-2})
TMP	0.17	0.19	120
DMP	0.05	0.11	90
DMEP	0.12	0.14	160
TTMP	—	0.02	1000
DTMP	—	0.06	180

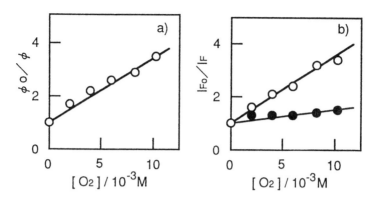

Figure 3. Stern-Volmer analysis of the quenching of the biaryl formation (a) and of the emission intensities (b) by oxygen. The ratio of the quantum yields in the absence to that in the presence of oxygen (a) and the ratios of IFs at λ =312 nm (monomer fluorescence, ●) and 370 nm (excimer fluorescence, ○) in the absence of oxygen to that in the presence of of oxygen were plotted against the oxygen concentration.

intensities at 312 and 370 nm were measured instead of the monomer fluorescence and the excimer fluorescence quantum yields, respectively. The ratios of the fluorescence intensity in the absence of O_2 (I_{Fo}) versus that in the presence of O_2 (I_F) were plotted against the oxygen concentration, as shown in Figure 3b. Each plot gave a good linear relationship with a slope of 50.7 M^{-1} (310 nm), and 247 M^{-1} (370 nm). Since the quenching rate determined for the excimer fluorescence agreed with the quenching rate of bisaryl formation, the bisaryl formation reaction must proceed through the intramolecular excimer.

<u>Single Photon Counting</u>   Lifetimes of the excited singlet state and the intramolecular excimer were measured in methanol by the aid of the single photon counting apparatus. Figure 4 shows a fluorescence decay profile of TMP. The decay profiles were analyzed using a program for three component processes. The lifetimes of the excited singlet state (τ_s) and the excimer (τ_{ex}) were about 4 ns and 11 ns, respectively, except for the τ_{ex} values of TTMP (31 ns) and DMP (23 ns), listed in Table I. The quenching of the emissions of TMP and DMP by oxygen was carried out in methanol. Because

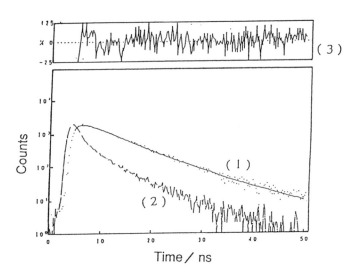

Fig. 4. Decay profiles of the fluorescence of DMP (1) and the lamp intensity (2) measured in the single photon counting in methanol. A fitting plot (3) is also shown.

the excimer fluorescence was quenched completely at a low oxygen concentration, only a quenching rate constant for the monomer fluorescence was determined to be 3.4×10^{10} M^{-1} s^{-1} for TMP and 2.5×10^{10} M^{-1} s^{-1} for DMP.

<u>Application of TAP to Photosensitive Materials</u>   To make a thin photosensitive film, the grained aluminum plate was spin-coated with solutions of a polymer and TAP or DAP. After drying under vacuum, the coated film (film thickness : ca. 3 μm) was heated at 100 ℃ (pre-baking) The irradiation was carried out using a low pressure Hg lamp. After photoirradiation, the film was heated at 100 ℃ (post-baking).   A tetrabutylammonium hydroxide aqueous solution was used as the developer. The unit operations are shown in Figure 5.

Because DAP is an acidic compound,[10] PSSt reacted with DAP without irradiation. Sensitivities (E) were determined as the minimum energy necessary to deprotect the polymer. The results are listed in Table II. The E value is inversely related to ϕ_{acid}. The sensitivity of TMP-PSSt film is the highest among the TAPs studied (E = 12 mJ cm^{-2}).

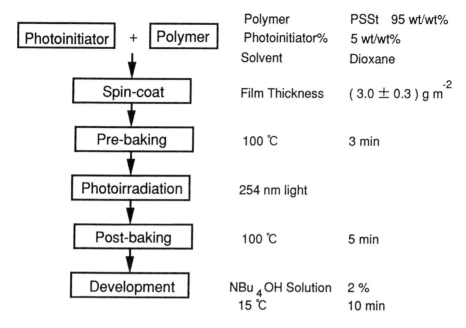

	PSSt	95 wt/wt%
Polymer		
Photoinitiator%	5 wt/wt%	
Solvent	Dioxane	

Film Thickness $(3.0 \pm 0.3)\, g\, m^{-2}$

100 °C   3 min

254 nm light

100 °C   5 min

NBu_4OH Solution   2 %
15 °C   10 min

Fig. 5. Materials and patterning conditions.

Conclusion

The photoreaction mechanism of TAP and DAP is concluded to be the following: the 254 nm irradiation of these compounds yields the intramolecular excimer by the way of the excited singlet state ($^1TAP^*$), as shown in eq. 4. The excimer yields the reaction products, a bisaryl and a phosphorus compound [Aryl-O-PO_3 (TAP) or HPO_4 (DAP)], as shown in eq. 5. The precursor of the reaction is believed to be the intramolecular excimer on the basis of the agreement of the quenching rate of the biaryl formation with that of the excimer emission. The lifetimes of the excited singlet state and the intra-molecular excimer for TAP were determined to be about 4 and 11 ns, respectively. The phosphorus compound yielded by the reaction may abstract hydrogen atoms from the solvent to give phosphoric acid monoaryl ester or phosphoric acid, which are strong and nonvolatile acids, as shown in eq. 6.

Through the application of TAPs and DAP to the photosensitive materials, the acid is generated in the polymer film by 254 nm light irradiation. Deprotection of the polymer, PSSt, is induced by the heat treatment, and finally, the pattern is formed in the non-irradiated polymer. In the case of DAP-PSSt film, since

DAPs are acidic compounds, the degradation proceeds without irradiation. The sensitivity of the thin TMP-PSSt film, the most sensitive film, is determined to be 12 mJ cm^{-2}. This material can resolve 0.6 μ m line and space patterns.

$$TAP \xrightarrow{h\nu} {}^1TAP^* \longrightarrow Excimer \quad (4)$$

$$Excimer \longrightarrow Aryl\text{-}Aryl + Aryl\text{-}O\text{-}PO_3 \quad (5)$$

$$Aryl\text{-}O\text{-}PO_3 \xrightarrow{Sovent} Aryl\text{-}O\text{-}PO(OH)_2 \quad (6)$$

Literature Cited

1) Thompson, L.F.; Willson, C.G.; Bowden, M.J. Edit., " Introduction to Microlithography, ACS Symposium Series 219 ", American Chemical Society, Wasington DC, (1983), P. 153.

2) Allen, N.S. Edit., " Photopolymerization and Photoimaging Science and Technology ", Elsevier Applied Science, London, (1989), p. 99.

3) Ueda, M.; Ito, H., *J. Synthetic Org. Chem., Jpn.* 1991, *49,* 437.

4) Shin, M.; Yamamoto, M.; Okamoto, Y.; Takamuku, S., *Phosphorus, Sulfur, and Silicon,* 1991, *60,* 1.

5) Naito, I.; Nakamura, Y.; Kinoshita, A.; Okamoto, Y.; Takamuku, S., *J. Imaging Sci. Eng.,* 1994, *38,* 49.

6) Canington, W.C., J.P. 63-292128 (1983).

7) Förster, T.; Kasper, K., *Z. Electrochem.,* 1955, *59,* 976.

8) Naito, I.; Tashiro, K.; Kinoshita, A.; Schnabel, W., *J. Photochem.,* 1983, *23,*73.

9) Murov, S.L.; Carmichael, I.; Hug, G.L., " Handbook of Photochemistry, second edition, Revised and Expanded ", Marcel Dekker, Inc., (1993) p. 291.

10) The pH values of DMP and DTMP in pure water (ca. 0.1 mM) were measued to be 3.11 and 3.46, respectively.

RECEIVED October 21, 1994

Chapter 12

Effect of Water on the Surface Insoluble Layer of Chemically Amplified Positive Resists

Jiro Nakamura, Hiroshi Ban, Yoshio Kawai, and Akinobu Tanaka

NTT LSI Laboratories, 3—1 Morinosato Wakamiya, Atsugi-shi, Kanagawa 243—01, Japan

The performance of a chemically amplified resist is deeply affected by water in the air. The resist sensitivity degrades and an insoluble surface layer thickens as an irradiated film is exposed to nitrogen stream containing more water. Thermal desorption spectroscopy (TDS) reveals that the generation of a strong acid promotes the absorption of atmospheric water by films. This absorbed water reduces the rate of an acid-catalyzed reaction, which is obtained by infrared (IR) spectroscopy. The effect of water on the acid-catalyzed reaction is explained in term of the acidity of hydrated protons. Its low acidity leads to a deterioration in the efficiency of the acid-catalyzed reaction.

Chemically amplified resists (1-4) have potential for use in deep-UV, electron-beam (EB), and X-ray lithographies. The concept for this resist system was developed in 1982, and is based on an acid-catalyzed reaction mechanism. A strong acid is generated from an acid generator upon exposure and acts as a catalyst for chemical amplification reactions during subsequent post-exposure baking (PEB). This reaction results in a change in film solubility during development. The reaction mechanisms produce chemically amplified resists with higher sensitivity than conventional ones. However, the reactions required for latent image formation are not completed during the exposure process, constituting a major difference between chemically amplified and conventional resist systems. This causes several problems, including acid diffusion (5-10) during PEB, and degradation (11-16) of the catalytic acid during the interval between exposure and PEB. Airborne contaminants are responsible for the formation of insoluble surface layers in positive resists which in turn gives rise to significant deterioration in sensitivity and resolution. The influence of a small amount of N-methylpyrollidone (NMP) on the formation of insoluble layers has already been reported by IBM groups (12-14). On the other hand, water has a higher basicity than other functional groups present in common resist materials and its content in ambient air is on the order of one percent; much higher than other basic compounds. Water can thus influence acid-catalyzed reactions if it is absorbed by resist films.

This paper describes the effects of water on the degradation of lithographic characteristics of a chemically amplified positive resist. We evaluate the changes in resist sensitivity during the interval between exposure and PEB, and measured the thickness of the insoluble surface layer. We also investigate the water uptake in the film before and after exposure, and the acid-catalyzed reaction that occurs after exposure to moisture. We also discuss the effect of water on acid-catalyzed reactions, focusing on proton transfer.

Experimental

We analyzed a chemically amplified resist (17) composed of partially t-butoxycarbonyl(t-Boc)-protected poly(hydroxystyrene) (PHS), t-Boc-protected bisphenol-A as a dissolution inhibitor, and di(4-t-butylphenyl)iodonium triflate as an acid generator for producing trifluoromethanesulfonic acid during exposure. Diethylene glycol dimethyl ether (diglyme) was used as the casting solvent, and a small amount of NMP was added to the resist material to enhance its stability against the delay time. The resist was spincoated on a Si wafer to a thickness of 1.0 μm for evaluating the lithographic characteristics and prebaked on a hot plate at 100°C for 120 s. The resist film was irradiated with an EB with an acceleration voltage of 30 kV and stored under various conditions. It was then post-exposure baked on a hot plate at 75°C for 120 s and developed in a 2.4% aqueous tetramethylammonium hydroxide solution for 60 s.

The experimental system used for controlling the atmospheric vapor content is shown in Figure 1. Nitrogen with a flow rate of 0.5-5 L/min was bubbled through water or organic compounds, mixed with a flow of dry nitrogen, and introduced into the reaction chamber. The concentration of vapors was controlled by varying the flow rates.

We used a cross-sectional development method, as shown in Figure 2, to evaluate the thickness of the insoluble surface layer. In this method, the exposure and PEB are performed in a conventional manner. Next, the film on the Si wafer is cut orthogonally to a line-and-space pattern to expose a cross-section, and is then developed. Finally, we measured the thickness of the resist above the lines and the linewidth.

Thermal desorption spectroscopy (TDS) was used to analyze the uptake of water in the films. TDS was carried out from room temperature to 300°C at a heating rate of 20°C/min after the base pressure was evacuated to less than 5×10^{-9} Torr. The spectrum of water ($M/e = 18$) was measured by a quadrupole mass analyzer. To evaluate the catalytic ability of acids after being stored in a nitrogen stream containing water, infrared (IR) spectroscopy measurements were taken both before exposure and after PEB using a Perkin Elmer 1760X spectrometer. The catalytic reaction was analyzed by following the peak at 1760 cm^{-1} assigned to carbonyl groups of t-Boc.

Results and Discussion

Sensitivity Stability.   Examples of resist pattern profiles are shown in Figure 3. Photomicrograph (a) shows a developed image when exposure was immediately followed by PEB and (b) shows a developed image of film stored in our laboratory for 60 min. A T-top shape appeared in the profile of the stored resist. It is difficult to identify the specific effects of each airborne contaminant in the ambient laboratory atmosphere. In this study, we distinguished their effects using the experimental system shown in Figure 1. The sensitivities of the resists stored in nitrogen containing various organic compounds or water are shown in Figure 4. Here, the sensitivity means the dose required to clear an exposed resist film of a 20x300-μm^2 pattern. The approximate vapor concentrations are shown in brackets. As expected, organic vapors with a high basicity greatly reduced the resist sensitivity while the influence of hydrocarbon compounds and ether compounds was negligibly small. Water was also

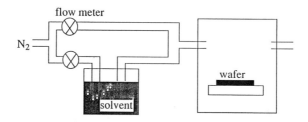

Figure 1. Experimental system for controlling atmosphere.

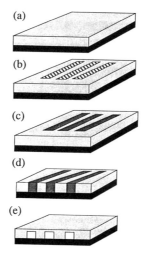

Figure 2. Sequence for the cross-sectional development method. (a) spin coat and prebake, (b) EB irradiation, (c) post-exposure bake (d) wafer cutting, (e) development.

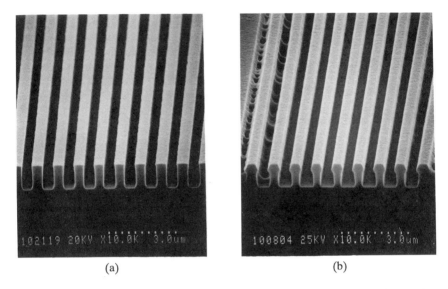

(a)   (b)

Figure 3. SEM photograph of replicated resist patterns. (a) no delay time and (b) delay time of 60 min. Pattern size was 0.5 μm.

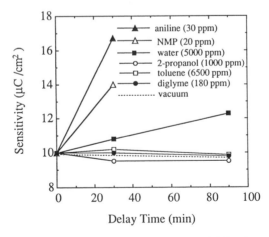

Figure 4. Effects of several kinds of solvents on sensitivity degradation. Delay time between exposure and PEB was 30 min. The evaluated pattern width was 20 μm.

observed to degrade the resist sensitivity. As reported by an IBM group (*12*), the change in performance of our resist was also negligibly small when the film was exposed to water before irradiation. On the other hand, the exposure to water after irradiation had a strong influence on the resist performance.

The basicity of water is weaker than that of NMP and aniline, but still stronger than that of the other functional groups present in the resist. Water incorporated in the film should influence the t-Boc group decomposition reaction. At identical concentrations, the effects of water may be smaller than those of organic compounds with high basicity. Nonetheless, the concentration of water in ambient air is on the order of one percent and is usually much higher than that of other basic compounds. As a result, water contributes significantly to the time delay problem.

The sensitivities of resist films stored in nitrogen containing various concentrations of water are shown in Figure 5. Converted values for relative humidity at 23°C are also shown for reference. The open and closed circles represent the sensitivity of films stored in a controlled nitrogen atmosphere and in air in our laboratory, respectively. The air in the laboratory means uncontrolled air and thus contains other airborne contaminants as well as water. The sensitivities became worse with higher humidity. After 30 min in a nitrogen stream at 40% humidity, the change in sensitivity is about 10%. The sensitivity degraded rapidly when the humidity exceeds 40%. This critical humidity, beyond which the sensitivity degrades rapidly, becomes lower with increasing delay time. The resist sensitivities of the films exposed to nitrogen streams containing water were not restored even after being stored in dry nitrogen or in a vacuum for 12 hours, presumably because protonated water is difficult to desorb from resist films. Moreover, the sensitivity of resist films exposed to the laboratory air was about the same as that of the films stored in a nitrogen flow at the same relative humidity. This confirms that atmospheric water plays a significant role in the time delay problem of our resist system. To improve the resist stability against the time delay, it is thus important to use an overcoat film to prevent water from penetrating the resist film. It has already been demonstrated that overcoat films such as poly(acrylic acid) and poly(α-methylstyrene) are able to improve the stability. (Nakamura, J., *Polymer Preprints*, in press)

Determination of the Depth of the Insoluble Surface Layer. The sensitivity degradation mentioned above results from a decrease in the dissolution rate at the film surface due to deactivation of catalytic acids by water. We used cross-sectional development to determine quantitatively the thickness of the insoluble surface layer. With this method, the thickness of this layer can be evaluated more accurately than with conventional development because the resist film is developed laterally. In addition, a change in the solubility of the bulk resist under an insoluble layer can also be observed. An example of the replicated pattern and evaluation parameters are shown in Figure 6. Here, two variables were measured. The thickness of the insoluble layer T is used for evaluating the influence of airborne contamination and the linewidth on the bottom W is used for the change in the linewidth caused by the dark reaction that occurs at room temperature during the interval between exposure and PEB. The dependence of insoluble layer thickness on the delay time between exposure and PEB is shown in Figure 7 as a function of relative humidity. The insoluble layer becomes thicker with increasing delay time. This indicates that water absorbed on the surface diffuses into the film during the delay time. The relationship between the thickness and the square root of the delay time is shown in Figure 8. The thickness was observed to be approximately proportional to the square root of the delay time. The relationship

$$C(x,t) = C_0 \{ 1 - erf[x/2(Dt)^{1/2}] \} \tag{1}$$

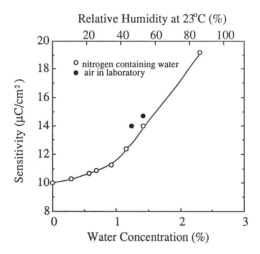

Figure 5. Dependence of sensitivity on water concentrations. The delay time between exposure and PEB was 30 min. The evaluated pattern was 20 μm wide.

T : Insoluble layer thickness
W : Linewidth

Figure 6. An example of the replicated pattern using cross-sectional development and evaluated parameters.

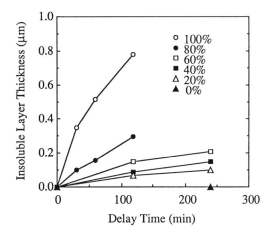

Figure 7. Dependence of the thickness of the insoluble surface layer on relative humidity.

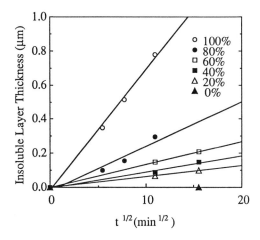

Figure 8. Thickness of the insoluble surface layer vs. (delay time)$^{1/2}$.

can be obtained from Fick's law if the water concentration C_0 on the surface of the film is assumed to be constant (*18*). Here, x is the depth from the surface, D is the diffusion constant of water, and t is the delay time. If the critical water concentration needed for the formation of an insoluble layer and D are kept constant, the argument of the error-function will also be constant. Therefore, x is proportional to the square root of t. The experimental results shown in Figure 8 thus indicate that the movement of water in resist films is governed by Fick's law.

The dependence of the formation rate of the insoluble layer, as represented by the slopes in Figure 8, on the relative humidity is shown in Figure 9. This rate increased with higher humidity. The linewidth at the bottom, however, increased only slightly with an increase in delay time, and was unaffected by humidity. For example, the variation for a 0.8-μm pattern was less than 10% over 180 min. This slight increase seems to be caused by dark reactions.

Water Absorption by Film.   The TDS spectra of water from the resist films after they were stored in 75% humidity for 24 hours are shown in Figure 10. Here, PHS containing 10 wt% of the acid generator was used as a model compound and the films were 100-nm thick. The solid line represents the result for a film exposed to deep-UV light and the dotted line that for an unexposed film. The amount of water absorbed in the exposed film is much larger than that in the unexposed film, presumably because generation of a strong acid by exposure results in an increase in the affinity for, and the absorption of water by the film. The vapor uptake in the film is controlled by solubility and diffusibility. Water is absorbed on the film surface and then diffuses into the bulk film. The film used in this study was very thin and the storage time was very long. As a result, the water absorption reached saturation. The difference between exposed and unexposed film was, therefore, caused by the solubility change. Moreover, the TDS results for the model compound make it clear that time delay between exposure and PEB influences the pattern profile significantly while the effects of delays between coating and exposure are less significant.

Decomposition of t-Boc.   Our resist system utilizes the change in the dissolution rate of a resist film in an alkaline solution caused by decomposition of the t-Boc group. We examined the degree of t-Boc decomposition in the exposed film after being stored under various conditions, to clarify the dissolution characteristics described above. The t-Boc conversion is shown in Figure 11 as a function of exposure dose. A 0.4-μm thick film was used to obtain information about the film's upper layer. The conversion was about 40% at an exposure dose of 10 μC/cm^2, which is the dose needed to clear the exposed film when PEB is at 75°C for 120 min. The dependence of t-Boc decomposition on delay time is shown in Figure 12. Here, the relative humidity is 40%, which is a typical condition in clean rooms, and the exposure dose was 14.4 μC/cm^2. At a relative humidity of 40%, the t-Boc conversion was greatly reduced with longer delay time. The rate of the acid-catalyzed reaction at a delay time of 300 min is about 20% lower than the initial value. As this is the average value for the 0.4-μm thick film, degradation in the vicinity of the film surface is more significant. On the other hand, the t-Boc deprotection for a film stored in vacuum slightly increased with an increase in delay time. This indicates that the reaction proceeds slowly at room temperature. The sensitivity change caused by this reaction was less than 10% at a delay time of 180 min. Dependence of t-Boc decomposition on relative humidity is shown in Figure 13. The delay time here was 30 min. The degree of deprotection decreased with higher relative humidity, to approximately one half at 90% humidity. This decrease in t-Boc decomposition causes a decrease in dissolution rate, resulting in a lower resist sensitivity at high humidity.

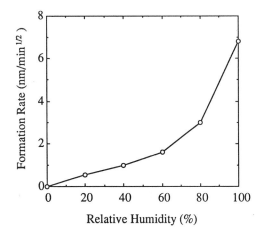

Figure 9. Dependence of the formation rate for the insoluble layer on relative humidity.

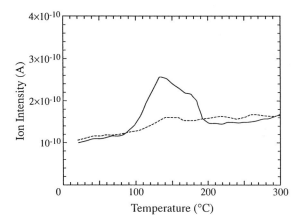

Figure 10. TDS spectra of water absorbed by film. The solid line represents the result for exposed film and the dotted line for unexposed film.

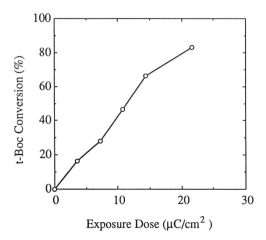

Figure 11. Dependence of t-Boc decomposition on exposure dose.

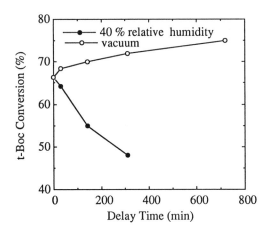

Figure 12. Dependence of t-Boc decomposition on delay time. Relative humidity was 40% and the exposure dose was 14.4 $\mu C/cm^2$.

Effect of Water on the Rate of t-Boc Decomposition. Here we discuss the effect of atmospheric water on the decomposition of t-Boc, focusing on proton transfer in the resist film. However, it should be pointed out that other factors may be involved, such as the appearance of another reaction path in the presence of water. Table I shows the pK_a values (19) of the major functional groups present in our resist system and of the materials used as an airborne contaminants in this study. The acid generators produce trifluoromethanesulfonic acid during exposure. This is more acidic than other functional groups present in the resist film. Therefore, trifluoromethanesulfonic acid releases a proton, which is transferred to the base polymer and/or residual casting solvent. The ether group of diglyme is the most basic one and therefore the proton is transferred to it in the absence of airborne impurities. However, water has a stronger basicity than the casting solvent and the proton is thus transferred to water when it is present. A proton transfer to t-Boc is required to decompose the t-Boc group. However, the acidity of protonated water is weaker than that of protonated casting solvents, thus leading to a deterioration in the efficiency of t-Boc decomposition. This results in a decrease in the reaction rate under high humidity. The same idea might be applied to other chemically amplified resist systems using acid-catalyzed reactions where water is not directly involved.

Table I. pK_a values for acids

Acid	Approximate pK_a (relative to water)
$Ar\!-\!OH_2^+$	-6.4
$R\!-\!\overset{+}{\underset{H}{O}}\!-\!R$	-3.5
H_3O^+	-1.74
$R\!-\!\overset{}{\underset{OH^+}{C}}\!-\!NH_2$	-0.5
$Ar\!-\!NH_3^+$	3-5

In general terms, the possibility exists that when any other impurities with stronger basicity than the base polymer and casting solvent are absorbed by the resist film, the acid-catalyzed reaction rate decreases to some extent, depending on the degree of their basicity. Actually, the results mentioned in Fig. 4 can be well explained by the pK_a values. The vapors with higher pK_a of protonated base have a stronger influence on the time delay problem. Moreover, this explanation in term of pK_a values indicates that the acidity of the protonated casting solvent becomes low, resulting in a decrease in initial sensitivity but an increase in the stability of resist performance during the interval between exposure and PEB. The resist used in this study contains a small amount of NMP as a casting solvent to enhance stability against the delay time. The addition of NMP increases the total basicity of the casting solvent which in turn improves the stability of resist performance against the delay time. A comparison of the sensitivity change between resists with and without addition of 0.4% NMP is shown in Figure 14. The addition of NMP reduced the initial sensitivity level from 3

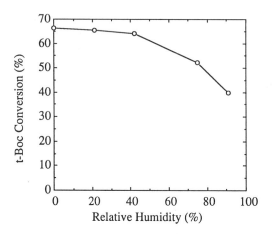

Figure 13. Dependence of t-Boc decomposition on relative humidity. Delay time was 30 min and the exposure dose was 14.4 μC/cm².

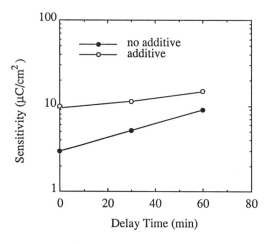

Figure 14. Effect of NMP addition to resist material on the time delay problem.

$\mu C/cm^2$ to 10 $\mu C/cm^2$. The sensitivity of the resist without NMP at delay time of 60 min degraded by approximately 3 times the initial value. In contrast, the sensitivity of the resist with NMP only degraded by about 30% at a delay time of 60 min. The addition of NMP as a casting solvent plays the same role as the use of a solvent with higher basicity.

Conclusion

We have investigated the influence of water and organic vapors in atmosphere on the lithographic performance of a chemically amplified positive resist. When an irradiated film was exposed to water and organic vapors with high basicity, the resist sensitivity became worse. On the other hand, hydrocarbon compounds and ether compounds had a negligibly small influence. The formation rate of an insoluble surface layer, which was determined by cross-sectional development, increased with increasing humidity. The thickness of this layer was approximately proportional to the square root of the time between exposure and PEB. This suggests that the water absorbed in the surface diffuses into the film, governed by Fick's law. TDS analysis suggests that the generation of catalytic acids promotes water uptake in the film. This is why the effect of time delay between exposure and PEB is more significant than the time delay between coating and exposure. Absorbed water decreased the rate of the deprotection reaction of t-Boc, because the low acidity of hydrated protons degrades the efficiency of proton transfer to t-Boc.

Acknowledgments

The authors wish to thank Tetsushi Sakai and Tadahito Matsuda for their advice and encouragement. They are also grateful to Norikuni Yabumoto for his TDS analysis and to Daisuke Takagi for his assistance in the experiment. They would also like to thank the staff of Project Team 2 at NTT LSI Laboratories for their contribution to this work.

Literature Cited

1) Ito, H.; Willson, C. G. *Polym. Eng. Sci.* **1983**, *23*, 1012.
2) Liu, H.; deGrandpre, M. P.; Feely, W. E. *J. Vac. Sci. & Technol.* **1988**, *B7*, 379.
3) Lingnau, J.; Dammel, R.; Theis, J. *Polym. Eng. Sci.* **1989**, *29*, 874.
4) Ban, H.; Nakamura, J.; Deguchi, K.; Tanaka, A. *J. Vac. Sci. & Technol.* **1991**, *B9*, 3387.
5) McKean, D. R.; Schaedeli, U.; MacDonald, S. A. In *Polymers in Microlithography;* Reichmanis, E.; MacDonald, S. A.; Iwayanagi, T., Eds.; ACS Symp. Ser. 412; ACS: Washington, D. C. 1989, 27-38.
6) Schlegel, L.; Ueno, T.; Hayashi, N.; Iwayanagi, T. *J. Vac. Sci. & Technol.* **1991**, *B9*, 278.
7) Nakamura, J.; Ban, H.; Deguchi, K.; Tanaka, A. *Jpn. J. Appl. Phys.* **1991**, *30*, 2619.
8) Nakamura, J.; Ban, H.; Tanaka, A. *Jpn. J. Appl. Phys.* **1992**, *31*, 4294.
9) Fedynyshyn, T.; Cronin, M.; Szmanda, C. *J. Vac. Sci. & Technol.* **1991**, *B9*, 3380.
10) Trefonas, P.; Allen, M. T. *Proc. SPIE* **1992**, *1672*, 74.
11) Nalamasu, O.; Cheng, M.; Kometani, J. M.; Vaidya, S., Reichmanis, E.; Thompson, L. F. *Proc. SPIE* **1990**, *1262*, 33.
12) MacDonald, S.; Clecak, N.; Wendt, R.; Willson, G.; Snyder, C.; Knors, C., Deyoe, N.; Maltabes, J.; Morrow, J.; McGuire, A.; Holmes, S. *Proc. SPIE* **1991** *1466*, 2.

13) Hinsberg, W.; MacDonald, S.; Clecak, N.; Snyder, C.; *Proc. SPIE* **1992**, *1672*, 24.

14) Ito, H.; England, W.; Sooriyakumaran, R.; Clecak, N.; Breyta, G.;Hinsberg, W.; Lee, H.; Yoon, D. *J. Photopolym. Sci. Technol.* **1993**, *6*, 547.

15) Kozawa, T.; Yoshida, Y.; Uesaka, M.; Tagawa, S. *Jpn. J. Appl. Phys.* **1992**, *31*, 4301.

16) Padmanaban, M.;Endo, H.; Inoguchi, Y.; Kinoshita, Y.; Kudo, T.; Masuda, S.; Nakajima, Y. *Proc. SPIE* **1992**, *1672*, 141.

17) Kawai, Y.; Tanaka, A.; Matsuda, T. *Jpn. J. Appl. Phys.* **1992**, *31*, 4316.

18) Crank, J. *The Mathematics of Diffusion* ; Clarendon Press: Oxford, 1975, Chaps. 2 and 3.

19) March, J. *Advanced Organic Chemistry* ; John Wiley & Sons: NY, 1992, Chap. 8.

RECEIVED September 13, 1994

Chapter 13

Thermal Properties of a Chemically Amplified Resist Resin

Koji Asakawa, Akinori Hongu, Naohiko Oyasato, and Makoto Nakase

Research and Development Center, Toshiba Corporation 1, Komukai Toshiba-cho, Saiwai-ku, Kawasaki 210, Japan

The thermal properties of a co-polymer of p-t-butoxycarbonyl-methoxystyrene and 4-hydroxystyrene (BCM-PHS) - which is used as chemical amplification resist resin - were investigated in terms of the thermal decomposition of the p-t-butoxycarbonylmethoxystyrene group (dissolution inhibitor) and the glass transition temperature of base polymer. Simple relationship between the inhibitor fraction, activation energy, glass transition temperature, etc. were derived. The thermal margins for chemical amplification were predicted from these results.

In typical acid-catalyzed chemical amplification resists, resist pattern formation comprises a series of processes, that is, acid is generated from a photo acid generator by exposure, and then decomposes the dissolution inhibitor which protects the resist from dissolving in the developer. This results in differential dissolution rates between the exposed and unexposed areas in the developer. Resists with superior performance are designed to have a greater differential dissolution rate between the exposed and unexposed parts, thus offering higher contrast.

Thermal decomposition of the inhibitor is unfavorable from the viewpoint of resist contrast, but it is necessary to understand this phenomenon in order to specify the resist processing conditions, such as the pre-exposure bake (pre-bake) or the post-exposure bake (PEB) conditions. It is generally thought that the inhibitors in not only the exposed areas but also the unexposed areas are decomposed by heat, thus degrading resist contrast and resolution.

It has been reported in previous studies (1-3) that the decomposition temperature increases with increasing inhibitor fraction. It was found that this increase is attributable to acid-catalyzed self decomposition of the resist, which is caused by the acid generated from the phenolic hydroxy group. This paper reports the results of investigation of thermal properties of positive chemical amplification resist resins, that consist of poly(4-hydroxystyrene) (PHS). In addition, a simple relationship between inhibitor fraction, activation energy, and thermal decomposition of the resist has been found.

0097–6156/94/0579–0165$08.00/0

Another factor, aside from thermal decomposition, that must be considered for thermal stability of resists is the glass transition temperature (Tg) of the base polymer, and to this end, the inhibitor fraction dependence of Tg was investigated for this co-polymer. This factor has a strong influence on the performances of the resist.

Experimental

Materials. A random co-polymer of p-t-butoxycarbonylmethoxystyrene and 4-hydroxystyrene (BCM-PHS) was used in this study. BCM-PHS was synthesized by reaction of t-butoxycarbonyl-α-bromoacetate with poly(4-hydroxystyrene) (PHS) (M_w = 5.1×10³) (2) in the presence of potassium carbonate and potassium iodide according to the literature. The t-butoxycarbonylmethoxy group of BCM, referred to as an inhibitor in this paper, makes this resist insoluble in an alkaline developer. This inhibitor decomposes into carboxylmethoxy group upon heating in the presence of acid (Scheme I).

Measurement of Dissolution rate.   After treatment to promote adhesion, the 1-acetoxy-2-ethoxyethane (ECA) solution of BCM-PHS was spin-coated on silicon wafers. The wafers were baked at various temperatures for various times according to the experimental requirements. They were then placed in a dissolution rate monitor (DRM) to measure the dissolution rate in 0.28 N aqueous tetramethylammonium hydroxide as an alkaline developer. The temperature of the developer was 23 °C.

TG Measurement.   The ECA solution of BCM-PHS was spin-coated onto silicon wafers and baked at 95 °C for 90 s. Then, about 5 mg of the film was removed from the wafer. The weight loss of the BCM-PHS was measured by thermogravimetry (TG) as it was heated at a certain rate from 25 to 250 °C.

DSC measurement.   Tg was measured by differential scanning calorimetry (DSC). The temperature was set to 70 °C for several minutes to remove the casting solvent. It was then reduced to -20 °C and measurements began at a heating rate of 5 °C/min up to 220 °C.

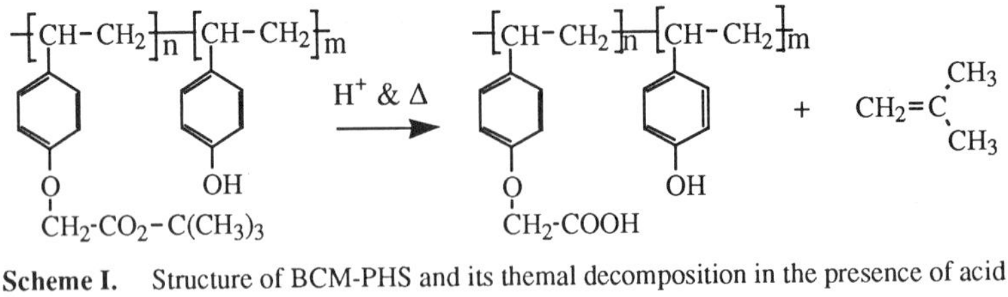

Scheme I.   Structure of BCM-PHS and its thermal decomposition in the presence of acid

Results and discussion

Time and temperature dependence of thermal decomposition.   The dissolution rate of a resist in an alkaline developer increases considerably because of changes in the polarity of the polymer if the dissolution inhibitor decomposes for some reason. The dissolution rate may even quadruple if the inhibitor fraction decreases from 25 to 20 mol%, which means only a 20 % decomposition of the inhibitor. This extremely sensitive inhibitor fraction dependence of the dissolution rate makes it easy to measure decomposition rates even when only a small amount of the inhibitor decomposes.

Figure 1 shows the change in dissolution rate after a certain baking time at a baking temperature of 95 °C. The dissolution rate increases with longer baking time. The dissolution rate doubled, for example, with a baking time of 4 hours. This can be converted into a inhibitor decomposition ratio if the relationship between dissolution rate and inhibitor fraction is known.

The dissolution rates of blends of pure PHS and BCM-PHS in the developer at 23.0 °C were measured to estimate the remaining inhibitor ratio after the baking process. The dissolution rate, D_R, appears to obey the empirical equation given below in the range of inhibitor fraction, C, from 5 to 30 mol% (8). This exponential relationship has not, however, been theoretically clarified for chemical amplification resists (5-7).

$$\log D_R = -0.12C + 2.58 . \tag{1}$$

Assuming that the co-polymer of BCM-PHS and the polymer blend of PHS and BCM-PHS have the same dissolution rates when both contain equal amounts of inhibitor. The dissolution rate can be thus converted into a baking time dependence of the thermal decomposition ratio with Equation (1), as shown on the right-hand axis of Figure 1. More inhibitor decomposed with a longer baking time. For example, a baking time of 4 hours resulted in 12 % inhibitor decomposition. It should be noted that thermal decomposition occurred at the low temperature of 95 °C. The half-life of the inhibitor decomposition can be estimated using the method of least-squares. It was found to be 8.1×10^4 s.

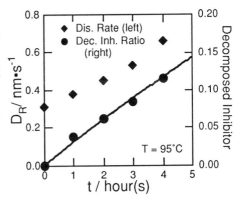

Figure 1. Baking time dependence of dissolution rate at a baking temperature of 95°C (left-hand axis). Developer was a 0.28N aqueous TMAH and its temperature was 23.0°C. Baking time dependence of the decomposed inhibitor at a baking temperature of 95°C (right-hand axis). The initial inhibitor-fraction was 26 mol%. It was converted using Equation (1).

Figure 2 shows the ratio of decomposed inhibitor versus the inverse of baking temperature, T^{-1}. The baking time was 800 s in this experiment. The estimation method was the same as that used in the time dependence experiments. Since decomposition of the inhibitor is considered to be a chemical reaction, the decomposition rate constant, k, is defined as:

$$k = (\ln2/t)\log(C_0/C), \qquad (2)$$

where C is the fraction of the inhibitor at a certain baking time, t, and C_0 is the initial fraction.

When the activation energy of thermal decomposition is Ea, k is expressed by the following Arrhenius equation:

$$k = k_0\exp(-Ea/RT), \qquad (3)$$

where R is the gas constant, T is the absolute temperature, and k_0 is the limit of decomposition rate at infinite temperature.

The inhibitor decomposition ratio was found to be proportional to the exponent of T^{-1}. This suggests that Equation (3) describes these phenomena well. Ea was calculated from the slope of the least-squares fit to the plots and was 159 kJ/mol. k_0 was also estimated to be 2.15×10^{17} s.

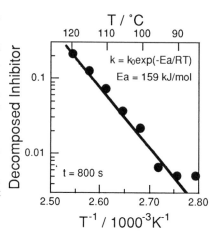

Figure 2. Baking temperature dependence of the decomposed inhibitor at a baking time of 800 s. The inhibitor-fraction was 26 mol%.

Measurement and simulation of TG. The weight loss of the resin was measured by TG from 25 to 250 °C at various heating rates. The experimental results for 4.5 °C/min are shown by the solid line in Figure 3. The vertical axis is the remaining inhibitor weight (a.u.) and the horizontal axis is temperature (°C). The horizontal axis also corresponds to time, because the heating rate was constant at 4.5 °C/min. The initial decomposition temperature, T_d, was defined as the point of intersection between the tangent from the lowest temperature and the tangent of the point where 50% of the inhibitor was decomposed. This definition is typically used in studies in the resist field. It should be noted, however, that T_d is not the temperature at which thermal decomposition is about to start. T_d was 138 °C for a heating rate of 4.5 °C/min from these results. Beyond this temperature, sample weights decreased rapidly because the inhibitors rapidly decomposed obeying Equation (3).

The weight loss of the resin during heating is equal to the total mass of decomposed inhibitor, which can be calculated by integrating Equation (3). This calculation is equivalent to simulating a TG measurement if carried out for a certain heating rate. The following difference equation is thus predicted:

$$W_{n+1} = W_n[1 - k_0\exp\{-Ea/RT(t)\}\Delta t], \qquad (4)$$

where W_n is the weight of the remaining inhibitor at a certain time and W_{n+1} is the inhibitor remaining after a very short time Δt.

The calculation results are shown in Figure 3. This calculation agrees very well with the experimental TG curves. T_d seems to be about 138 °C from these results. This curve was, however, obtained using the difference Equation (4) which is based on the assumption that Equation (3) is appropriate. Therefore, it is possible to derive the activation energy using the TG curve from Equation (4).

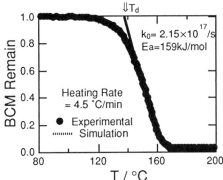

Figure 3. Comparison between the experimental results of TG and the calculated curve of equation (4) (heating rate : 4.5°C).

Inhibitor fraction dependence of thermal decomposition. Figure 4 shows the simulated TG curve calculated using Equation (4) at various Ea when the heating rate is 4.5 °C/min. The curves shifted to higher temperatures as Ea increased and thus T_d also increased as Ea increased. The experimental T_d data are shown as a function of the inhibitor fraction on the left-hand axis of Figure 5. This indicates that T_d increases as the inhibitor fraction becomes larger.

The relationship between Ea and C can be derived from these experimental data and by curve-fitting to thermal decomposition. The results, which were derived from the calculation using Equation (4) and definition T_d is shown on the right-hand axis of Figure 5. Ea for C of 100 mol% was estimated to be approximately 190 kJ/mol by extrapolating to an inhibitor fraction of 100 mol%, although the theoretical relationship between the C and the Ea has yet to be clarified.

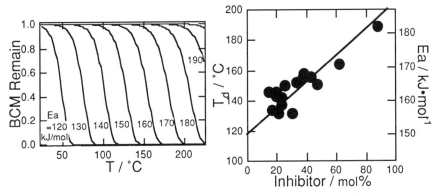

Figure 4. Activation energy dependence of the thermal decomposition temperature calculated from equation (4) at a heating rate of 4.5°C/min.

Figure 5. Inhibitor concentration dependence of thermal decomposition temperature (left-hand axis) and activation energy (right-hand axis).

Estimation of thermal decomposition time.
The thermal decomposition rates of these resins during the baking process can be predicted from the activation energy as estimated in this study. Figure 6 shows the half-life for inhibitor decomposition, τ, during baking, as an example of a thermal decomposition rate where half of the inhibitor is decomposed by heat. These values are calculated from the following equation:

$$\tau = \ln 2/k, \qquad (5)$$

where k can be calculated from Equation (3).

Figure 6 indicates that the half-life for thermal decomposition of the inhibitor decreases rapidly as the baking temperature increases. The inhibitor fraction dependence of activation energy of the unexposed areas

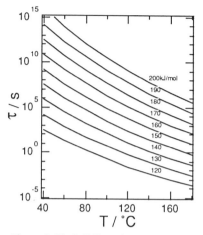

Figure 6. The half-lives of the inhibitor decomposition at various activation energies.

can be evaluated from Figure 6. The half-life becomes longer as the inhibitor fraction increases because of the greater activation energy. Figure 6 also shows that the ratio of decomposition rates between greater and smaller inhibitor fractions decreases at higher baking temperatures.

Although thermal decomposition cannot be avoided, it can be made small enough to be neglected, meaning that its dissolution rate in the developer hardly changes. Assuming that 1% of the inhibitor decomposes, the dissolution rate is then approximately 1.3 times greater. This value is considered small enough to not affect resist performance. The remaining inhibitor ratio, W(t)/W(t=0), can be described as follows:

$$W(t)/W(t=0) = 0.5^{t/\tau}. \qquad (6)$$

where t is the baking time and τ is the half-life which is a function of baking temperature and activation energy. Therefore, if the half-lives were about 7000 s, only 1 % of the inhibitor would decompose during a bake of 100 s. The dissolution rate becomes considerably greater at half-lives below 7000 s. This temperature is thus thought to be the upper limit for the baking process. For example, an Ea of 159 kJ/mol, which corresponds to an inhibitor fraction of 26 mol%, must be baked below 120 °C.

The Ea value of the exposed area of this resist has already been measured in a previous study (8). It was found to be 117 kJ/mol when the resist contained 1 wt% of triphenylsulphonium trifluoromethanesulphonate (TPS-OTf) as a photo-acid generator (PAG). The half-lives of the inhibitor on exposed areas are predicted to fall just below the curve of 120 kJ/mol in Figure 4. The difference in decomposition rate between the exposed and unexposed areas tends to decrease at higher baking temperatures and the contrast becomes inferior. From this viewpoint, a lower temperature post-exposure bake (PEB) effectively gives better resist performance.

Mechanism for thermal decomposition of the inhibitor. The mechanism by which the inhibitor thermally decomposes can be simply and uniquely described in terms of the activation energy, both for the exposed and unexposed areas, during the baking process. The thermal decomposition of a resist is caused by a small number of protons generated by phenolic hydroxy groups; more protons are generated as the ratio of the hydroxy group increases (decreasing ratio of inhibitor). Increasing the proton concentration reduces the activation energy for the decomposition of inhibitors. The thermal decomposition temperature of resists thus decreases rapidly with decreasing inhibitor fraction. For example, these protons reduce the activation energy for thermal decomposition from 190 to 159 kJ/mol when the inhibitor fraction decreases from 100 to 26 mol%. Since the phenolic hydroxy group is a weak acid and generates only a small number of protons, this indicates that chemical amplification resists are extremely sensitive to acidic contamination, such as by organic acid.

Furthermore, the activation energy of the exposed areas is drastically reduced when considerably more protons are generated by PAG: it decreases to 117 kJ/mol under the influence of acid generated from exposed PAG (8). Unexposed PAG does not reduce the activation energy. In chemical amplification resists, an acid acts as a catalyst, which means that one definite role of the acid is to reduce the activation energy. Good resists have a large differential activation energy between the exposed and unexposed areas for higher contrast. A drastic reduction in the activation energy is thus an essential condition for a good PAG.

Dissolution rate at a certain baking time. A lower PEB temperature is not suitable for the process because it would entail on extremely long PEB time although this would effectively improve resist contrast. In spite of this, decomposition rates at a certain limited PEB time can be estimated from Equation (3). Dissolution rates in the developer should be found to clarify this relationship, since resist patterns are formed by the difference in the dissolution rates between the exposed and unexposed areas.

Figure 7 shows the PEB temperature dependence of the dissolution rate for individual activation energies as derived using Equations (1) and (3). The PEB time here was fixed at 100 s. This clearly indicates the temperature margin for the baking process. For example, the PEB temperature must be lower than 140 °C for resists whose Ea for the unexposed areas is 160 kJ/mol. It must be higher than 70 °C for resists with an Ea of 120 kJ/mol for the exposed areas. These margins might be expected to be somewhat narrower in practice than this prediction because of other uncertainties such as exposure defocus, non-uniformity of temperature on the hot-plate, etc.

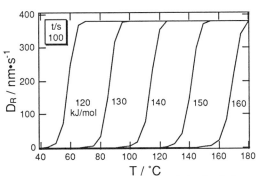

Figure 7 Temperature dependence of the dissolution rate with the developer at various activation energies. Baking time was set to be 100 s.

Relationship between Tg and inhibitor fraction.   Another thermal stability factor besides thermal decomposition is the Tg of the base polymer. Tg is defined as the temperature at which the polymer changes from a glassy to a rubber state. On a molecular level, the polymer backbones are fixed in a glassy state but can move in a rubber state. The resist pattern would be inferior if the resist was heated to a temperature higher than Tg after the exposure because of the mutual diffusion of the polymer coil of the base resin.

The DSC curves are shown in Figure 8. The heating rate was 5 °C/min. The Tg of PHS, which forms the base polymer of the resin, was 146 °C from this measurement. This temperature is considerably lower than that given in the literature (181 °C) (9). This is thought to be due to the extremely low molecular weight (5.1×10^3) of this polymer (10). At inhibitor fractions below 30 mol%, the Tg values of the resin are close to the thermal decomposition temperature, as shown in Figure 8. This makes it difficult to distinguish the Tg peak from the DSC curve.

There is a well-known relationship between Tg and the volume or weight fraction of the polymer when co-polymers or blended polymers are well mixed at the molecular level, that is, they are in a homogeneous state. If Tg_A is the Tg of polymer A and Tg_B is the Tg of polymer B, the Tg of co-polymer or polymer blend in a homogeneous state is described as follows:

$$Tg = w_A Tg_A + (1-w_A) Tg_B , \qquad\qquad (7)$$

where w_A is the weight fraction of polymer A. Tg decreases and the thermal decomposition temperature increases as the inhibitor fraction increases in this system. Therefore, it becomes easy to distinguish the Tg peak from its thermal decomposition curve.

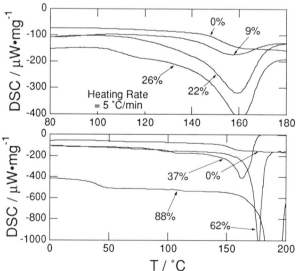

Figure 8. DSC curves of BCM-PHS at various inhibitor-fractions at a heating rate of 5.0 °C/min.

The Tg values of inhibitor fractions 88, 62, and 37 mol% were estimated to be 33, 84, and 97 °C from the DSC curves, respectively. The molar fraction of the inhibitor must be converted to a weight fraction to use Equation (7). Figure 9 shows Tg versus weight fraction of the inhibitor. The Tg of a smaller fraction of the inhibitor can be roughly predicted from this figure. Exact Tg values of smaller inhibitor fractions (26, 22, and 9 mol%) could be experimentally estimated to be 111, 120, and 131 °C, respectively, from the approximately predicted Tg and experimental DSC curves.

Since polymer coils move into a rubber state, the resist pattern would be inferior if the baking temperature was set higher than Tg. Figure 9 shows that Tg decreases as the inhibitor fraction increases. Resist patterns have been known to become inferior with increasing inhibitor fraction. Thermal decomposition is negligible for inhibitor fractions of more than 20 mol% at a baking temperature of 110 °C because stability against thermal decomposition improves with increasing inhibitor fraction. The polymer behaves as a viscous liquid and its volume increases when it enters the rubber state. As the volume of the thin film increases, stress occurs in the film especially in the direction parallel to the surface, because it can escape in the perpendicular direction but a large strain occurs in the parallel direction. This is one reason for the degraded resist pattern, sometimes observed at high PEB temperatures. Tg is, therefore, another upper limit for the PEB temperature when it is lower than the decomposition temperature.

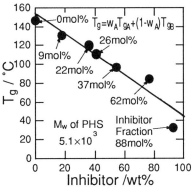

Figure 9. Inhibitor fraction dependence of glass transition temperature. The horizontal axis is weight fraction. Corresponding molar fractions are shown in the figure. Weight average degree of polymerization was 41.

Determination of thermal process of chemical amplification resists   The upper and lower limit baking temperatures for this type of resist can easily be estimated from what has been elucidated. The lower limit pre-bake temperature is determined by its relationship to the remaining amount of solvent (*11*). It has a large influence on the diffusion range of acid and causes a troublesome time-delay effect. If the pre-bake temperature is too low, the diffusion range of the acid becomes too long for a resist profile to be formed. The authors have already reported that there is a certain pre-bake temperature above which only a small amount of solvent remains. For example, this pre-bake temperature was 110 °C for ECA (*11*). There is a belief that the pre-bake temperature should be higher than Tg because the base resin shrinks and the free volume is reduced (*12*). If this were true, the lower limit for the pre-bake temperature would be the Tg of the base resin. The upper limit for the pre-bake temperature is derived from the thermal decomposition of the inhibitor, as discussed earlier.

The PEB temperature margin can be derived from Figure 7. Tg may also be considered for the upper limit of PEB temperature. Tg was predicted from its relationship with inhibitor fraction shown in Figure 9. The lower temperature for the thermal decomposition of the unexposed area or Tg is thus the predicted upper limit for the PEB temperature.

From these considerations, there remain only very few alternatives for the baking conditions of this type of resist (Scheme II). The pre-bake condition should be higher than the temperature determined by the remaining solvent and Tg. The upper limit should be derived from the thermal decomposition. The PEB condition is derived from the thermal decomposition of the exposed areas at the lower limit. Thermal decomposition of the unexposed areas or the glass transition has to equal the upper limit temperature.

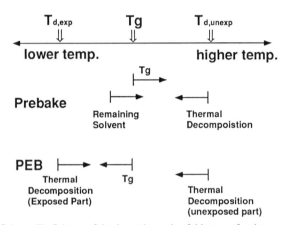

Scheme II. Scheme of the thermal margin of this type of resists.

Summary

The thermal properties of chemical amplification resists are discussed in this paper. These phenomena obey very simple relationships with the activation energy, the acid concentration, and so on. Decomposition of the inhibitor is simply and uniquely described by the activation energy in chemical amplification resist systems. This activation energy of inhibitor decomposition decreases with increasing acid concentration in both the exposed and unexposed areas.

Once the relationships between activation energy, acid concentration, exposure dose, inhibitor fraction, etc. have been derived, it is very easy to predict the thermal decomposition rates. The Tg is also an important parameter related to the thermal stability of resists, and it is not difficult to estimate Tg as a function of inhibitor fraction. Therefore, the upper and lower limits for pre-bake or PEB can be derived. Lithographic experiments should be thus carried out only within these limits. In addition, resist patterns can also be predicted using computer simulations, thereby avoiding at least some of the time-consuming and troublesome experiments (13).

Acknowledgment

The authors wish to thank Mr. T. Ushirogouchi and Dr. T. Naito for their helpful discussions. They also express their thanks to Mr. Y. Onishi and Mr. K. Sato for useful information about resist properties.

Literature Cited

[1] H. Ito; J. Polym. Sci., **24**, 2971 (1986)
[2] Y. Onishi, N. Oyasato, H. Niki, R. H. Hayase, Y. Kobayashi, K. Sato, M. Miyamura; J. Photopolym. Sci. Technol., **5**, 47 (1992)
[3] A. S. Gozdz, J. A. Shelburne III; Proc. SPIE, **1672**, 184 (1992)
[4] F. Houlihan, F. Bouchard, J. M. J. Frechet, C. G. Willson; Can. J. Chem., **63**, 153 (1985)
[5] D. J. Kim, W. G. Oldham, A. R. Neureuther; IEEE Trans., **ED-31**, 1730 (1984)
[6] C. A. Mack; J.Electrochem.Soc.Solid State Sci. Technol., **134**, 148 (1987)
[7] Y. Hirai, M. Sasago, M. Endo, K. Tsuji, Y. Mano; IEEE Trans. **CAD-6**, 403 (1987)
[8] K. Asakawa; J. Photopolym. Sci. Technol., **6**, 505 (1993)
[9] A. C. Puleo, N. Muruganandam, D. R. Paul; J. Polym. Sci. Part B, Phys., **27**, 2385 (1989)
[10] H. Hatakeyama, S. Hirose, T. Hatakeyama; ACS Symp. Ser., **397**, 205 (1989)
[11] K. Asakawa, T. Ushirogouchi, M. Nakase; J. Photopolym.Sci.Technol., **7**, 497, (1994)
[12] H. Ito, W. P. England, N. J. Clecak, G. Breyta, H. Lee, D. Y. Yoon, R. Sooriyakumaram, W. D. Hinsberg; Proc. SPIE, **1925**, 65 (1993)
[13] A. Hongu, K. Asakawa, T. Ushirogouchi, H. Wakabayashi, S. Saito, M. Nakakase; Proc. Polym. Microelctro. 93, p85 (1993)

RECEIVED September 13, 1994

Chapter 14

Modeling and Simulation of Chemically Amplified Resist Systems

**Akinori Hongu, Koji Asakawa, Tohru Ushirogouchi,
Hiromitsu Wakabayashi, Satoshi Saito, and Makoto Nakase**

**Research and Development Center, Toshiba Corporation 1, Komukai
Toshiba-cho, Saiwai-ku, Kawasaki 210, Japan**

A novel computer simulation program for chemical amplification
resist systems have been created accounting for the diffusions of
acid and trapping substance, and the reaction caused by the acid to
simulate the post-exposure bake (PEB) process of chemical
amplification resist systems. Some reaction rate constants those
are needed for the simulation have been estimated by experimental
data. The model for the PEB process is that the dissolution
inhibitor is decomposed by the acid, the acid and the trapping agent
deactivate each other, and the acid and the trapping agent diffuse
according to their concentration gradients. The dissolution rate of
the resist is assumed to be a function of the amount of the
remaining dissolution inhibitor. Resist profiles obtained by
simulation have the same tendency with actual resist profiles
obtained by experiment. It is confirmed that the model, the
algorithm, and the constants used in the simulation system are
acceptable for simulating chemical amplification resist systems.

A number of models have been created to understand the mechanism of novolac-
diazoquinone type positive-tone photoresist systems and computer simulation
programs are commercially available today. Several of them have provided
successful results. However, few modeling or simulation studies of chemical
amplification resist systems have been reported.

Conventional simulation programs *(1)* for positive-tone photoresist systems
derive the amount of photo-induced destruction of the dissolution inhibitor
(diazoquinone) according to the exposure light intensity distribution, and convert
the concentration of the remaining dissolution inhibitor into the dissolution rate.
On the contrary, in chemical amplification positive-tone resist systems, there is

another important factor, such as acid diffusion during post-exposure bake (PEB), since the amount of dissolution inhibitor decomposition is determined by the degree of the acid catalyzed reaction. This factor is intrinsically important for chemical amplification resist systems. Furthermore, if any basic substance exists in the resist, or if such a substance enters the resist from the atmosphere via the resist surface, the acid may be trapped and deactivated by the basic substance. In such a system, the developing rate of the resist film after PEB depends on the acid diffusion, the dissolution inhibitor destruction caused by the acid, and the acid trapping by the basic substance, whereas the exposed UV light intensity distribution is only a factor that defines the initial concentration of the acid.

Model and Algorithm

The model and the algorithm reported here account for the diffusions of the acid and the trapping substance, and the reaction caused by the acid to simulate the PEB process of chemical amplification resist systems. The model can be written in the form of differential equations of the concentrations of the resist components which affect the reactions, accounting for their mutual reaction, self reaction, diffusion, and trapping.

Exposure. The model for the exposure process is that the acid generation rate is proportional to the exposed light intensity. Those can be written as follows:

$$- \partial \, C_{PAG}(r,t)/ \partial \, t \quad = \quad \phi \cdot \varepsilon_{PAG} \cdot \ln(10) \cdot I(r) \cdot \lambda /(h \cdot c) \cdot C_{PAG}(r,t), \qquad (1)$$

where $C_{PAG}(r,t)$ is the concentration of the photo acid generator (PAG) at position vector r and exposure time t, ϕ is the quantum yield of acid generation, ε_{PAG} is the molar extinction coefficient of the PAG, $I(r)$ is the intensity of the exposed light intensity, λ is the wavelength of the exposed light, h is Planck's constant, and c is the light velocity. This differential equation has a solution,

$$C_{PAG}(r,t) \quad = C_{PAG}(r,0) \cdot \exp(- \phi \cdot \varepsilon_{PAG} \cdot \ln(10) \cdot I(r) \cdot \lambda /(h \cdot c) \cdot t). \qquad (2)$$

The authors used this equation to calculate the concentration of the PAG after the exposure. The concentration of acid C_{acid} after exposure is given by

$$C_{acid}(r,t) \quad = C_{acid}(r,0) + \{C_{PAG}(r,0) - C_{PAG}(r,t)\}. \qquad (3)$$

The distribution of the exposed light intensity, $I(r)$, was calculated according to a conventional method *(2-4)*

PEB. The model for the PEB process is that the dissolution inhibitor is decomposed by the acid, the acid and the trapping agent deactivate each other, and the acid and the trapping agent diffuse according to their concentration gradients. These can be written as follows:

$$\partial \, Cinh(r,t)/ \partial \, t \;\; = \; - \; k_1 \cdot Cinh(r,t) \cdot Cacid(r,t)^n \; - \; k_2 \cdot Cinh(r,t), \qquad (4)$$

$$\partial \, Cacid(r,t)/ \partial \, t \;\; = \; - \; k_3 \cdot Cacid(r,t) \cdot Ctrap(r,t) \; + \; Dacid \cdot \nabla^2 Cacid(r,t), \qquad (5)$$

$$\partial \, Ctrap(r,t)/ \partial \, t \;\; = \; - \; k_3 \cdot Cacid(r,t) \cdot Ctrap(r,t) \; + \; Dtrap \cdot \nabla^2 Ctrap(r,t), \qquad (6)$$

where $Cinh$, $Cacid$, and $Ctrap$ are the concentrations of the dissolution inhibitor, the acid, and the trapping agent, respectively, n is a constant for the acid concentration exponent, t is the PEB time, k_1 and k_3 are the rate constants of the reaction between the dissolution inhibitor and the acid, and of the reaction of the acid and the trapping agent, respectively, k_2 is the rate constant of the self decomposition of the dissolution inhibitor, $Dacid$ and $Dtrap$ are the diffusion coefficients of the acid and the trapping agent, and ∇ is nabla.

It should be noted that it has become possible by taking the trapping agent into account to evaluate the effect of basic additives like amine and/or basic impurities diffusing from the atmosphere through the resist surface or from the substrate.

The above differential equations were transformed to the following difference equations to carry out simulations on a digital computer system:

$$Cinh(r, \, t+\Delta t) \;\; = \; - \; k_1 \cdot Cinh(r,t) \cdot Cacid(r,t)^n \cdot \Delta t \qquad (4')$$
$$\qquad\qquad\qquad\;\; - \; k_2 \cdot Cinh(r,t) \cdot \Delta t,$$

$$Cacid(r, \, t+\Delta t) \;\; = \; - \; k_3 \cdot Cacid(r,t) \cdot Ctrap(r,t) \cdot \Delta t \qquad (5')$$
$$\qquad\qquad\qquad\;\; + \; Dacid \cdot \nabla^2 Cacid(r,t) \cdot \Delta t,$$

$$Ctrap(r, \, t+\Delta t) \;\; = \; - \; k_3 \cdot Cacid(r,t) \cdot Ctrap(r,t) \cdot \Delta t \qquad (6')$$
$$\qquad\qquad\qquad\;\; + \; Dtrap \cdot \nabla^2 Ctrap(r,t) \cdot \Delta t,$$

where Δt is assumed to be a very small time fraction.

∇^2 of both $Cacid$ and $Ctrap$ is calculated by

$$\nabla^2 C \;\; = \; \{C(x+\Delta x,y,z) + C(x-\Delta x,y,z) -2 \cdot C(x,y,z)\}/(\Delta x)^2 \qquad (7)$$
$$\qquad\; + \; \{C(x,y+\Delta y,z) + C(x,y-\Delta y,z) -2 \cdot C(x,y,z)\}/(\Delta y)^2$$
$$\qquad\; + \; \{C(x,y,z+\Delta z) + C(x,y,z-\Delta z) -2 \cdot C(x,y,z)\}/(\Delta z)^2,$$

where x, y, and z are the Cartesian coordinates of r.

Development. The dissolution rate $R(r)$ of the resist is assumed to be a function of the amount of the remaining dissolution inhibitor as follows:

$$R(r) \quad = f\{ \, Cinh(r) \, \}, \tag{8}$$

where $Cinh(r)$ is the amount of the remaining dissolution inhibitor at position vector r after PEB. The rate determining function f is defined from experimental data.

The minimum duration time, $td(x)$, which is the time between the start of the development and end point of the dissolution of the resist at the position, x, can be described as follows:

$$td(x) \quad = min\{ \, t(p) \, \}, \tag{9}$$
$$t(p) \quad = \int_p R(r) \, dr, \tag{10}$$

where $t(p)$ is the duration time through a pass, p, from the surface to position x, and $min\{ \, t(p) \, \}$ is the minimum value among all $t(p)$s.

Constants

To carry out the simulation using the above mentioned model, the quantum yield and the molar extinction coefficient of PAG, the reaction rate constants, the diffusion coefficients, and the dissolution rate function should be known. The KrF excimer resist was used for the measurements of the constants, which consists of poly (4–hydroxystyrene) partially protected with t–butoxycarbonylmethyl group as a base polymer and triphenylsulfonium trifluoromethylsulfonate as a photo acid generator.

Quantum Yield of PAG. The quantum yield can be derived from the comparison of the exposure dose and the concentration of the generated acid. The measurements of the concentration of the generated acid was made as follows:

The resist was coated on a silicone wafer,
exposed to a certain amount of exposure dose,
and dissolved into an organic solvent which contains a pH indicator,
and the absorbance at 602.5 nm was measured.

Ethylcellosolve acetate was used as the solvent, and tetrabromophenol blue was used as the pH indicator. The amount of the generated acid was estimated using a calibration curve between the acid concentration and the absorbance at 602.5 nm.

Figure 1 shows the experimental data for acid generation. It is clearly seen in the figure that the generated acid increases with increasing the exposure dose, and reaches the amount of the initial PAG concentration of

$2.73 \times 10^{-7} \mu$ m^{-3}. By fitting a theoretical curve, equations 2 and 3, to these experimental data using the method of least squares the quantum yield was estimated to be 0.27. That is, 27% of the photon absorbed by PAG results in acid generation.

Reaction Rate Constant. To determine the reaction rate of the decomposition of the dissolution inhibitor catalyzed by the acid, k_1 in equations 4 and 4', the activation energy, E_a, of the reaction was measured as follows:

The resist was coated on a silicon wafer,
exposed to a certain amount of exposure dose,
and scraped from the wafer,
its weight loss was measured by thermo gravinometry (TG),
and a TG simulation curve was fitted to the measured TG data (5)

TG simulation curve is given by the following difference equation:

$$W(t) \quad = W(t+\Delta t) \times [1 - K_0 \exp\{-E_a / R \cdot T(t)\} \Delta t], \tag{11}$$

where W is the weight, K_0 is the decomposition rate at the infinite temperature, R is the gas constant, and T is the absolute temperature. Figure 2 shows one example of the TG data for both simulated and measured. In the case of this figure, the activation energy was estimated to be 1.45×10^{-19} J.

Therefore, the activation energy should be converted to the reaction rate to make the simulation become easier. The reaction rate, k_1, and the exponent, n, as functions of the temperature can be obtained by fitting equation 4 using the method of least squares to a series of activation energy data for various concentrations of the acid. As a result, the k_1 and n were estimated to be 32.37 nm^3/s and 0.733, respectively.

Diffusion Coefficient. It is useful to use the relationship that the diffusion coefficient is proportional to the square of the diffusion range. The PEB temperature dependence of the diffusion coefficient of the acid can be estimated by using the PEB temperature dependence of the diffusion range of the acid, shown in Figure 3 (7) The diffusion range for PEB temperature of 120 ℃ was estimated to be 1.76 nm^2/sec.

Dissolution Rate Function. Figure 4 shows the experimental data of the dissolution rate as a function of the inhibitor content. The dissolution rate can be described by the following equation:

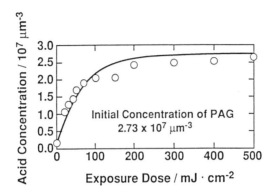

Figure 1. Acid Condition vs. Exposure Dose.
Experimental data (O) and fitted line.

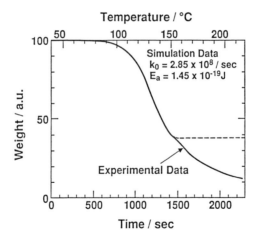

Figure 2. TG Data of Exposed Resist.
Experimental data (solid line) and the fitted simulation data (broken line).

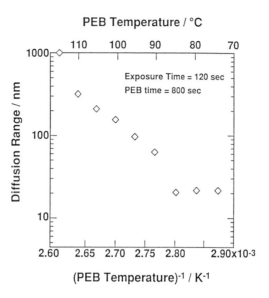

Figure 3. PEB Temperature Dependence of Diffusion Range.

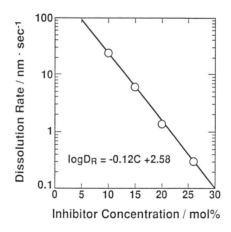

Figure 4. Dissolution Rate vs. Inhibitor Content.
Experimental data (O) and fitted line.

$$\log[\ R(r)/(\text{nm sec}^{-1})\ ]\ = 2.58 - 0.12 \times [\ C_{inh}(r)/\text{mol\%}\ ], \tag{12}$$

where $R(r)$ is the dissolution rate at position vector r, and $C_{inh}(r)$ is the amount of the remaining dissolution inhibitor at position vector r after PEB.

Results and Discussion

Figure 5 shows an example of simulated resist profiles in comparison with actually

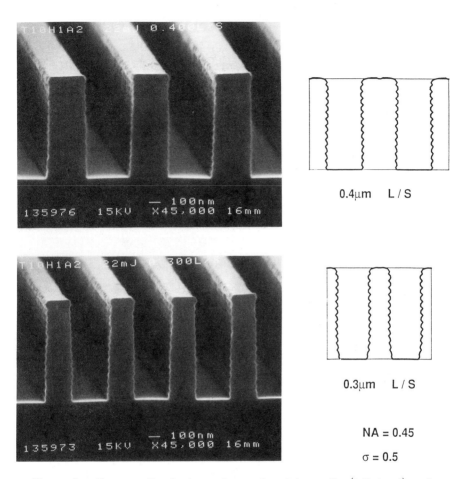

Figure 5.   Cross-sectional views of experimental results (left hand) and simulation results (right hand) using line width as a parameter.

obtained resist profiles, where the exposure was made by using KrF excimer laser stepper equipped with 0.45 NA projection lens and C_{trap} was assumed to be zero. A good agreement between the simulation and experiment can be seen.

Figure 6 shows the resist profile considering the trapping agent diffuses from the atmosphere, where assumed values of D_{trap} and C_{trap} were used for the simulation. The so called T-top profile and its dependences on D_{trap} and C_{trap} can be successfully simulated.

Conclusion

Resist profiles obtained by simulation have the same tendency with actual resist profiles obtained by experiment. It can be said, then, that the model, the

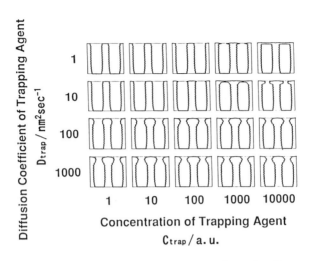

Figure 6. Cross-sectional views of simulation results, using C_{trap} and D_{trap} as parameters.

algorithm, and the constants used in the simulation system are acceptable for simulating chemical amplification resist systems.

The simulator will become a useful tool for developing chemical amplification resist systems, as it can predict the resist profile much faster than by experimental trials.

Literature Cited

[1] Mack, C. A. *Proc. SPIE 538 Optical Microlithography IV*.1992, pp.207–220.
[2] Lin, B. J. *IEEE Trans. Elec. Dev.*1980, ED–27(5), pp.931–938.
[3] Subramanian, S. *Appl. Opt.*1981, 20(10), pp.1854–1857.
[4] Widmann, D. W. *ibid.*1975, 14(4), pp.931–934.
[5] Asakawa, K.; Hongu, A.; Oyasato, N.; Nakase, M. *Proc. Polym. for Microelectronics -Science and Technology- '93*.1993, pp.78–81.
[6] Allen, M. T. *Proc. SPIE 1672 Advances in Resist Technology and Processing IX*, 1992, 74 93.
[7] Asakawa, K.; Sakai, K.; Ushirogouchi, T.; Hongu, A.; Nakase, M. *Extended Abstracts (The 54th Autumn Meeting); The Japan Soc. of Appl. Phys.*1993, (2), 546.

RECEIVED September 13, 1994

Chapter 15

Surface Imaging Using Photoinduced Acid-Catalyzed Formation of Polysiloxanes at Air–Polymer Interface

Masamitsu Shirai, Tomonobu Sumino, and Masahiro Tsunooka

Department of Applied Chemistry, College of Engineering, University of Osaka Prefecture, Sakai, Osaka 593, Japan

Polysiloxane formation by a chemical vapor deposition (CVD) method on the UV-irradiated surface of polymer films was studied. Polymers which have both 4-vinylbiphenyl (4-VB) units and 1,2,3,4-tetrahydro-1-naphthylideneamino p-styrenesulfonate (NISS) units were synthesized. When the irradiated surface of the polymer films was exposed to the vapor of alkoxysilanes at 30 °C, polysiloxane networks were formed on the surface. No polysiloxane networks were formed in the unirradiated areas. The thickness of the polysiloxane layer which is formed beneath the film surface was reduced by the existence of 4-VB unit in the polymers because the 4-VB units worked as a UV absorbent. The photoinduced acid-catalyzed crosslinking of polymers containing NISS and glycidyl methacrylate effectively decreased the thickness of the polysiloxane network layer beneath the film surface because of the reduced diffusible depth of alkoxysilane vapor into the film. The polymer films which were irradiated and subsequently exposed to the vapor of alkoxysilanes showed a good resistance to etching using an oxygen plasma.

In the microlithographic process the use of deep UV light to provide higher resolution causes new problems due to decreased depth of focus (DOF) and increased substrate reflectance. DOF has more influence by the high numerical aperture of the lens. Surface imaging has been proposed to alleviate these problems. One approach to accomplish the surface imaging is the selective modification of the irradiated or unirradiated polymer surface using organic or inorganic chemicals which show resistance to etching using an oxygen plasma. A bisarylazide/isoprene crosslinkable polymer system was irradiated with UV light and, in a subsequent step, treated with the vapor of inorganic halide such as $SnCl_4$, $SiCl_4$, or $(CH_3)_2SiCl_2$ (1, 2). Reaction products containing Si or Sn could be formed at the near surface of the polymer films. Pattern development for microlithography was achieved using oxygen reactive ion etching (O_2 RIE). The gas-phase silylation of a proprietary

0097–6156/94/0579–0185$08.00/0

diazonaphthoquinone/phenolic matrix resin (3) or polymer films (4) bearing photochemically formed phenolic -OH groups has been studied for the surface imaging.

The selective formation of polysiloxanes or metal oxides at the irradiated surface of polymer films is an important approach to designing surface imaging systems. Follett and co-workers (5) have reported the plasma-developable electron-beam resists. The essential feature involved selective diffusion of $(CH_3)_2SiCl_2$ into the irradiated areas of poly(methyl methacrylate), followed by hydrolysis of the chlorosilane by exposure to water vapor, resulting in the formation of an interpenetrating network of polysiloxanes. O_2 RIE yielded a negative tone image. Taylor and co-workers (6-9) reported a photoresist based on the surface oxidation of chlorine-containing polystyrenes using 248.4- or 193-nm light and subsequent treatment with gaseous $TiCl_4$.

Recently we have reported photoinduced acid-catalyzed formation of polysiloxanes at the irradiated polymer surface by a chemical vapor deposition (CVD) method using alkoxysilane vapor (10-12). The methodology of our system for the surface imaging is shown in   Scheme I. Upon irradiation at 254 nm the surface of polymers having photoacid generating units becomes hydrophilic because of the formation of acid. Water sorption from the atmosphere occurs at the top surface of the irradiated films.   When the irradiated surface is exposed to the vapor of alkoxysilanes, polysiloxane networks are formed at the near surface of the polymer films.   The polysiloxane network is not formed at unirradiated areas because the photochemically-formed acids are necessary for the polysiloxane formation by hydrolysis and subsequent polycondensation reactions of alkoxysilanes.   This system gives a negative tone image by O_2 RIE.   Since gaseous alkoxysilane diffuses into the polymer film, the hydrolysis and subsequent polycondensation reactions occur beneath the film surface and at the film-air interface. The maximum thickness of the polysiloxane network layer can be determined by the light penetrating depth which depends on the molar extinction coefficient of the photoacid generating units. For surface imaging systems, it is preferred that polysiloxane network can be formed at the top surface of the polymer films. In this paper we report two methods to limit the thickness of the polysiloxane network formation to the near surface of polymer films. (i) Polymers containing 1,2,3,4-tetrahydro-1-naphthylideneamino p-styrenesulfonate (NISS) as a photoacid generating unit and 4-vinylbiphenyl units (4-VB) were synthesized (Scheme II). 4-Vinylbiphenyl units are photochemically stable and strongly absorb the light at 254 nm by which NISS units can be photolyzed. Thus, the introduction of the 4-VB units into the photoacid generating polymers can reduce the thickness of the surface layer where the acid can be photochemically formed.   (ii) Photocrosslinkable polymers containing NISS and glycidyl methacrylate were prepared (Scheme III).   The thickness of the polysiloxane layer at the irradiated surface was effectively reduced because of the decreased diffusible depth of alkoxysilane vapor into the crosslinked film.

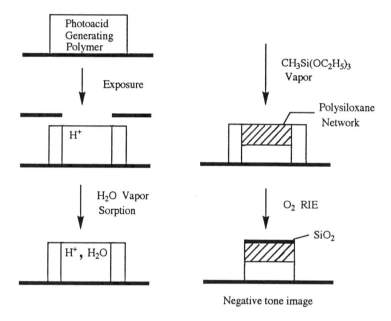

Scheme I. Surface imaging using photoinduced acid-catalyzed formation of polysiloxanes on polymers

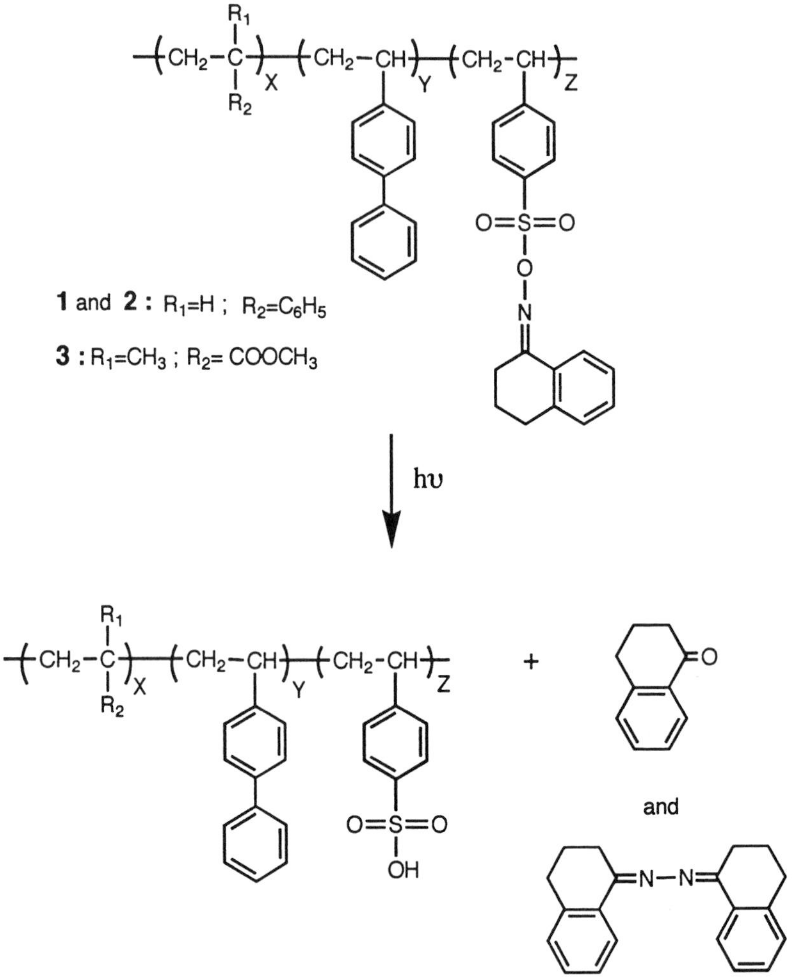

Scheme II. Structures and photochemical reactions of polymers 1, 2, and 3

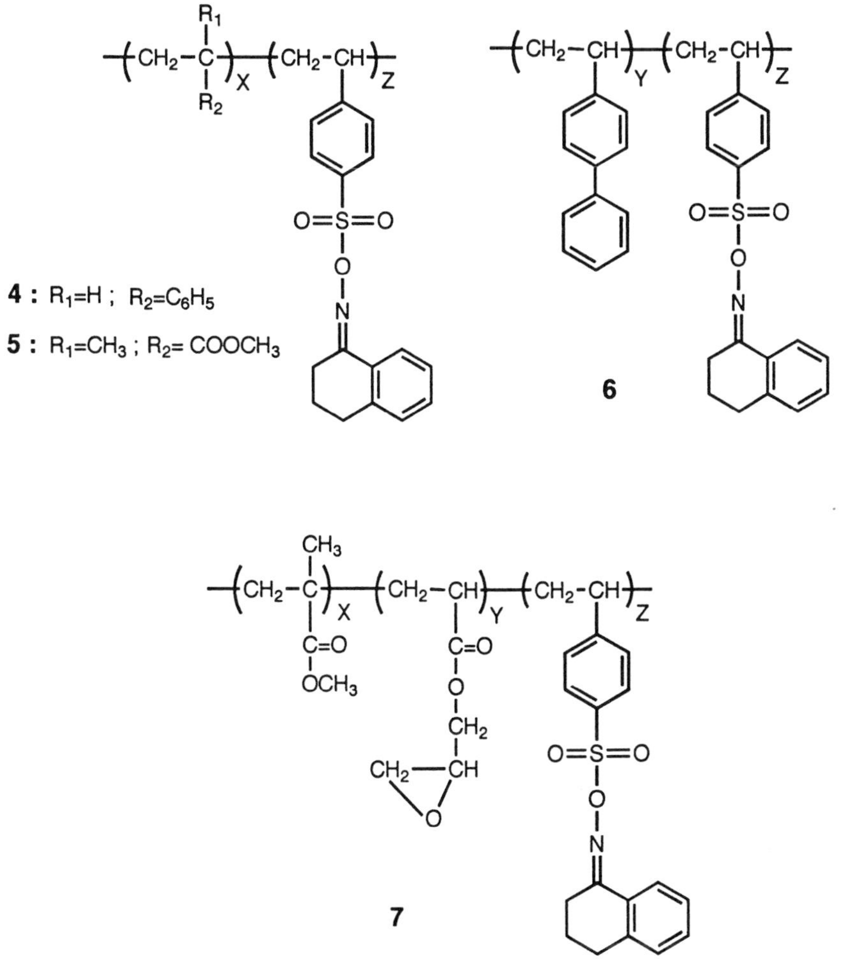

4 : R_1=H ; R_2=C_6H_5

5 : R_1=CH_3 ; R_2= COOCH$_3$

6

7

Scheme III. Structures of polymers 4, 5, 6, and 7

EXPERIMENTAL SECTION

Materials

Syntheses of 1,2,3,4-tetrahydro-1-naphthylideneamino p-styrenesulfonate (NISS) (10) and 1,2,3,4-tetrahydro-1-naphthylideneamino p-toluenesulfonate (NITS) (13) were reported in detail elsewhere. 4-Vinylbiphenyl (4-VB) was reagent grade and used after recrystallization from methanol. Methyltriethoxysilane (MTEOS) and methyltrimethoxysilane (MTMOS) were reagent grade and used without further purification. Styrene (St), methyl methacrylate (MMA), and glycidyl methacrylate (GMA) were distilled before use.

Preparation of polymers

Terpolymers (1-3), NISS-co-styrene (4), NISS-co-methyl methacrylate (5), and NISS-co-4-VB (6) were prepared by the photochemically-initiated copolymerization of corresponding monomers with 2,2'-azobis(isobutyronitrile) (AIBN) as an initiator at 29 °C by irradiation with light at wavelengths above 350 nm. The flux was 4 mW/cm^2. The photodecomposition of NISS did not occur by the light at wavelengths above 350 nm. The concentrations of total monomer and AIBN in benzene were usually 4.5 and 1.6×10^{-2} mol/L, respectively. The sample solution was degassed under vacuum by repeating freeze-thaw cycles before polymerization. The contents of NISS units and 4-VB units in the polymers 1, 2, and 3 were determined by measuring the absorbances at 254 and 300 nm in CH_2Cl_2. The molar extinction coefficient of the NISS units was estimated to be equal to that of the model compound NITS, ε being 15300 L/mol· cm at 254 nm and 2270 L/mol·cm at 300 nm in CH_2Cl_2 at room temperature. The molar extinction coefficient of the 4-VB units was estimated to be equal to that of the radically polymerized poly(4-vinylbiphenyl) (Mn=9×10^4), ε being 25900 L/mol·cm at 254 nm and 1390 L/mol·cm at 300 nm in CH_2Cl_2 at room temperature. The composition of the copolymers 4, 5, and 6 was determined by measuring absorbance at 254 nm in CH_2Cl_2. The composition of the polymer 7 was determined from NMR spectrum and absorbance at 254 nm. Polymerization conditions and characteristics of the polymers are shown in Table I. Although the polymers could be obtained by the conventional thermally initiated copolymerization of the corresponding monomers with AIBN, they showed wide molecular weight distributions (Mw/Mn > 3.5). Structures of the polymers used in this study are shown in Schemes II and III.

Water sorption

A laboratory-constructed piezoelectric apparatus (quartz crystal microbalance) was used to measure water sorption in the polymer films. The AT-cut quartz crystal with gold electrodes (Webster Electronics, WW1476) had a resonance frequency of 10.000 MHz. With this crystal, a frequency shift of 1 Hz corresponded to a mass change of 0.84 ng. The frequency change is linearly related to the mass sorbed on the quartz plate (14, 15).

Polymers were deposited onto the quartz crystal (1.2 cm diameter) by casting from chloroform solution. The area coated with the polymer film was usually 0.19

Table I. Polymerization Conditions and Polymer Properties[a]

Polymer	Monomer in Feed					Polymerization Time (h)	Conversion (%)	Mn X10^{-4}	Mw/Mn	Composition[b,c] (mol %)			Tg[d] (°C)
	NISS (g)	4-VB (g)	MMA (mL)	St (mL)	GMA (mL)					X	Y	Z	
1[b]	0.28	0.23			0.25	7	33	6.4	2.0	65	22	13	-[e]
2[b]	0.49	0.25			0.52	13	28	6.6	2.3	50	23	27	-[e]
3[b]	1.96	0.70	0.54			10	48	1.8	2.4	60	29	11	-[e]
4[c]	0.70			2.21		31	37	2.3	2.9	79		21	125
5[c]	0.41		3.0			15	47	11.5	2.0	93		7	114
6[c]	0.10	0.62				9	18	9.2	1.8		74	26	-[e]
7[c]	0.50		2.12		1.22	11	45	26.3	15.0	45	43	12	109

[a] [Total monomer]=4.5 mol/L, [AIBN]=1.6X10^{-2} mol/L. Benzene was used as a solvent. [b] See Scheme II. [c] See Scheme III. [d] Glass transition temperature. [e] Distinct Tg was not observed.

cm^2.   The thickness of the film on quartz crystal was adjusted by changing concentration of the polymer solution.   The quartz crystal was placed in the middle of the sealed glass vessel which had a quartz window for UV irradiation.   A saturated NaBr solution was placed at the bottom of the vessel to control its humidity at a constant temperature.   Irradiation of polymer films on the quartz crystal through the quartz window of the vessel was carried out with 254-nm light. The intensity of the incident light determined with a chemical actinometer (potassium ferrioxalate) (16) was 0.1 mJ/cm^2 ·sec at 254 nm.

Deposition of polysiloxane

The polymer films were prepared on glass plates (8.8 X 50 mm) by casting from chloroform solutions and drying under vacuum at room temperature.   Sample films (8.8 X 22 mm) on the glass plates were obtained by removing both edges of the film. The polymer weight on the glass plate was usually 2 X 10^{-4} g and area was 1.94 cm^2. If the polymer is assumed to have a density of 1 g/cm^3, the film thickness is ≈1 μm. After exposure with 254-nm light, the glass plate coated with polymer film was placed at the center of a 500 mL glass vessel which had gas-inlet and -outlet valves.   Fifty mL of water or salt solution was placed at the bottom of the vessel to adjust the relative humidity in the vessel and equilibrated for 10 min prior to introduction of the vapor of alkoxysilanes.   During the polysiloxane network formation nitrogen gas (50 mL/min) flowed through a bubbler which contained liquid alkoxysilanes.   The bubbler and reaction vessel were placed in a thermostatic oven at 30 °C. The amounts of polysiloxanes formed at the near surface of the polymer films were determined from the difference between the weight of the sample plate before and after exposure to the vapor of alkoxysilanes (a conventional gravimetric method).

Etching with an oxygen plasma

Oxygen plasma etching was carried out at room temperature using a laboratory-constructed apparatus where the oxygen plasma was generated using two parallel electrodes and RF power supplies.   The typical etching conditions were as follows: 20W power (13.56 MHz), power density of 1.0 W/cm^2, 125 mTorr, and oxygen flow of 1 sccm.

RESULTS AND DISCUSSION

Irradiation with 254-nm light cleaves -O-N= bonds in the NISS units.   Subsequent abstraction of hydrogen atoms from residual solvent in the polymer film or from polymer molecules leads to the formation of p-styrenesulfonic acid units, tetralone, and tetralone azine as shown in Scheme II (13). The quantum yield for the photolysis of the NISS units incorporated into poly(methyl methacrylate) was reported to be 0.33 for 254-nm irradiation in air (10).   Absorption spectra of NITS and poly(4-vinylbiphenyl) (PVB) are shown in Figure 1.   Both compounds showed the absorption maximum at 254 nm and the molar extinction coefficients of NITS and PVB at 254 nm were 15300 and 25900 L/mol·cm, respectively. The photoreactivity of the NISS units was not affected by the 4-VB units incorporated in the polymer chain.   Thus, the 4-VB units did not work as neither sensitizer nor quencher for the photolysis of the NISS units.   Furthermore, the 4-VB units were photochemically stable.   Thus the 4-VB units in the polymers worked as light absorbing units.

When the polymer films were irradiated with 254-nm light, water sorption occurred because the hydrophobic polymer surface became hydrophilic due to the formation of acid.   Water sorption began immediately upon irradiation.   It increased with irradiation time and gradually saturated.   The sorbed water could be removed, when the sample film was placed in a dry nitrogen atmosphere.   Figure 2 shows the relationship between the thickness of the irradiated 1 and 4 films (area=0.19 cm^2) on the quartz crystal and water sorbed.   Water sorption was determined after the equilibrium was established.   The water sorption for both films increased with increasing the thickness and a saturation phenomenon was observed for the 4 film above the thickness of around 50 nm.   Thus the water sorption occurred at the top surface of the film.   The 1 film did not show a distinct saturation below the thickness of 100 nm.   The 4-VB units may allow deeper diffusion of water. Although the NISS unit fraction in 1 was lower than that in 4, the amounts of water sorbed by the irradiated 1 film was higher than those by the irradiated 4.   It has been reported that water sorption increases with the NISS unit concentration in polymers, if the irradiation dose is the same (10).   Furthermore, it has been reported that hydrophobic nature of the polymers reduces the water sorption ability.   The introduction of the 4-VB units in polymers did not increase the hydrophobic nature of the polymer, because the water sorption ability of the irradiated 4 and 6 was almost the same.   The 4-VB units in the terpolymers may cause the phase separation at the irradiated surface and the hydrophilic p-styrenesulfonic acid units may assemble at the film-air interface. The water sorption ability of the irradiated 3 film bearing the 4-VB units was higher than that of the irradiated 5 film having no 4-VB units.   The saturation phenomenon observed for the water sorption of the irradiated films was not due to the limited light penetration, since the absorbances of the films on quartz crystal were estimated to be < 0.2 at 254 nm.   Upon photolysis of the NISS units, 1-tetralone and 1-tetralone azine, which have poor affinity for water, could be formed in addition to the acid.   In the previous paper, we have reported that the aggregation of 1-tetralone and/or 1-tetralone azine acts as a "hydrophobic barrier" to prevent further diffusion of water molecules (10).

In the presence of strong acid such as sulfonic acid and water, hydrolysis and subsequent polycondensation reactions of MTEOS lead to the formation of polysiloxane networks, which is known as the sol-gel process for the silica glass formation (17, 18).   When the irradiated 1-7 films were exposed to the vapor of

$$n\ CH_3Si(OC_2H_5)_3\ +\ 1.5n\ H_2O\ \xrightarrow{H^+}\ Polysiloxanes + 3n\ C_2H_5OH$$

MTEOS at 30 °C, polysiloxane networks were formed in the near surface region of the films.   The formation of the polysiloxane network was confirmed by FTIR spectroscopy (Figure 3).   The spectrum was measured by an attenuated total reflection (ATR) method.   The new peaks were observed at 1000-1200 (Si-O-Si)

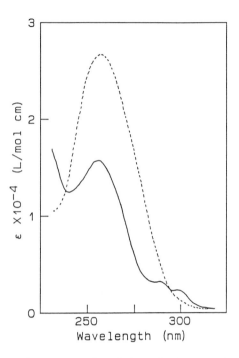

Figure 1. Absorption spectra of 1,2,3,4-tetrahydro-1-naphthylideneamino p-toluenesulfonate (solid line) and poly(4-vinylbiphenyl) (dotted line) in CH$_2$Cl$_2$ at room temperature.

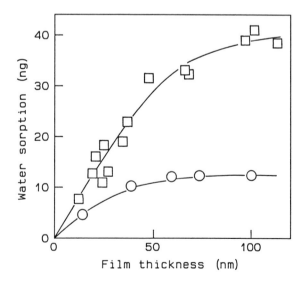

Figure 2. Water sorption at 58% RH into 1 (□) and 4 (○) films irradiated with 46 mJ/cm^2 of 254-nm light at 30 °C.

and 1272 cm⁻¹ (Si-CH₃) for the films irradiated and subsequently exposed to the vapor of MTEOS. The polysiloxane network formation was limited to the surface because the shallow light penetration confined the acid formation at the air interface even if water was present in the bulk of the film. No formation of polysiloxanes was observed at the unirradiated areas of the films because the p-styrenesulfonic acid units formed photochemically were essential for the formation of polysiloxane networks. The unreacted alkoxysilanes in the bulk of the film completely escaped after the CVD treatment.

Figure 4 shows the relationship between the amounts of polysiloxane network formed at the irradiated surface of **1** and **3** films and CVD treatment time. Although the NISS fractions in **1** and **3** were almost the same, the water sorption ability of **3** was higher than that of **1** because **3** is less hydrophobic than **1**. For both films, the formation of polysiloxanes linearly increased with increasing CVD treatment time. The rate of polysiloxane formation for **3** was 4 times higher than that for **1**.   An induction period (ca. 5 min) for the polysiloxane formation was observed.   This period was needed to fill the reaction chamber with the vapor of alkoxysilanes.

Figure 5 shows the effect of film thickness on the formation of polysiloxane networks at the near surface of **2** and **4**. The exposure dose and the CVD treatment conditions were the same for **2** and **4** films. The formation of polysiloxanes increased with increasing film thickness and showed a saturation at the thicknesses above ca. 0.35 and 1.35 μm for **2** and **4**, respectively. These values mean the penetrating depth of the polysiloxane networks and this roughly corresponds to the penetration depth of 254-nm light which can induce the photolysis of the NISS units. The absorbances of **2** (thickness=0.35 μm) and **4** (thickness=1.35 μm) films at 254 nm were calculated to be 3.5 and 3.8 at 254 nm, respectively. The saturation phenomenon was not due to the diffusion depth of the alkoxysilane vapor during the CVD treatment time, because the saturated point did not depend on the CVD treatment time under the present experimental conditions. Thus, the thickness of the interpenetrating polysiloxane network layer could be reduced by introducing the 4-VB units in polymer chains. Although the NISS units fraction in **2** was about 30% higher than that in **4**, the amount of polysiloxane formed at the surface of the **2** film was about twice as much as that of the **4** film, suggesting that introduction of 4-VB in polymers could enhance the polysiloxane formation at the irradiated surface. This is because the introduction of the 4-VB units in polymers can enhance the water sorption ability as shown in Figure 2.

Figure 6 shows the effect of film thickness on the formation of polysiloxane at the irradiated surface of **3** and **5** films. It was found that the thickness of the interpenetrating polysiloxane networks was 0.3 and 1.2 μm for **3** and **5**, respectively. The 4-VB units introduced in the polymer chain effectively reduced the light-penetrating depth and enhanced the depth of the formation of polysiloxane networks as observed for the irradiated **2** and **4** films.

The depth of the polysiloxane network layer observed for **2**, **3**, **4**, and **5** films was beyond the depth of the water sorption layer of the irradiated polymers (see Figure 2). The ethanol liberated during the hydrolysis of MTEOS in the film may help the diffusion of the water in the film by destroying the "hydrophobic barrier" of the photochemically formed 1-tetralone and/or 1-tetralone azine as discussed before (10).

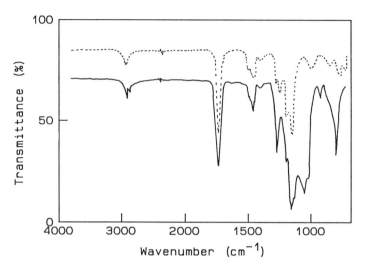

Figure 3. FTIR spectra of the irradiated **5** film before (dotted line) and after (solid line) CVD treatment using the vapor of MTEOS at 30 °C.   Spectra were measured by an attenuated total reflectance (ATR) method using Ge prism.

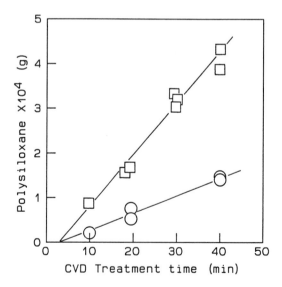

Figure 4. Polysiloxane formation at the irradiated surface of **1** (○) and **3** (□) films using the vapor of MTEOS at 30 °C.   Irradiation dose:180 mJ/cm^2. Relative humidity: 100%.

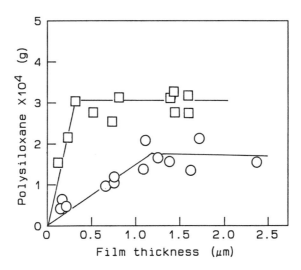

Figure 5. Effect of film thickness on the polysiloxane formation using the vapor of MTEOS at 30 °C. Polymer: (□) **2**, (○) **4**. Irradiation dose: 180 mJ/cm². CVD treatment: 30 min. Relative humidity: 100%

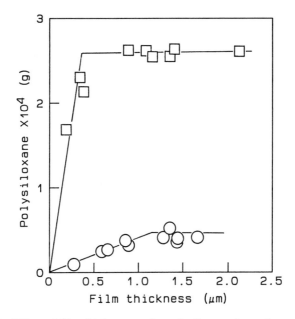

Figure 6. Effect of film thickness on the polysiloxane formation using the vapor of MTEOS at 30 °C. Polymer: (□) **3**, (○) **5**. Irradiation dose: 180 mJ/cm². CVD treatment: 30 min. Relative humidity: 100%.

It has been reported that the polymers bearing glycidyl methacrylate and NISS are crosslinked upon irradiation (19, 20). The photochemically formed acid initiates the polymerization of epoxy groups in the polymer chain. The degree of crosslinking increased on heating the irradiated films (19, 20). When the surface of the irradiated **7** film was exposed to the vapor of MTMOS, polysiloxane networks were formed at the surface, because the photochemically formed acid remained in the film after the crosslinking reaction. Figure 7 shows the effect of film thickness on the polysiloxane formation at the irradiated surface of the **7** film. The formation of polysiloxane networks increased with increasing film thickness and showed a saturation at the thickness above ca. 0.95 and 0.3 μm for the irradiated **7** film before and after heating at 80 °C, respectively. The heating after irradiation largely reduced the depth of the polysiloxane network layer beneath the film surface. This may be due to the decreased diffusible depth of MTMOS vapor into the crosslinked film. Although the photoinduced crosslinking of the film decreased the amount of polysiloxanes formed at the surface, a good resistance to the etching with an oxygen plasma was observed.

Figure 8 shows the relationship between thickness loss of **1** and **4** films and etch time using an oxygen plasma. Under the present etching conditions, the etching speeds for **1** and **4** films were 0.068 and 0.064 μm/min and they were almost half of the etching rate for poly(methyl methacrylate) film. The etching profile was dependent upon CVD treatment time of the sample films. The etching speed for the **4** film which was irradiated and subsequently exposed to the vapor of MTEOS can be divided into three different regions that occur in sequence: the initial rapid-rate region, the very-slow-rate region, and the rapid-rate region, the latter being almost equal to the etching rate for the **4** film. The period during which the etch resistance to an oxygen plasma could be observed increased with CVD treatment time of the irradiated films. The initial rapid-rate region corresponds to the etching of the polymer surface containing polysiloxane networks to form a SiO_2-containing layer acting as an etch barrier to an oxygen plasma. In the very-slow-rate region the SiO_2-containing layer acted as a good etch barrier, and the etching rate was 21 times slower than that of the unreacted **4** film. The latter rapid-rate region corresponded to the etching of the bulk of **4** layer after the removal of the SiO_2-containing layer by oxygen plasma etching. For the **1** film irradiated and subsequently exposed to the vapor of MTEOS, the initial rapid-rate region was not observed because the **1** film containing the 4-VB units can form the denser polysiloxane networks at the top surface. The etching speed at the slow-rate region was 26 times slower than that of the **1** film.

CONCLUSIONS

Photosensitive polymers containing 1,2,3,4-tetrahydro-1-naphthylideneamino p-styrenesulfonate (NISS) and 4-vinylbiphenyl (4-VB) were prepared. Upon irradiation with 254-nm light the NISS units were converted to p-styrenesulfonic acid units. The 4-VB units were photochemically stable and worked as an absorbent of 254-nm light. When the irradiated polymer films were exposed in the presence of moisture to the vapor of methyltriethoxysilane at 30 °C, polysiloxane networks were formed at the surface of the films. No polysiloxane networks were formed at

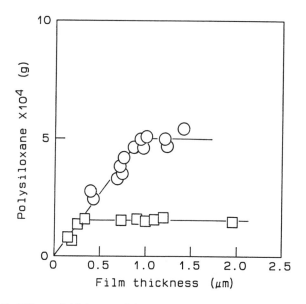

Figure 7. Effect of thickness of the **7** film on the polysiloxane formation using the vapor of MTMOS at 30 °C. Postexposure bake: (○) 0; (□) 3 min at 80 °C. Irradiation dose: 180 mJ/cm². CVD treatment time: 10 min. Relative humidity: 95%.

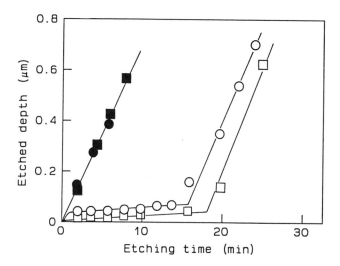

Figure 8. Oxygen plasma etching of **1** (square symbol) and **4** (circular symbol). Irradiation dose: 180 mJ/cm². CVD treatment: (■ , ●) 0, (○) 23, (□) 43 min. Relative humidity during CVD treatment: 58% for **1** and 95% for **4**.

unirradiated polymer surface. The depth of polysiloxane layer formed beneath the film surface was reduced by introducing the 4-VB units into polymer chains. The photocrosslinking of polymers bearing NISS and glycidyl methacrylate effectively reduced the depth of the polysiloxane layer beneath the film surface. The polymer films obtained after irradiation and subsequent CVD treatment with the vapor of alkoxysilanes showed a good etch resistance to an oxygen plasma.

REFERENCES

1. Taylor, G. N.; Stillwagon, L. E.; Venkatesan, T. *J. Electrochem. Soc.* **1984**, 131, 1658.
2. Wolf, T. M.; Taylor, G. N.;Venkatesan, T.; Kraetsch, R. T. *J. Electrochem. So* **1984**, 131, 1664.
3. Coopmans, F.; Roland, B. *Proc. SPIE* **1986**, 631, 34.
4. MacDonald, S. A.; Ito, H.; Hiraoka, H.; Willson, C. G. *Proceedings of SPE Regional Technical Conference on Photopolymers*: NE York Mid-Hudson Section; Society of Plastic Engineers: Ellenville, New York, 1985; p 177.
5. Follett, D.; Weiss, K.; Moore, J. A.; Steckl, A. J.;Liu, W. T. *The Electrochemical Society Extended Abstracts* ; The Electrochemical Society: Pennington, NJ, 1982; Vol. 82-2, Abstract 201, p 321.
6. Stillwagon, L. E.; Vasile, M. J.; Baiocchi, F. A.; Silverman, P. J.; Taylor, G. N. *Microelectron. Eng.* **1987**, 6, 381.
7. Taylor, G. N.; Nalamasu, O.; Stillwagon, L. E. *Microelectron. Eng.* **1989**, 9, 513.
8. Nalamasu, O.; Taylor,G. N. *Proc. SPIE* **1989**, 1086, 186.
9. Taylor, G. N.; Nalamasu, O.; Hutton, R. S. *Polym. News* **1990**, 15, 268.
10. Shirai, M.; Hayashi, M.; Tsunooka, M. *Macromolecules,* **1992**, 25, 195.
11. Shirai, M.; Kinoshita, T.; Sumino, T.; Miwa, T.; Tsunooka, M. *Chem. Mater.* **1993**, 5, 98.
12. Shirai, M.; Sumino, T.; Tsunooka, M. *Eur. Polym. J.* **1993**, 29, 831.
13. Shirai, M.; Masuda, T.; Ishida, H.; Tsunooka, M. *Eur. Polym. J.* **1985**, 21, 781.
14. Sauerbrey, G. *Z. Phys.* **1959**, 155, 206.
15. Alder, J. F.; McCallum, J. J. *Analyst* **1983**, 108, 1291.
16. Murov, S. L. *Handbook of Photochemistry* ; Dekker: New York, **1973**.
17. Bradley, D. C. *Chem. Rev.* **1989**, 89, 1317.
18. Hench, L. L.; West, J. K. *Chem. Rev.* **1990**, 90, 33.
19. Shirai, M.; Masuda, T.; Ishida, H.; Tsunooka, M.; Tanaka, M. *Eur. Polym.. J.* **1985**, 21, 781.
20. Shirai, M.; Wakinaka, S.; Ishida, H.; Tsunooka,M.; Tanaka, M. *J. Polym. Sci.:Part C: Polym. Lett.,* **1986**, 24, 119.

RECEIVED September 13, 1994

Chapter 16

Surface Imaging for Applications to Sub-0.35-μm Lithography

Ki-Ho Baik[1] and Luc Van den hove

Interuniversity Microelectronics Center (IMEC), Kapeldreef 75, B–3001 Leuven, Belgium

New silylation solutions for liquid phase silylation (LPS) for the positive tone diffusion enhanced silylated resist (DESIRE) process are presented. The silylation solution consists of the silylating agent bis(dimethylamino)dimethylsilane (B[DMA]DS) with the resist solvent propylene glycol methyl ether acetate (PGMEA) and n-decane as a safer solvent. The silylation process and the composition of the silylation solution have been optimized using statistically designed experiments. Characterization of the process has been performed using fourier transform infrared spectroscopy (FTIR), and rutherford backscattering spectroscopy (RBS). Several techniques to reduce proximity effects are proposed. The lithographic performance using this process has been evaluated for both conventional masks and phase shifting masks. A cost analysis for LPS, based on the costs of the chemical consumption, is presented.

Lithography is one of the most important processing steps of semiconductor manufacturing, not only because it has allowed the required continuous scaling of feature dimension, but also because it is one of the most frequently used steps in the semiconductor fabrication process (main cost of manufacturing and investment). The transition towards higher numerical apertures (NA) and shorter wavelengths has lengthened the lifetime of optical lithography. The first generation of 64 M DRAM devices has been realized using i-line lithography (λ=365 nm) using steppers with variable NA and partial coherence factors (1). The demand for more dense ULSI devices has driven lithography into the sub-0.35 micron regime. The main technological challenges of advanced lithography are very high resolution (< 0.25 μm) with a high aspect ratio, wider process latitudes, small linewidth variation over topography, the reduction of proximity effects, and very precise overlay accuracy (< 90 nm) while maintaining high throughput.

---

[1]Current address: Semiconductor Research and Development Laboratory, Hyundai Electronics Industries Company, Ltd., San 136–1, Ami-ri, Bubal-eub, Ichon-kun, Kyoungki-do, 467–860 Korea

The resolution limitation has been alleviated by increasing NA and decreasing the wavelength. Higher NA lenses extend the limit of resolution at the expense of a rapid loss of depth of focus (DOF). Additionally, each transition towards a shorter wavelength and/or higher NA requires a new stepper, which is a considerable investment (double investment for every new generation device). The transition towards shorter wavelengths each time requires developments of all aspects of the technology. The most serious problem for shorter wavelength lithography (DUV) is linewidth control over topography due to interference effects caused by reflections of topographic features and non-uniform reflectivity on the multiple film layers over topography.

Several approaches to sub-0.35 μm lithography have been suggested in order to solve these problems, such as Deep UV lithography, phase shifting mask (2), off-axis illumination (3), and surface imaging (4). Each technique has its advantages, drawbacks, and limitations. DUV lithography can improve the resolution and process latitudes. Phase shifting masks (PSM) enhance the resolution and DOF using the same exposure tools. The main drawbacks of PSM are mask design and fabrication difficulties, mask inspection and repair of defects, and limited applicability. However, the weak PSM (such as Rim, half tone) can be applied to real devices without difficulties of layout at the expense of resolution enhancement. The off-axis illuminations techniques have received a lot attention because the resolution limit and process latitudes for specific feature sizes improve and because it is easier to apply. The disadvantages using this technique are degradation of 45° patterns, the reduction of DOF for larger dimensions, linearity, and feature distortion problems. Device production, however, additionally requires good control of C.D.(Critical Dimension) over topography, as well as a high resolution and wide process latitudes. While maintaining single layer process simplicity, the DESIRE process has been suggested as an attractive solution to cope not only with limitations of resolution and process latitudes, but also with linewidth variations due to reflections over steps (see Fig. 1). However, this technique has until now not received a wide acceptance as a production worthy process because stepper and resist improvements have allowed single wet developable resists to keep pace with industry needs, and because of the stringent requirements on suitable silylation and dry development equipment that have good uniformity for 200 mm wafers and good reproducibility and have to be available from a reliable supplier. However, for quarter micron lithography, surface imaging may become a mandatory process for patterning on topographic layers.

Experimental

The wafers were coated with a newly formulated resist which is a positive working resist (from JSR Electronics) at 4500 RPM for 30 s and soft baked on a hot plate at 135 °C for 60 s, resulting in a resist thickness of 1.1 μm. Exposures were carried out on an ASM-L PAS 5000/70 (NA 0.42) deep-UV (λ=248 nm) stepper. Liquid phase silylation was carried out in a modified MTI track at room temperature using B[DMA]DS. The liquid phase silylation solution consists of the silylating agent B[DMA]DS with PGMEA a resist solvent that can act as a diffusion promoter and n-decane as the solvent. It is safer than xylene. After liquid phase silylation, the wafers were rinsed with n-decane, and subsequently baked on a hot plate at 90°C for 60 s before the dry development. The dry development was performed on an MRC MIE (Magnetron enhanced reactive Ion Etching) 720 etcher using a 2 step process. Linewidths were measured using a low voltage Hitachi S-6100 scanning electron microscope (SEM).

Results and discussion

Liquid phase silylation   In the negative DESIRE process hexamethyldisilazane (HMDS) has typically been used in the gas phase as the silylating agent, although an alternative silylating agent, 1,1,3,3-tetramethyldisilazane (TMDS) has also been studied (5-7). Such silylating agents have demonstrated some interesting advantages over HMDS, such as reduced vertical and lateral swelling, improved surface roughness, increased process latitude, and low temperature utilization. However, this technique requires a stringent control of the silylation track and dry development equipment.

Liquid phase silylation (8-9) using B[DMA]DS with N-methyl-2-pyrrolidone (NMP) in xylene with Plasmask 200-G for i-line exposures exhibited important advantages over gas phase silylation, such as (i) an improved silicon contrast, resulting in high resolution (0.25 μm lines/spaces) (see Fig. 2) and wide focus latitudes (DOF>1.8 μm for 0.3 μm l/s) as well as exposure latitudes using conventional transmission masks (i-line, 0.48 NA). This performance results from an improved silylation contrast as revealed by SEM cross-sections after staining techniques (larger α angle)(Fig. 3) and FTIR measurements (S-shaped response)(see Fig. 4)., (ii) extremely high Si incorporation (up to 25 wt. %), resulting in largely improved dry development selectivities (see Fig. 5).

Liquid phase silylation (LPS) was found to be useful for i-line and DUV (248 nm) exposures. Liquid phase silylation for i-line exposures with Plasmask 200-G results in negative tone images. For DUV exposures, crosslinking is induced in the exposed areas leading to positive tone images (10).

In the case of the DUV exposures, the positive tone DESIRE process (11) has been presented using a newly formulated resist (which consists of a novolac based resin with a photo crosslinker). DUV exposures convert the photo-crosslinker to a crosslinked network in the exposed regions. This DUV induced crosslinking inhibits the Si-diffusion in the exposed areas. The liquid phase silylation is performed at room temperature. Dry development results in positive tone images. The liquid phase silylation for both i-line and DUV exposures used the B[DMA]DS with NMP as a diffusion promoter and xylene as a solvent. The positive tone DESIRE process using LPS has exhibited several advantages, such as an improved silylation contrast, an increased Si concentration, the use of room temperature silylation, a simpler process without post-exposure baking steps, etc. This room temperature wet silylation has been performed on existing tracks with minor modifications in puddle mode. In this way, the requirements for silylation and dry development equipment are significantly relaxed.

Resist   The resist chemistry for the negative tone DESIRE process, using both gas phase and liquid phase silylation, has been discussed in earlier publications (5,8). A base polymer carrying phenolic OH groups has reactive sites to bind the trimethyl or dimethyl silyl groups from the silylating agents during the silylation step. Diazoquinone compounds are used to afford the required photo-selectivity. Unexposed diazoquinone results in thermal crosslinking during the pre-silylation bake, which limits the permeation of the silylating agents in the unexposed parts of the resist. In the exposed parts the photo-decomposed diazoquinone can not crosslink any more resulting in unrestricted permeation of the silylating agents in these exposed parts followed by reaction with the phenolic OH groups. This then will result in a negative tone pattern after dry development.

In order to obtain a positive tone image, crosslinking should take place in the exposed areas. Therefore resist compositions were used consisting of phenolic resins (to supply reactive OH groups) and photo crosslinkers, such as bisazides. These photo crosslinkers will crosslink the phenolic resin upon exposure resulting in restriction of the permeability of the silylating agent in these exposed parts. After dry development this will result in a positive tone image. Since in this case crosslinking

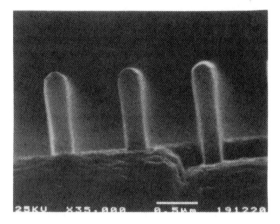

Figure 1. SEM micrograph of 0.3 μm isolated lines over 400 nm poly topography after LPS (B[DMA]DS)(i-line, 0.48 NA).

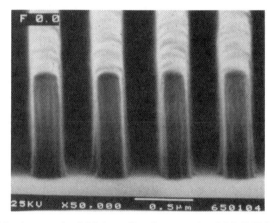

Figure 2. SEM micrograph of 0.26 μm lines/spaces after LPS (B[DMA]DS)(i-line, 0.48 NA).

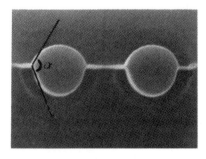

Figure 3. SEM micrograph of silylated profile using liquid phase silylation (B[DMA]DS)(i-line, 0.48 NA).

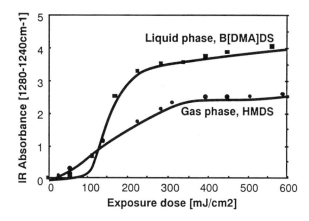

Figure 4. Comparison of silylation using gas phase [HMDS] and liquid phase silylation (B[DMA]DS).

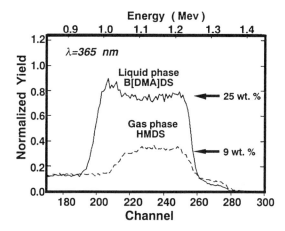

Figure 5. RBS measurements of comparison of silylation gas phase [HMDS] and liquid phase silylation (B[DMA]DS).

takes place during the exposure, no pre silylation bake is necessary. The choice of resin and photo-crosslinker results in crosslinking taking place at a reasonable exposure dose to ensure a sufficient reduction in permeability of the silylating agents.

Silylation solution   In addition to the lithographic performance of a process, safety and cost issues must also be considered in order to implement it into a manufacturing environment. We have investigated the application of new safer solvents and a feasibility study of liquid phase silylation on a track. We have investigated the silylation composition using statistically designed experiments.

In earlier studies, for liquid phase silylation NMP was used as a diffusion promoter and xylene as a solvent. The NMP concentration had a strong influence on the diffusion behaviour of the silylating agent. Therefore the NMP concentration is very critical for the process. Because of safety reasons, xylene is not desirable for high volume production. We have been investigating safer solvents as an alternative for xylene. The PGMEA that is used as a safe resist solvent has been considered as a replacement for NMP. For the this experiment we have used the newly formulated resist and the positive tone DESIRE process at 248 nm. Table I summarizes the chemical specifications of the solvents that have been investigated.

Table I. solvents

Solvent name	formula	M.W.	B.P. [°C]	M.P. [°C]	Vapor pressure m torr (20°C)
1. n-decane	$CH_3(CH_2)_8CH_3$	142.29	174	- 30	1.3
2. n-heptane	$CH_3(CH_2)_5CH_3$	100.21	98	-91	40
3. xylene	C_8H_{10}	106.17	137	-48	8.3

A relatively high soft bake temperature (140 °C for 60 s) has been used compared to conventional resist processing conditions (11). This temperature is needed to prevent dissolution of the resist by the silylation solution. This dissolution is mainly due to the solvent. We have investigated alternative aprotic (which do not react with hydroxyl groups) solvents from the viewpoint of safety, cost, dissolution of the resist, compatibility with the PGMEA, good carrier of the silylating agent. N-decane is an attractive alternative candidate. It results in less dissolution at lower soft bake temperature (130°C). The lower soft bake temperature is more desirable, because it can result in higher selectivity compared to that of the higher soft bake temperature. The thickness loss is 200 Å/min for n-decane compared to 850 Å/min for xylene at 130°C soft bake temperature. However, n-decane (30$/l) is slightly more expensive compared to xylene (20$/l). N-heptane has also shown good results as an alternative safer solvent due to its reduced dissolution rate (350 Å/min at 130°C) and its safety. We have concentrated on n-decane as a safer solvent for the rest of the study, because n-decane has less dissolution rate of the resist than n-heptane. The soft bake temperature is fixed at 135 °C for 60 s, as a compromise between dissolution and silylation selectivity.

In order to optimize the silylation process, the silylation solution concentration have been evaluated using statistically designed experiments. The five most important variables were determined with a screening experimental design. The five variables chosen are as follows: (1) the silylating agent (B[DMA]DS) concentration (2.5 % to 25 % per volume), (2) the PGMEA concentration (35 to 50 % per volume), (3) the silylation time (50 s to 80 s) at the room temperature, (4) the soft bake temperature is fixed at 135°C for 60 s, (5) for the dry development the same conditions as described above were used. Table II shows the first experimental design runs using a fixed silylation time of 60 s. The variations of the input parameter had however only minor effect on the results. These silylation experiments were performed on a track in puddle mode at room temperature. FTIR analysis of these wafers shows that the concentration of PGMEA had an influence on the diffusion rate. The results of the table indicate that the concentration of the silylation agent does not have a major influence on the silylation contrast. In a previous study (7,8), we have reported that a higher B[DMA]DS concentration in the silylation solution results in a higher atomic Si concentration after silylation as indicated by the RBS measurement in the case of liquid phase silylation for *i*-line exposures with Plasmask 200-G. A more detailed study will be performed for the positive tone process using RBS and FTIR measurements.

In the case of the lower PGMEA concentration (35%) and silylation time (60 s), not enough Si has penetrated into the unexposed areas. Therefore we have set up a second experimental design with as variables: silylation time and lower silylation concentrations (below 10%). Table II shows that for the case of lower silylation concentration (10%), these is still enough Si to diffuse into the unexposed regions. Table III summarize the results and experimental runs for 3 variables. These results indicate that the 2.5 % silylation concentration is still acceptable for the silylation process. The concentration of the silylation agents is a very important factor for the cost of the chemicals, because the silylation agent is the most expensive chemical in the silylation mixtures. Below 5 % the silylation agent concentration becomes irrelevant for the total cost. This will be discussed in detail in the next section. The results of Table III indicate that the longer silylation time can compensate the lower PGMEA concentration and the silylation concentration. The composition of the silylation mixture seems to be less critical for mixtures based on n-decane and PGMEA as compared to those based on NMP. These experimental results indicate that the lithographic performance is influenced less by the silylation concentration and the PGMEA concentration. This relative lower sensitivity is very attractive for implementation into manufacturing. The resolution can be even more enhanced by using the highest partial coherence factors (σ=0.77), since the liquid phase silylation requires lower image contrast. A more detailed study on the process latitudes by changing the silylation composition will be presented elsewhere.

Characterization of the silylation   Characterization of the silylation reaction and mechanism have been performed using FTIR and RBS. FTIR spectroscopy has been used for the quantitative determination of the Si incorporation. The integrated absorbance of 1300-1230 cm^{-1} (Si-C bond deformation) was chosen as a quantitative measure for the built-in Si.

Figure 6 shows the influence of the silylation agent concentration. This plot indicates that the optimum silylation concentration is 5-10 % and by increasing the silylation concentration above 15 % the Si content decreases again. Because by increasing the concentrations of the silylating agent, the resist will become insoluble and impenetrable and retard Si diffusion (12). The concentration of the silylating agent is significant from the viewpoint of cost. Therefore 5-10 % is the optimum concentration of silylating agent. This plot also shows that the exposed area does not contain any Si.

RBS spectroscopy was used to determine the Si atomic concentration and the Si-depth profile. Figure 7 shows that the Si concentration (6 atomic % or 18 wt. %) is larger than compared to that of the gas phase silylation in the negative DESIRE process (3 atomic % or 9 wt. %). Figure 8 shows RBS measurements with various silylation agent concentrations using this process. This plot indicates that the Si atomic concentration is constant for silylation agent concentrations above 5 %. The exposed areas do not contain any Si, because the exposed areas cross link during the

Table II

Runs of the experimental design for 2 variables for 60 s silylation time

Run No	B[DMA]DS [%]	PGMEA [%]	Sily. Time [s]	Exposure dose [mJ/cm2]	Resolution limit [μm]	
					T	P
1	10	35	60	under sily.	...	...
2.	10	50	60	65	0.22	0.18
3.	18	50	60	70	0.22	0.18
4.	18	42	60	80	0.24	0.2
5.	25	35	60	under sily.	...	...
6.	10	42	60	80	0.24	0.2
7.	25	42	60	80	0.24	0.2
8.	18	35	60	under sil.	...	...
9.	25	45	60	80	0.24	0.18

Table III

Runs of the experimental design for 3 variables

Run No	B[DMA]DS [%]	PGMEA [%]	Sily. Time [s]	Exposure dose [mJ/cm2]	Resolution limit [μm]	
					T	P
1	10	45	65	62	0.24	0.2
2.	18	42	65	62	0.22	0.18
3.	25	35	75	64	0.24	0.2
4.	5	50	50	106	0.22	0.2
5.	5	45	50	74	0.22	0.18
6.	2.5	45	65	64	0.24	0.2
7.	2.5	45	70	64	0.24	0.2

T means conventional transmission mask
P means Levenson type phase-shifting mask
under sil. means under silylated

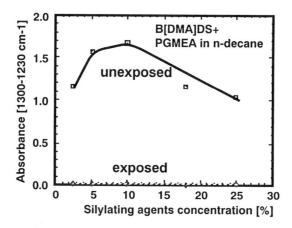

Figure 6. Si content as measured by FTIR versus silylating agents concentration (2.5% to 25 % B[DMA]DS, PGMEA in n-decane) at room temperature.

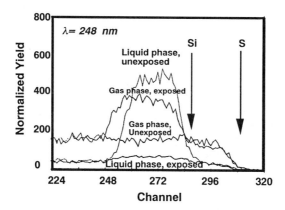

Figure 7. RBS measurements of comparison of negative tone DESIRE process using gas phase silylation (TMDS) and positive tone DESIRE process using LPS (B[DMA]DS). The arrows indicated the surface positions of the elements. S means sulfur.

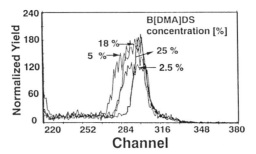

Figure 8. RBS measurements for positive tone DESIRE process using LPS (B[DMA]DS) silylation with the various concentrations of a silylating agent at room temperatures.

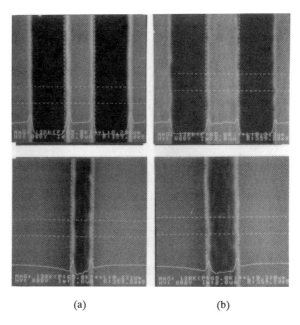

(a)   (b)

Figure 9. SEM micrographs of the various processes at DUV exposures (λ=248 nm, 0.42 NA with σ=0.58) using a transmission mask: (a) dry development with Plasmask 305-U (GPS), (b) dry development after LPS with newly formulated resists.

exposure. These results indicate that the optimum silylation agent concentrations are between 5-10 % and the positive tone DESIRE process using the new safer solvents n-decane and PGMEA improves the silylation contrast and dry development selectivity.

Proximity effects   For high contrast processes such as the DESIRE process operating close to the resolution edge, local proximity effects are typically observed, showing up as small differences in linewidth between inner and outer lines of periodical structures and isolated lines. As feature size is reduced below 0.35 µm, it becomes difficult to find an optimum single exposure dose that is satisfactory for different size and geometries (dense lines/spaces (l/s), isolated line and spaces, contact windows). For implementation on real devices, the proximity effect should be reduced. The proximity effect can be reduced by using a silylation process with high contrast and/or by optimizing the dry development process conditions (resulting in minimal microloading effect) such as low operating pressures (for obtaining anisotropic etching) and high plasma densities (high etch rate and selectivity). This can be obtained using systems such as helicon, TCP (Transformer Coupled Plasma), HRe⁻, ECR,...

In a previous study (9), we have demonstrated the reduction of the proximity effects using negative tone liquid phase silylation for *i*-line exposures with Plasmask 200-G, since the exposure latitude is increased by a factor of 2 as a result of the higher silylation contrast and the higher Si concentration. For this positive tone process using LPS, we can predict that the proximity effects can be even more reduced according to the aerial images simulations. For the simulated aerial images of the positive tone resists, the isolated lines are wider than the dense lines. For developed images the linewidth differences can be minimized since the difference resulting from aerial image can be compensated by the loss during dry development. Proximity effects are increasing relatively for features smaller than 0.4 µm using DUV wavelength (λ=248 nm, N.A. 0.42).

Figure 9 shows that the reduction of proximity effects for this positive process is superior to that of conventional negative tone wet development scheme and that of the negative tone DESIRE process using gas phase silylation. Both negative tone processes based on wet and dry developments (gas phase silylation) suffer from proximity effects. Figure 9 indicates that the positive tone process reduces the local proximity effects since the difference in aerial image is compensated during development.

In the case of 0.35 µm patterns for the positive tone process, the isolated lines are wider than dense lines due to the aerial images. In this higher contrast process, isolated lines of the final developed images are wider than the dense lines above 0.35 µm patterns because this process replicates the aerial image effect more faithfully than lower contrast ones. By approaching the resolution limit the aerial images degrade and the loss of the silylated areas increases. Therefore the differences of the linewidths can be minimized by the compensation of these two factors. Figure 9 shows that the difference of linewidths for 0.3 µm patterns is considerably small. This small difference of the linewidth can be accepted for real device implementation for sub-0.35 µm process. For 0.25 µm pattern (edge of resolution limits) larger differences have been observed.

The developed images are determined by the combination of the aerial images (positive tone vs. negative tone, partial coherence, N.A. of the lens, and masks types) and process conditions (silylation process and dry development process). In this paper, we have been investigating proximity effects which depend on the silylation process and the optical effects. The dry development conditions might be one of the most significant factors contributing to the reduction of proximity effects and the differences of resist profiles.

We have found that the combination of off-axis illumination techniques with an attenuated PSM reduced the proximity effect between dense l/s and isolated lines even for 0.25 μm feature sizes (13). This is because off-axis illumination improves the imaging for dense structures while an attenuating PSM improves the isolated features.

Lithographic performance

Resolution limits   We have evaluated the limit of resolution using both conventional transmission masks and phase shifting masks. Under optimized conditions, 0.22 μm and 0.24 μm patterns (l/s) (see Fig. 10) have been obtained with conventional masks , and 0.18 μm patterns (l/s) (see Fig. 11) with a phase-shifting mask on an ASM-L PAS 5000/70 DUV stepper (λ=248 nm, 0.42 NA). Figure 12 shows 0.1 μm isolated lines and 0.15 μm isolated spaces using a transmission mask. In optical lithography a positive tone resist process is strongly recommended for printing contact holes because a dark field mask can be used. In the case of a negative tone process, a light field has to be used, resulting in an inferior aerial image for contact holes and a higher sensitivity for defects. A dark field mask provides a higher aerial image quality than a light field one, especially for out of focus conditions. Figure 13 shows the cross-sectional SEM micrograph of 0.25 μm contact holes with an attenuated phase shifting mask.

Topography   As the wavelength is reduced, the control of C.D. variations over topography is becoming more difficult due to the higher reflectivity of most materials at shorter wavelengths. Therefore, top surface imaging techniques have received more attention not only to obtain the high resolution and the wider process latitudes, but also to control linewidth variations over topography. Figure 14 shows 0.3 μm poly-silicon lines over topography in real device applications. These SEM micrographs show the perfect linewidth control over topography.

Linearity   The linearity of the positive tone silylation process is one of most critical issues. Non-constant silylation depths for various linewidths have been reported to result in bad linear performance (14). We have investigated the linearity of this new positive DESIRE process using a conventional transmission mask. Very good linearity has been observed (see Fig 15). Over-silylation on the larger areas can be avoided by using optimum process and resist conditions (such as soft bake conditions, optimum resist design, and high silylation contrast process).

Economic issues   It is obvious that the cost of a process step is one of the most important issues to be addressed before it may be considered for manufacturing (15). The cost comparison for the process should be based on the cost of the chemicals, estimation of the equipment costs for the silylation and dry development equipment (throughput, maintenance costs, reliability). A full cost of ownership is not considered for this study. We estimated the costs of the chemical consumption including the resist cost for DUV lithography. Table IV lists the costs of the chemicals based on liter quantities. Table V summarizes estimates of the cost of the chemical consumption including the resist cost. The costs per wafer are larger for the liquid phase silylation, because of the considerably small amount of chemical consumption in gas phase silylation. The table shows that liquid phase silylation is roughly 10 times as expensive as gas phase silylation. However this cost is only 10% of the chemical costs. If you compare only the chemical cost, the costs of the dry development process (5 dollars per wafer) are significantly cheaper than for the wet development case (more than 12 dollars per wafer), because of the extremely high cost of commercially available DUV resists (10 times or 15 times as expensive as *i*-

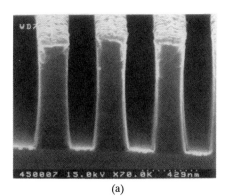

(a)   (b)

Figure 10. SEM micrographs of (a) 0.22 µm l/s (b) 0.24 µm l/s (λ=248 nm, 0.42 NA) after LPS (B[DMA]DS) using a conventional mask.

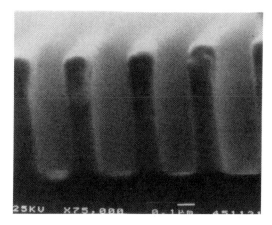

Figure 11. SEM micrograph of 0.18 µm l/s using a PSM (248 nm, 0.42 NA).

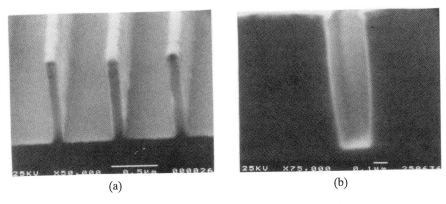

(a)   (b)

Figure 12. SEM micrographs of; (a) 0.1 μm isolated lines, (b) 0.15 μm isolated spaces using a transmission mask .

Figure 13. SEM micrograph of 0.25 μm contact hole (248 nm, 0.42 NA) obtained with positive tone DESIRE process using an attenuated PSM.

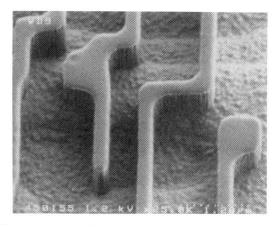

Figure 14. SEM micrograph of 0.3 μm poly patterns after LPS at DUV exposures (0.42 NA) over poly-silicon topography.

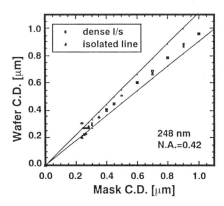

Figure 15. Experimental linearity using positive tone DESIRE process.

Table IV

Costs of silylation solution chemicals and resists

Products	prices/ liter [$]	Products	prices/ liter [$]
1. HMCTS	1100	1. IX 500 (wet)	184
2. B[DMA]DS	370	2. XP89131(wet)	2113
3. TMDS	650	3. TAR	225
4. Xylene	20	4. ARC	725
5. PGMEA	14	5. Plasmask (dry)	800
6. N-Decane	30		

based on catalog prices in liter quantities
HMCTS (Hexamethyl cyclotrisilazane), IX500 (*i*-line resist from JSR Electronics),
XP89131 (DUV resist from Shipley)

Table V

Cost comparison for the process for DUV lithography
(wet vs. dry development and gas and liquid phase silylation)

process	Resist cost/wafer (8", 6 ml)	silylation solution (8", 8 ml)	Total chemical costs ($/wafer)
1. wet dev. (DUV)	12.68	...	12.68
2. wet + ARC	17.03	...	17.03
3. dry (GPS)	4.80	0.065*	4.87
4. dry (LPS)	4.80	0.612**	5.41

* Based on the 0.1 ml consumption per wafer at the silylation track.
** Based on composition of the silylation solution; B[DMA]DS: PGMEA: n-decane=5:45:50
   The consumption is based on the 8 ml per wafer for the silylation plus 10 ml (n-decane) solvent
   consumption for the rinse step after silylation on the 8 inch wafer.
   costs for the liquid phase silylation (B[DMA]DS: $ 0.148+ PGMEA; $ 0.056+ n-decane;$ 0.408)

line resist). The ARC (Anti Reflection Coating) or TAR (Top Anti-Reflection) process can be more expensive. These rough estimations show that liquid phase silylation is more expensive than gas phase silylation when only the cost of the silylation chemicals is considered. However, the silylation chemicals contribute to only a small part of the total cost.

In the case of dry development, a dry etch system has to be taken into account for the dry development. In the case of liquid phase silylation, the silylation can be performed on a conventional track (maybe slightly modified wet development track), whereas for gas phase silylation, a more expensive vacuum based silylation unit is needed. A detailed estimation of the contribution of equipment depreciation to the total cost per wafer is difficult to make at this moment. However, liquid phase silylation can be attractive for DUV lithography also from the economic point of view.

Conclusions

In this paper, new safer solvents have been investigated for application to the positive tone DESIRE process using LPS in DUV lithography. The proposed liquid phase silylation solution consists of B[DMA]DS with n-decane and PGMEA. The new silylation solutions have been shown to exhibit several advantages over xylene and NMP, such as, (i) safety, (ii) reduced dissolution rate of the resist, (iii) relaxation sensitivity to the silylation solution. We have investigated and characterized the silylation process and proximity effects in the positive tone DESIRE process. The characterization of the silylation process has been performed using FTIR and RBS spectroscopy. A higher Si atomic concentration has been observed for the positive tone DESIRE process compared to that of negative tone DESIRE process using gas phase silylation (TMDS). We have studied the reduction of the proximity effects. Several solutions have been proposed such as (i) the higher contrast silylation process (LPS), (ii) the optical effects (positive tone, high N.A., and partial coherence factor, ..), (iii) optimum dry development conditions. This new positive tone DESIRE process is considerably simpler as well as safer compared to the process with xylene and NMP and results in a reduction of the proximity effect and in improved sensitivity as compared to gas phase silylation. A perfect linewidth control over topography in real device applications has been demonstrated. The cost of the process has been estimated based on the consumption of the chemicals including the resist cost and the silylation cost. Liquid phase silylation can be an attractive solution not only based on lithographic performance (resolution and process latitudes) but also based on cost effectiveness in DUV lithography.

Acknowledgements

The authors would like to thank B. Roland for resist formulations and useful discussion, K. Ronse for PSM, A.M. Goethals for some useful discussion, and Prof. G. Declerck, Dr. S. H. Choi, and Dr. K.H. Oh (Hyundai Electronics) for their encouragement and interest in this work. Part of this work has been sponsored by the Sony corporation.

[1]   K. Yamanaka , H. Iwasaki, H. Nozue, and K. Kasama, Proc. **SPIE 1927**, 310 (1993).
[2]   M. D. Leveson, N.S. Viswanathan, and R.A. Simpson, IEE Trans. ED-29, 1828 (1982).
[3]   M. Noguchi, M. Muraki, Y. Iwasaki, and A. Suzuki, Proc. **SPIE 1674**, 392 (1992).
[4]   F. Coopmans and B. Roland, Proc. **SPIE 631**, 34 (1986).
[5]   Ki-Ho Baik, L. Van den hove, A.M. Goethals, M.Op de Beeck, and B. Roland, J. Vac. Sci. Technol. **B 8(6)**, 1481 (1990).

[6]   A.M. Goethals, K.H. Baik, L. Van den hove, and S. Tedesco, Proc. **SPIE 1466**, 604 (1991).
[7]   Ki-Ho Baik, R. Jonckheere, A. Seabra, and L. Van den hove, Proc. **ME 91**, vol. 17, 269, (1992).
[8]   Ki-Ho Baik, L. Van den hove, and B. Roland, J. Vac. Sci. Technol. **B 9(6)**, 3399 (1991).
[9]   Ki-Ho Baik, K. Ronse, L. Van den hove, and B. Roland, Proc. **SPIE 1672**, 362 (1992).
[10]   Ki-Ho Baik, "Fundamental study of silylation for application to sub-0.35μm lithography", doctoral dissertation, Katholieke Universiteit Leuven, Ch IV (1992).
[11]   Ki-Ho Baik, K. Ronse, L. Van den hove, and B. Roland, Proc. **SPIE 1925**, 302(1993).
[12]   J. M. Shaw, M. Hatzakis, E. D. Babich, J. R. Paraszczak, D. F. Witman, and K. J. Stewart, J. Vac. Sci. Technol. **B 7(6)**, 1709 (1989).
[13]   K. Ronse, R. Pforr, K. H. Baik, R. Jonckheere, and L. Van den hove, Proc. **ME 93**, 133 (1994).
[14]   G.S. Calabrese, E.K. Pavelchek, P.W. Freeman, J.F. Bohland, S.K. Jones, and B.W. Dudley, Proc. **ME 92**, 231 (1993).
[15]   M. A. Hartney, R. R. Kunz, L. M. Eriksen, and D.C. La Tulipe, Proc. **SPIE 1925**, 270 (1993).

RECEIVED September 13, 1994

INSULATING POLYMERS

Chapter 17

Molecular Design of Epoxy Resins for Microelectronics Packaging

Masashi Kaji

Research and Development Laboratories, Nippon Steel Chemical
Company, Ltd., 46–80 Nakabaru, Sakinohama, Tobata-ku,
Kitakyusyu-shi, Fukuoka 804, Japan

The relation between the structure and properties of epoxy resins were
investigated to achieve the followings, (1) improved toughness, (2)
low moisture absorption, (3) increased heat resistance, (4) low thermal
expansion, and (5) decreased viscosity of epoxy resins for high filler
loading. It was found that introduction of a rigid group, such as 4, 4'-
biphenyl or 2, 6-naphthalene moiety, was an effective way to improve
fracture toughness, and the naphthalene based resins were effective for
lowering moisture absorption, increasing glass transition temperature,
and lowering thermal expansion. Several epoxy resins of the bisphenol
type with lower melting viscosity were synthesized and used as
molding compounds for IC packaging. Crack resistance of the packages
using these resins was markedly superior owing to high filler loading
due to their lower viscosity.

Epoxy resins are extensively used as transfer molding compounds for plastic IC and
LSI packaging. In this field, the package size has been steadily increasing while the
thickness has been decreasing. Also the mounting method has changed from insertion
mounting to surface mounting. These trends require development of new high-
performance epoxy resins to overcome the problem of package cracking.

Package cracking is due to moisture in the package (1). Moisture condensed
between the bottom side of a die pad and a packaging material is expanded at the
soldering temperatures of 215 to 260 °C. The resulting high pressure inside makes the
package swell and start cracking at the edge of the die pad.

The cracking mechanism indicates that there are several effective approaches for
the base resin to overcome the problem of package cracking. They are: lowering of
moisture content, increasing of toughness, increasing of grass transition temperature
(Tg) in order to increase mechanical strength at soldering temperature, lowering of
thermal expansion, and lowering of viscosity for high filler loading.

EXPERIMENTAL

Phenol novolac was   used as a curing agent   and   triphenylphosphine as a curing accelerator. Epoxy molding compounds obtained by compounding were molded at 150 ℃ for 3 minutes, followed by post curing at 180 ℃ for 16 hours.

The conditions for   characterization of the cured products are   described in the previous paper (*2*). Cross-linking density was determined by using the kinetic theory of rubber elasticity (*3*).

RESULTS AND DISCUSSION

Structural Modification of Epoxy Resins to Improve Toughness

Although epoxy resins have good mechanical properties, their poor fracture resistance is one of their serious disadvantages. Resins having improved toughness show resistance against the yield caused by package swelling. There have been many studies on improving the fracture resistance of cured products by incorporation of elastomeric materials. Butadiene-acrylonitrile copolymers (*4*) or siloxane oligomers (*5*) were incorporated in the base resin in order to form a finely dispersed sea-island structure. But these methods often lead to lower Tg and deteriorate mechanical strength. Influences of cross-linking density and rigidness of the resin backbone were investigated to improve the fracture toughness of the cured products without addition of elastomeric materials.

Influence of Cross-linking Density. In general, the fracture toughness of the cured product increases with   decreasing cross-linking density of the network. However this causes Tg decrease. We can make a compromise between fracture toughness and Tg for the conventional resin systems (*6*).

In order to increase Tg without increasing cross-linking density, bulky groups were incorporated into the resin structure (Fig. 1). An increase in the bulkiness increased Tg, which seemed to depend on the restricted molecular mobility of the resins. For the resin having the fluorene moiety (No. 5 in Fig. 1), low values for tan δ were observed in the vicinity of 100 ℃ by dynamic mechanical analysis (*7*), which were attributed to the restricted molecular mobility of the phenyl ring of the resin backbone. However, the fracture toughness decreased with increasing bulkiness (Fig. 2). The above compromise can still be reached by controlling cross-linking density based on the introduction of steric hindrance into the resin backbone.

Influence of Rigidness of Resin Backbone. We examined next the effect of the rigidness of the resin backbone. Fig. 3 shows the relationship between cross-linking density and fracture toughness for the bisphenol type resins. The fracture toughness of the cured product from the resin having biphenyl structure (No. 6) increased significantly without decreasing cross-linking density, with Tg maintained at   more than 170 °C. But, the cured product from the 2, 2'-disubstituted biphenyl isomer did not give higher fracture toughness, which gave a similar relationship as seen in the conventional resin systems.

In the naphthalenediol based resin systems, an extraordinary behavior was observed (Fig. 4). The cured product from the 2, 6-isomer gave a significantly higher fracture toughness than those from the other isomers.

The above results seem to indicate that the cured products from these resins having a symmetric and rigid backbone, such as 4, 4'-biphenyl and 2,6-naphthalene structure, have a different network structure compared with those from the conventional resin systems.

For explaining these behaviors, we proposed a network model for the cured products (Fig. 5). The cured products from the resins having a symmetric and rigid backbone will have a larger free volume fraction than those from the conventional resins, because the former have looser molecular packing due to steric hindrance resulted from their rigid backbone. The larger free volume fraction seem to cause large plastic deformation at the crack tip.

Dynamic mechanical analyses revealed that the storage modulus of the cured product of the diglycidyl ether of 4, 4'-dihydroxybiphenyl was smaller than that from the 2, 2'-biphenyl isomer at the glassy state. And the specific gravity of the cured product of diglycidyl ether of 4,4'-dihydroxybiphenyl was smaller than that from the 2, 2'-biphenyl isomer which had a flexible bending backbone (Table I). These observations presumably could be explained by the larger free volume fraction of the cured products from the resins having rigid backbone.

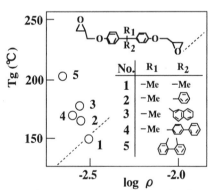

Fig. 1. Tg vs. cross-linking density (log ρ). Dotted line shows the relationship for the conventional resins.

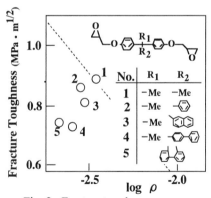

Fig. 2. Fracture toughness vs. cross-linking density (log ρ). Dotted line shows the relationship for the conventional resins.

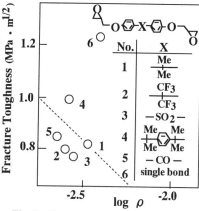

Fig. 3. Fracture toughness vs. cross-linking density (log ρ). Dotted line shows the relationship for the conventional resins.

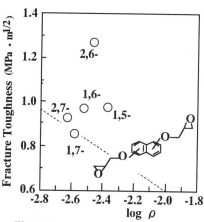

Fig. 4. Fracture toughness vs. cross-linking density (log ρ). Dotted line shows the relationship for the conventional resins.

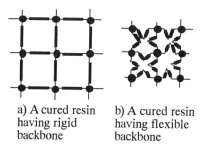

a) A cured resin having rigid backbone

b) A cured resin having flexible backbone

Fig. 5. Network models of cured resins.
● ; crosslinking point
▬ ; resin backbone

Table I. Specific Gravities of Epoxy Cured Products

No.	epoxy resin	specific gravity (g/cc)
1	4, 4'-isomer	1.18
2	2, 2'-isomer	1.20

Lowering of Moisture Absorption

Moisture absorption is one of the most critical factors causing the package cracking. Packages are sometimes supplied from venders in dry shipping containers, or they may be baked before soldering. These treatments have no industrial merit (8). As the amount of water absorbed in a package depends on the properties of the base resins used, lowering of moisture absorption is a strong demand for epoxy resins.

Introduction of Polynuclear Aromatics. It is reported that moisture content increases with increasing cross-linking density for the o-cresol novolac epoxy resin system (9). For the conventional resin system, lowering of moisture absorption limit by decreasing cross-linking density would then leads to a decrease in Tg. The amount of water has a close relation to the content of the functional group, i.e., phenolic hydroxyl groups in a curing agent used (10).   An increase in hydrophobicity of a resin is considered to be a good approach to decrease moisture absorption of a cured product. Introduction of condensed polynuclear aromatic moieties, e.g., naphthalene and anthracene rings, and so on, is an effective approach to lower moisture absorption without decreasing Tg. Synthesis and properties of the naphthalene based epoxy resin are reported (11). This resin was prepared by replacing the o-cresol moiety with the 1-naphthol moiety in the o-cresol novolac epoxy resin.   Although the cured product showed improved heat and humidity resistance,   the resin had a higher melting viscosity, which spoils moldability as a molding compound. No description about the relationship between the structure and properties of the cured product is given.

Several kinds of the naphthalene based epoxy resins were   synthesized through condensation reaction of   naphthalene compounds with condensing agents in the presence of acid catalyst,   followed by glycidyl etherification with epichlorohydrin (scheme 1),   and the relationships between the structure and properties of the cured products were studied.

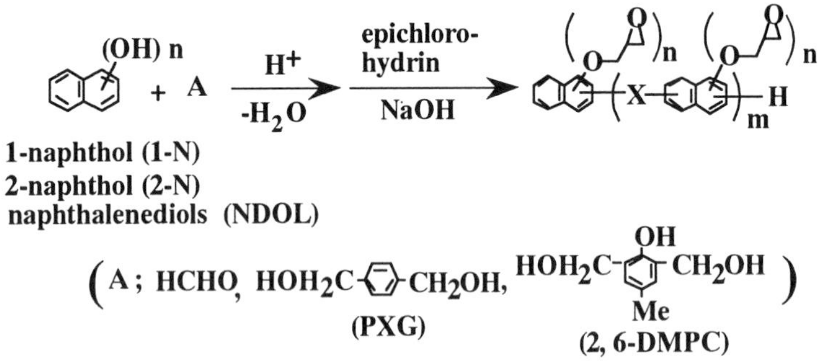

1-naphthol (1-N)
2-naphthol (2-N)
naphthalenediols (NDOL)

scheme 1

Table II.   Properties of Naphthalene Based Epoxy Resins

No.	naphthalene compound	condensing agent	epoxy equiv.	softening point (°C)	melting[a] viscosity
i	1-N	HCHO	244	98	10.5
ii	1-N	PXG	281	83	3.4
iii	1-N	2,6-DMPC	236	91	6.5
iv	2-N	HCHO	238	-	-
v	2-N	PXG	277	88	6.0
vi	2-N	2,6-DMPC	219	89	6.0
vii	1,6-NDOL	HCHO	155	59	1.0
viii	1,6-NDOL	PXG	166	69	2.0
ix	1,6-NDOL	2,6-DMPC	165	75	4.0
x	1,7-NDOL	PXG	173	79	5.0
xi	2,7-NDOL	PXG	166	72	3.5

[a]poise at 150°C

In general, moisture absorption increases with an increase in Tg. In other words, lowering of moisture absorption spoils the heat resistance of a cured product. This relationship was also noticed for the naphthalene based resins, but was less pronounced for these resin systems than for the phenyl ring based resin systems (Fig. 6). Especially, the resins having *p*-xylylene moiety (No. ii, v, viii, and x) gave more satisfactory results.

The lower moisture absorption in the naphthalene based resins corresponds to the lower content of the functional group, i.e., epoxy group, in the resins, through replacement of the phenyl ring by the naphthalene ring.

Fig. 7 shows the relationship of moisture content vs. epoxy equivalent which corresponds to the functionality of resins. No difference was noticed between the naphthalene based resins and the phenyl ring based resins, indicating that moisture absorption corresponds to the functionality of resins.

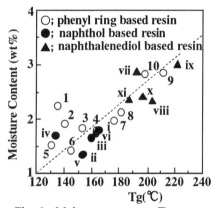

Fig. 6. Moisture content vs. Tg. Numbers of the points refer to resins listed in Table II and III.

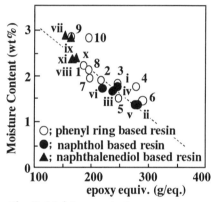

Fig. 7. Moisture content vs. epoxy-equivalent. Numbers of the points refer to resins listed in Table II and III.

Table III.   Phenyl Ring Based Epoxy Resins

Increasing of Glass Transition Temperature

Increasing of flexural strength at higher temperatures promises one direction for improving package crack resistance. As the mechanical properties of polymers change dramatically at Tg, where the polymers change from the glassy states to the rubbery states, higher Tg should be sought in order to hold a higher flexural strength at higher temperature. Suzuki et. al. reported that the flexural strength at 215 ℃ for the cured products of the epoxy resins depended on Tg, and also showed that the polyimide based molding compound having higher Tg (e.g., 250 ℃) exhibited good crack resistance (*12*) .

Naphthalene Based Epoxy Resins. Generally, an increase in Tg leads to an increase in the moisture content and a decrease of fracture resistance. Polyfunctional epoxy resins (No. 9 and 10 in Table III), for example, are known to have a higher Tg, they exhibit higher moisture absorption and lower fracture resistance. These properties are disadvantageous for improving package cracking.

Introduction of condensed polynuclear aromatic moieties, such as naphthalene ring, is an effective way to increase Tg while keeping lower moisture content as discussed above. Fig. 8 shows the relationship between cross-linking density and Tg. Tg increased with an increase in cross-linking density. The naphthalene based resins had higher values for Tg compared with the phenyl ring based resins, indicating that the Tg of the resins depends on their condensed aromatic structure. Especially, the naphthalenediol based resins having *p*-xylylene moiety (No. viii , x, and xi) gave markedly higher Tg while keeping lower moisture absorption and higher fracture toughness (Fig. 6 and 9). Furthermore, the Tg was raised further by post curing at higher temperatures (e.g., 230 ℃), and the cured products gave higher flexural strength and modulus at 215 ℃, correlating to their Tg (Table IV).

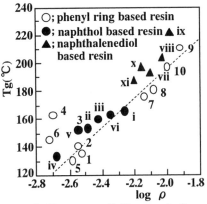

Fig. 8. Tg vs. cross-linking density(log ρ). Numbers of the points refer to resins listed in Table II and III.

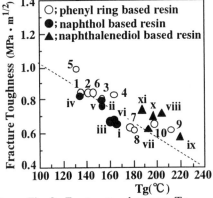

Fig. 9. Fracture toughness vs. Tg. Numbers of the points refer to resins listed in Table II and III.

Table IV.   Physical Properties of Epoxy Cured Products at 215°C

No.	epoxy resin	Tg(°C)	flexural strength (kgf/mm^2)	flexural modulus (kgf/mm^2)
1	No 7. in Table III	177	1.7	210
2	No v. in Table II	144	0.7	90
3	No viii. in Table II	227	4.9	830

SiO$_2$ content ; 75 wt%

Lowering of Thermal Expansion

Package cracking results from volume expansion of the water absorbed at the interface between the bottom side of a die pad and a packaging material. Increasing adhesion strength at the interface is an effective approach to lower the amount of water absorbed at the interface, leading to an improved crack resistance. As delamination at the interface is caused by thermal stress due to the difference in thermal expansion between a die pad and a packaging material, matching of thermal expansion of the cured product is required.

Influence of Cross-linking Density.   Fig. 10 shows the relationship between cross-linking density and CTE for the naphthalene based resins, in which the bottom points correspond to the grassy states and the upper points correspond to the rubbery states. Ochi et. al. reported that the cured product of the resin from the diglycidyl ether of 1, 6-naphthalenediol had a smaller coefficient of thermal expansion (CTE) (*13*). In our study, CTE for the naphthalene based resins were also smaller than those for the phenyl ring based resins. However, the CTE at the glassy state in the naphthalene based resins increased with increasing cross-linking density and reached the same value as the CTE for the phenyl ring based resins in the cured products having the Tg of around 200 °C.

Ogata et. al. reported on the thermal expansion behaviors in the *o*-cresol novolac epoxy resin (*9*), in which an increase in CTE with increasing cross-linking density was explained in terms of an increase in free volume fraction. However, the above extraordinary behavior of the naphthalene based resins has so far been inexplicable.

We proposed a network model as shown in Fig. 11 and the following equation:

$$\alpha = \mathbf{a} \cdot \log \rho + \mathbf{b} \tag{1}$$

where α is CTE, and ρ is cross-linking density, in which CTE can be expressed as a function of cross-linking density. In this equation, $\mathbf{a}$ and $\mathbf{b}$ are constants; $\mathbf{a}$ is associated with the vibration of cross-linking point and $\mathbf{b}$ with the vibration of resin backbone. Smaller CTE values in the naphthalene based resins are considered to correspond to smaller vibration members of the backbone having condensed nuclear aromatic structure, i.e., naphthalene; the naphthalene based resins have smaller constant $\mathbf{b}$. An increase in CTE with cross-linking density is explained by the increased contribution of the vibration of cross-linking points. Naphthalene based resins have a large value for constant **a**, which correspond to increased transmission of the vibration of cross-linking points through their rigid backbone. The CTE behavior in the rubbery state also can be explained in a similar manner.

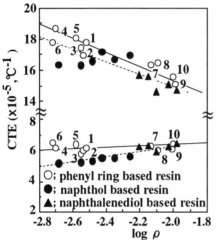

O; phenyl ring based resin
●; naphthol based resin
▲; naphthalenediol based resin

Fig. 10. Coefficient of linear thermal expansion(CTE) vs. cross-linking density(log ρ). Numbers of the points refer to resins listed in Table III.

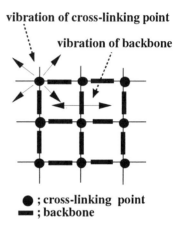

● ; cross-linking point
▬ ; backbone

Fig.11. Network model of cured product.

Low Viscosity Epoxy Resins

High filler loading is a very effective way to improve crack resistance as it leads to a decrease in moisture content, increase toughness and mechanical strength, and also to increase adhesion strength between a packaging material and a chip by reducing CTE difference between both materials. Lowering of the melting viscosity of the base resins is essential for high filler loading.

Although the *o*-cresol novolac epoxy resin has been widely used in the field of IC plastic packaging material, a limitation exists for lowering the melting viscosity while keeping a solid state at ordinary temperatures. Therefore, new types of resin should be designed for high filler loading. The biphenyl type epoxy resin is a promising material as a low viscosity base resin for the future use (*14*), and shows one of the directions for developing new low- viscosity epoxy resins.

Syntheses. For designing low-viscosity epoxy resins which exist as a solid at ordinary temperatures, it is necessary to consider their molecular weight and symmetry in molecular structure. We investigated bisphenol type epoxy resins, whose structures are depicted in Table V.

Table VI shows the influence of various connecting groups on the properties of the resins. The resins having ether or sulfide linkage (No. 1 and 2 in Table VI) gave crystalline products, which had lower viscosity and suitable melting point as base resins for transfer molding compounds . On the other hand, the resins having ketone or sulfone linkage (No. 3 and 4 in Table VI) gave higher viscosity and melting point, because of their strong molecular interaction due to their higher polarity.

Table V. Structures of Bisphenol Type Epoxy Resins Studied

X;	R1, R2, R3;
$-O-$	$-H$
$-S-$	$-Me$
$-CO-$	$-^iPr$
$-SO_2-$	$-^tBu$
$-CH_2-$	phenyl
$\underset{Me}{\overset{Me}{|}}$	
Me, Me / Me, Me	

Table VI. Properties of Bisphenol Type Epoxy Resins

No.	X in (1)	melting point (℃)	viscosity [a] (cps)
1	$-O-$	81	35
2	$-S-$	52	36
3	$-CO-$	139	79
4	$-SO_2-$	165-170	80
5	$-CH_2-$	-	36
6	$\underset{Me}{\overset{Me}{|}}$	-	44
7	Me, Me / Me, Me	64	56
8	biphenyl type [b]	105	59

[a] 30wt% in m-cresol at 25 ℃
[b] Diglycidyl ether of 3,3',5,5'-tetra-methyl -4,4'-dihydroxybiphenyl

(1)

Next, the influence of introduction of alkyl substituent groups on the aromatic rings was studied. Although the resin having methylene linkage (No. 5 in Table VI) had a lower viscosity, it gave a liquid product at ordinary temperatures. However, some resins having alkyl substituents gave crystalline products having suitable melting points. Through this study, we found several kinds of crystalline low viscosity epoxy resins as shown iv Table VII.

Moldability.   Moldability of the resins was evaluated by compounding with   78.5 wt.% of silica. Fig. 12 shows the relationship between gelation time and spiral flow at 175 ℃. The resins having ether or methylene linkage (No. 1 and 2) gave excellent moldability due to their lower viscosity.

Table VII. Structures of Epoxy Resins Selected

No.	structure (G; glycidyl group)	melting point (℃)	viscosity [a] (cps)
1	GO–⬡–O–⬡–OG	79	36
2	Me, Me GO–⬡–CH$_2$–⬡–OG Me, Me	78	45
3	GO–⬡–S–⬡–OG	49	36
4	^{t}Bu, Me GO–⬡–S–⬡–OG Me, ^{t}Bu	121	72
5	Me, Me GO–⬡–⬡–⬡–OG Me, Me	65	56
6	Me, Me, Me, Me GO–⬡–⬡–⬡–OG Me, Me	89	65
7	Me, Me GO–⬡–⬡–OG Me, Me	105	59

[a] 30wt% in m-cresol at 25 ℃

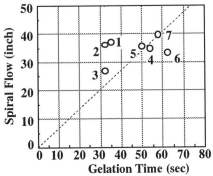

Fig. 12. Spiral Flow vs. gelation time.
Numbers of the points refer to the resins
in Table VII.

Properties of Cured Products. Moisture content of the cured products is shown in Fig. 13. The resins having lower viscosity had higher moisture absorption. This depends on their higher functionality due to their smaller molecular weight. In general, moisture absorption increases with raising Tg, which corresponds to cross-linking density as discussed before. The dotted line in Fig. 14 shows the relationship. The resins having ether or sulfide linkage (No. 1 and 3 in Table VII) had higher moisture content compared with the other resins because of higher polarity of the oxygen and sulfur atoms in the former resins.

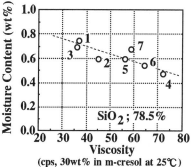

Fig. 13. Moisture content vs. viscosity. Numbers of the points refer to the resins in Table VII. Moisture absorption condition; 133 ° C, 3atm, 96 hours.

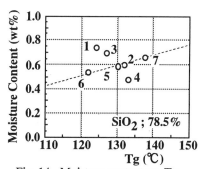

Fig. 14. Moisture content vs. Tg. Numbers of the points refer to the resins in Table VII. Moisture absorption condition; 133 ° C, 3atm, 96 hours.

Application to IC Packaging. The ether type resin (No. 1 in Table VII), which showed lower viscosity and good moldability, was used as a molding compound for IC packaging and evaluated in terms of crack resistance.

Higher filler content (over 90 wt.%) was achieved by use of the combination of the ether type resin and spherical silica, which exhibited the same level of moldability as that when the biphenyl type resin(No. 7 in Table VII) was used. Although the cured product of the ether type resin had higher moisture content than that of the biphenyl type resin with the same filler content (Fig. 13 and 14), their molding compounds gave opposite results on moisture absorption due to higher filler loading in the ether type resin system. And, in addition, the CTE of the ether type resin is lower as shown in Table VIII.

Table VIII. Physical Properties of Epoxy Cured Products

No.	epoxy resin in Table VII	filler content (wt%)	moisture absorption [a] (wt%)	CTE (x 10^{-5}, $°C^{-1}$, <Tg)
1	No. 7	88.5	0.22	1.10
2	No. 1	90	0.20	0.90
3	No. 1	91	0.17	0.75

[a] 85°C, 85%R.H., 100 hours

Fig. 15a shows the rate of package damaged by cracking after moisture absorption, followed by solder dipping at 260 °C for 10 seconds. All of the packages obtained from the biphenyl type resin system containing 84 wt.% of silica were damaged after

48 hours. However, no crack was observed in the ether type resin system containing 90 wt.% of silica, even after 96 hours.

It was also confirmed that a molding compound having a higher filler content gave higher adhesion strength at the interface between the bottom side of a die pad and a packaging material. Fig. 15b shows the proportion of the area of the bottom side of a die pad delaminated by solder dipping at 260 °C for 10 seconds after moisture absorption. Most parts of the area of the bottom side of a die pad were delaminated from the beginning of the test for the biphenyl type resin system containing 84 wt.% of silica, while 80 % of the area for the ether type resin system containing 90 wt.% of silica were still attached even after 96 hours.

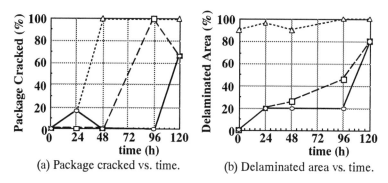

(a) Package cracked vs. time.   (b) Delaminated area vs. time.

Fig. 15. Package cracking test results.

moisture absorption condition; 85 °C, 85% R.H.
package type ; QFP-80pin
··△··· biphenyl type resin (SiO$_2$; 84 wt%)
─□─ ether type resin (SiO$_2$; 88.5 wt%)
─○─ ether type resin (SiO$_2$; 90 wt%)

CONCLUSION

In order to delineate promising direction for designing new epoxy resins, we examined several approaches to improve toughness, lower moisture content, increase heat resistance, and lower thermal expansion of the cured products of epoxy resins, and also a method for decreasing their melting viscosity in order to achieve higher filler loading.

The study revealed the following:
(1) Introduction of a rigid and symmetric groups, such as 4, 4'-biphenyl or 2, 6-naphthalene moiety, into the resin backbone was an effective way to improve fracture toughness. The increased fracture toughness seems due to increased plastic deformation at the crack tip, and is attributed to larger free volume fraction.
(2) The naphthalene based resins were effective for lowering moisture content, increasing Tg, and lowering thermal expansion. Especially, the resins having aralkyl linkages showed well-balanced properties against moisture, heat, and fracture.

(3) Several kinds of the bisphenol type epoxy resins having lower melting viscosity and existing a solid state at ordinary temperatures were synthesized as molding compounds for IC packaging. Packages made from them exhibited a greatly improved crack resistance owing to higher filler loading.

LITERATURE CITED

1. Iko, K.; Nakamura, Y.; Yamaguchi, M.; Imamura, N. *IEEE Electrical Insulation Magazine*, **1990**, *6(4)*, 25
2. Kaji, M.; Aramaki, T.; Nakahara, K. *J.Thermosetting Plastic, Japan*, **1993**, *14(2)*, 71
3. Kinjo, N.; Ogata, M.; Numata, S.; Yokoyama, T. *J. Thermosetting Plastic, Japan*, **1985**, *6(3)*, 31
4. Riew, C. K.; Rowe, E.H.; Siebert, A.R. *Am.Chem.Soc.Adv.Chem.* **1976**, Ser.,*154*, 326
5. Yorkgitis, E.M.; Eiss, Jr. N.S.; Tran, C.; Wilkes, G.L.; McGrath, J. E. *Advances in Polymer Science*, **1985**,*72*, 79
6. Kaji, M.; Yamada, Y. *IUPAC International Symposium, Preprints*, **1989**, 123
7. Kaji, M.;*The Japan Society of Epoxy Resin Technology, Lecture Preprint*, **1992**, Dec.
8. Nakagawa, O.; Shimamoto, H.; Ueda, T.; Shimomura, K.; Hata, T.; Tachikawa, T.; Fukushima, J.; Banjo, T.; Yamamoto, I. *J.Electronic Materials*, **1989**, *18(5)*, 633
9. Ogata, M.; Kinjo, N.; Kawata, T. *J. Applied Polymer Science*, **1993**, *48*, 583
10. Kaji, M.; Aramaki, T.; Nakahara, K. *J. Thermosetting Plastic, Japan*, **1993**, *14(4)*, 189
11. USP 4,551,508
12. Suzuki, H.; Sashima, H.; Kawata, T.; Ichimura, S. *Hitachi Technical Reports*, **1992**, *19*, 23
13. Ochi, K.; Tuboi, T.; Kageyama, H.; Shinpo, M. *J. Adhesion Society of Japan*, **1989**, *25*, 222
14. Mogi, N.; Yasuda, H. *IEEE*, **1992**, 1023

RECEIVED September 13, 1994

Chapter 18

Uniaxial and In-Plane Molecular Orientation of Polyimides and Their Precursor as Studied by Absorption Dichroism of Perylenebisimide Dye

M. Hasegawa, T. Matano, Y. Shindo, and T. Sugimura

Department of Chemistry, Faculty of Science, Toho University, 2–2–1 Miyama, Funabashi, Chiba 274, Japan

Molecular Orientation of a uniaxially stretched poly(amic acid) (PAA) films and the corresponding polyimide (PI) film cured thermally has been examined by measuring the dichroic ratio of a rigid-rod dye (perylenebisimide) dispersed molecularly in their films. The dichroic spectra of this dye at an incidence angle was also applied to evaluate the in-plane molecular orientation of PAA films cast on a glass plate (undrawn) and the thermally imidized PIs. For uniaxially stretched samples, the dichroic spectra showed that semi-rigid PI(BPDA/PDA) chains oriented spontaneously toward the stretching direction during thermal imidization of the uniaxially drawn PAA films (DR = 50 %). High Young's modulus of the uniaxially oriented PI(BPDA/PDA) film ($\sim$60 GPa) was rationalized in terms of the considerably large orientation factor (f = 0.6). On the other hand, flexible PI(BPDA/ODA) showed no spontaneous molecular orientation during thermal imidization, corresponding to the fact that Young's modulus of PI(BPDA/ODA) is nearly independent of a draw ratio. Thermal imidization of PAA(BPDA/PDA) cast on a substrate enhanced markedly the in-plane orientation of the polymer chains, no spontaneous in-plane orientation was observed for BPDA/ODA system. The spontaneous in-plane orientation behavior became marked as the film thickness decreased. Free-cured PI(BPDA/PDA) showed a degree of the in-plane orientation much smaller than the PI film imidized on substrate.

Physical properties of aromatic polyimides (PI) such as the thermal expansion coefficient (TEC) (1) and Young's modulus (2) have been widely known to be strongly affected by molecular orientation of PI chains. This paper is focused on spontaneous uniaxial and in-plane orientation of PI chains induced by thermal imidization of poly(amic acid) (PAA).

The uniaxial molecular orientation can be qualitatively estimated by the optical birefringence measurements. For quantitative analysis, the infrared absorption dichroism for a specific group in polymer chains and the uv-vis absorption dichroism for rigid-rod dyes introduced covalently or mechanically are available. Several research groups showed that PI chains orient in the film plane by using X-ray method (3-5) and the optical birefringence (n_o, n_e) measurement.(3, 6, 7) In the present paper, we describe the molecular orientation of PAA and PI chains determined by the absorption dichroism of perylenebisimide dye dispersed in PAA and PI films and focus on the spontaneous molecular orientation occurring in the course of thermal imidization of PAA films.(8) Biphenyl type PI is known to show the intrinsic absorption dichroism at 290 nm.(9) But, since the absorption at 290 nm is too strong, it is not available for thicker PAA and PI films (the use of the intrinsic dichroism is limited to PI films less than about 0.2 μm in thickness). Hence, the dichroic spectrum measurement of the dyes is powerful and convenient way to PAA and PI films ranging from several μm to hundreds μm thick.

EXPERIMENTAL

Chemical formulae of PAAs, PIs, and the dye are shown in Figure 1. Dye 1 and 2 were used for evaluation of the uniaxial and in-plane orientation, respectively. The dyes were dissolved rapidly in N-methyl-2-pyrrolidone (NMP) at 160°C, then the solution was cooled to room temperature and vigorously mixed with the N,N-dimethylacetamide (DMAc) solution of PAA (10 wt %). The dye-containing PAA solution was defoamed under reduced pressure and cast on a glass plate at 65°C. The concentration of the dye 1 and 2 in PAA films were ca. 4×10^{-4} M and 3×10^{-3} M, respectively. Due to this procedure, the dyes were dispersed molecularly in the films. The visible absorption spectrum of the dye-containing films was quite similar to that of the DMAc solution except for a slight spectral shift, and the dyes were fluorescent in PI films as well as in the DMAc solution.

For the uniaxial orientation measurement, the PAA specimens containing the dye were uniaxially stretched at room temperature. The drawn PAA films fixed with a frame were thermally imidized at 250°C for 2 h. Herman's orientation factor F of the drawn PAA and PI films given by

$$F = \frac{3\langle\cos^2\Theta\rangle - 1}{2} = \frac{D_0 + 2}{D_0 - 1} \cdot \frac{D - 1}{D + 2} \qquad (1)$$

was determined from the dichroic absorption spectra of the dye measured at incidence angle = 0 by using an uv-vis spectrophotometer equipped with polarizers, where Θ is the angle between the polymer chain axis and the drawing direction, D (= $A_\parallel /A_\perp$) is the dichroic ratio at the incidence angle = 0 which is the ratio of absorbances at the peak wavelength (near 535 nm) for incident lights polarized parallel ($A_\parallel$) and perpendicular ($A_\perp$) to the stretching direction, and D_0 is the dichroic ratio for perfect uniaxial orientation and defined as $D_0 = 2\cot^2\beta$; β is the angle between polymer chain axis and the transition dipole of the dye. The f is the orientation factor for the dye itself. The orientation factor for polymer chains, F, changes within the range, 0 (isotropic) $\leq$ F $\leq$ 1 (perfect uniaxial orientation). If $\beta = 0°$, F = f = (D-1)/(D+2).

For the in-plane orientation measurement, the PAA solution was cast on a glass plate, and the cast film was then thermally imidized on the substrate or without substrate (free-standing cure). All the as-cast films were perfectly isotropic in the film plane. The in-plane orientation of PAA and PI chains were estimated by measuring the dichroic spectra at an incidence angle as shown in Figure 2. In order to measure the dichroic spectra at a constant refraction angle α, the incidence angle θ for both P-and S-polarized lights were adjusted on the basis of the refractive indices (the ordinary (n_o) and the extraordinary (n_e)) of PAA and PI films measured by using an Abbe's refractometer. θ was calculated from the following equation:

$$(\text{Snell's law}) \quad n \sin\alpha = \sin\theta$$
$$n = n_o \text{ (for S-polarized light)}$$
$$n^{-2} = n_o^{-2}\cos^2\alpha + n_e^{-2}\sin^2\alpha \quad \text{(for P-polarized light)} \quad (2)$$

We defined IP=$(1-R)/(1-R_0)$ as the degree of in-plane orientation, where R $(=A_p/A_s$, A: absorbance at the peak wavelength at 535 nm) is the dichroic ratio, R_0 $(=\cos^2\alpha)$ is the dichroic ratio for the perfect in-plane orientation, and 0 (3D random) $\leq$ IP $\leq$ 1 (2D random distribution).

RESULTS AND DISCUSSION

Uniaxial Orientation. Prior to use the perylenebisimide dye, we examined whether the dye reflects polymer chain orientation by comparing with 1,6-diphenyl-1,3,5-hexatriene (DPH) which is known to align parallel to poly(vinyl chloride) (PVC) chains.(10) The measurement of the dichroic orientation factor, f, for the perylenebisimide dye and DPH in PVC film as a function of draw ratio showed that the f values for the perylenebisimide dye are only sightly lower than that for DPH. Accordingly, it is most likely expected that the rigid-rod perylenebisimide dye orients nearly parallel to PAA and PI chain segments ($\beta \approx 0$).

Figure 3 shows the orientation factors of PAA and PI chains for (a) BPDA/PDA, (b) BPDA/ODA systems, and (c) Young's moduli of PI(BPDA/PDA) film as a function of draw ratio of the corresponding PAA specimens. For the PAA(BPDA/PDA) system, stretching of about 50 % caused only a low degree of molecular orientation ($f \approx 0.1$). Surprisingly, thermal imidization of the slightly drawn of PAA film enhanced markedly the orientation factor up to ca. 0.6, thus indicating that the spontaneous molecular orientation toward the stretching direction occurred during thermal imidization. The orientation factor increased linearly with a draw ratio. On the other hand, for a flexible BPDA/ODA system (Figure 3b), no spontaneous orientation was observed during thermal imidization. On the contrary, the f values decreased slightly by thermal imidization. Thus, it was found that PI chain linearity (rigidity) is one of the most important factor to the spontaneous uniaxial orientation behavior. What is the driving force for the spontaneous orientation behavior induced by thermal imidization remains unsolved.

As shown in Figure 3c, the effect of stretching on Young's moduli is rationalized in terms of the measured orientation factors for both systems. Annealing at 330°C

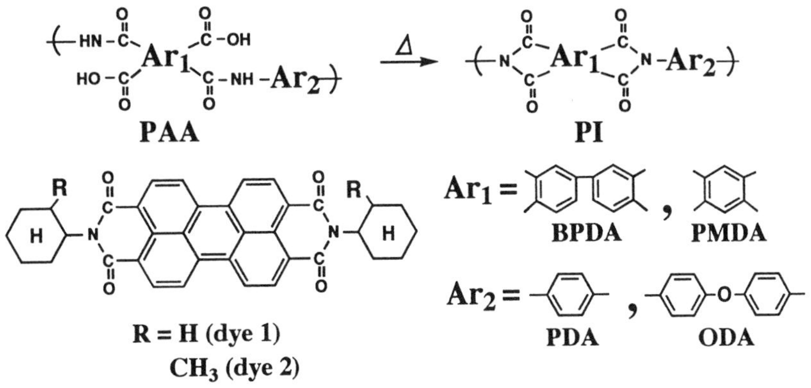

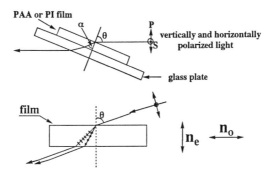

PAA   PI

Ar₁ = , BPDA PMDA

Ar₂ = , PDA ODA

R = H (dye 1)
CH₃ (dye 2)

Figure 1. Chemical formulae of PAA, PI and perylenebisimide dyes.

PAA or PI film
α
θ
P
vertically and horizontally
polarized light
S
glass plate

film
θ
n_e n_o

Figure 2. Schematic diagram for the in-plane orientation measurement.

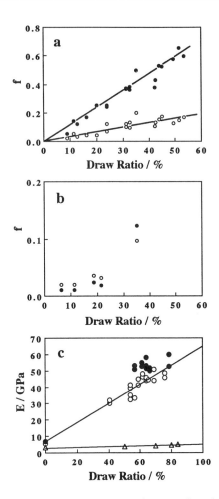

Figure 3. The orientation factors for PAA and PI as a function of draw ratio for (a) BPDA/PDA and (b) BPDA/ODA systems. (c) Young's modulus of PI films cured at 250°C (○, △) and annealed at 330°C (●). (Reproduced with permission from ref. 8. Copyright 1994 John Wiley & Sons, Inc.)

increased somewhat Young's modulus of the oriented PI(BPDA/PDA) film. In fact, the orientation factor didn't change by the same annealing, suggesting that the increase in Young's modulus is attributed to an increase in crystallinity or intermolecular interaction such as charge-transfer interaction.(11)

In-plane Orientation. Figure 4 illustrates the optical birefringence Δn (=n_o-n_e) of PAA and PI films as a function of film thickness for the BPDA/PDA system. The value of Δn decreased with increasing thickness for both PAA and PI. The fact that Δn for the PI film is much larger than that for the PAA film gives the impression that the degree of in-plane orientation for the former is also far larger than that for the latter. But in fact, quantitative comparison of the in-plane orientation between different polymers is

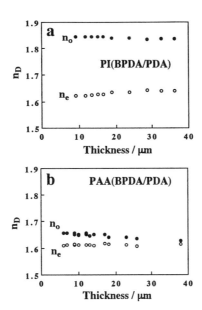

Figure 4. The ordinary (n_o) and extraordinary (n_e) refractive indices of (a) PAA(BPDA/PDA) and (b) PI(BPDA/PDA) films.

generally difficult if their intrinsic birefringence Δn_0 are unknown. In the present study, connecting the optical information (refractive indices) with spectroscopic data (the dichroic ratio of the perylenebisimide dye measured at an incidence angle) makes quantitative evaluation of the degree of in-plane orientation for various PAAs and PIs possible.

Figure 5 shows the effect of film thickness on the in-plane orientation for three systems. For the BPDA/PDA system (Figure 5a), the IP value for PAA film decreased as the film becomes thick. Even for the flexible PAA chain, the small extent of the in-plane orientation was observed. It should be noted that thermal imidization of the PAA on a glass plate enhanced markedly the in-plane orientation. On the other hand, no spontaneous in-plane orientation was observed for the flexible BPDA/ODA system as illustrated in Figure 5b. These are very similar to the phenomenon observed in the uniaxially drawn system described above (Figure 3a and b). For PMDA/ODA system in Figure 5c, which has shown to align in the film plane by the X-ray method (3,4), it was found that the degree of in-plane orientation is not so high compared with the BPDA/PDA system.

Figure 6 shows the comparison between the samples cured on glass plate and cured in free-standing. Although the sample cured in free-standing showed the IP value much smaller than that of the PI film cured on glass plate, the spontaneous in-plane orientation by thermal imidization was evidently observed even for the PI films cures in free-standing. Furthermore, we found that TEC varies in inverse proportion to IP for PI(BPDA/PDA).

Thus, this method is available for the evaluation of the uniaxial and in-plane molecular orientation for various PAA and PI films of several μm to hundreds μm in thickness.

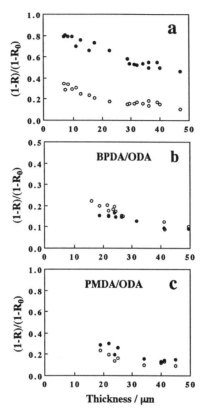

Figure 5. Thickness dependence of the in-plane orientation of PAA (○) and PI (●) chains for (a) BPDA/PDA, (b) BPDA/ODA, and (c) PMDA/ODA systems.

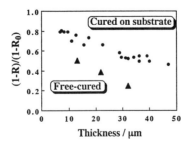

Figure 6. Thickness dependence of the in-plane orientation for PI(BPDA/PDA) cured on glass plate (●) and cured in free-standing (▲).

Literature Cited.
1. Numata, S.; Fujisaki, K.; Kinjo, N. *Polymer* , **1987**, 28, 2282.
2. Kochi, M.; Uruji, T.; Iizuka, T.; Mita, I.; Yokota, R. *J. Polym. Sci.: C*, **1987**, 25, 441.
3. Russel, T. P.; Gugger, H.; Swalen, J. D. *J. Polym. Sci. Phys. Ed.*, **1983**, 21, 1745.
4. Takahashi, N.; Yoon, D. Y.; Parrish, W. *Macromolecules*, **1984**, 17, 2583.
5. Jou, J. H.; Huang, P. T.; Chen, H. C.; Liao, C. N. *Polymer*, **1992**, 33, 967.
6. Herminghaus, S.; Boese, D.; Yoon, D. Y.; Smith, B. A. *Appl. Phys. Lett.*, **1991**, 59, 1043.
7. Ando, S. In *Recent Advances in Polyimides 1992*; Yokota, R, Ed; Raytech Co., Tokyo, **1993**, pp 63.
8. Hasegawa, M.; Shindo, Y.; Sugimura, T.; Yokota, R.; Kochi, M.; Mita, I. *J. Polym. Sci.: B*, in press.
9. Nishikata, Y.; Konishi, T.; Morikawa, A.; Kakimoto, M; Imai, Y. *Polym. J.*, **1988**, 20, 269.
10. Neuert, R.; Springer, H.; Hinrichsen, G. *Progr. Colloid Polym. Sci.*, **1985**, 71, 134.
11. Hasegawa, M.; Kochi, M.; Mita, I.; Yokota, R. *Eur. Polym. J.*, **1989**, 25, 349.

RECEIVED September 13, 1994

Chapter 19

Novel Photosensitive Polyimide Precursor Based on Polyisoimide Using Nifedipine as a Dissolution Inhibitor

Amane Mochizuki[1], Tadashi Teranishi[1], Mitsuru Ueda[1], and Toshihiko Omote[2]

[1]Department of Materials Science and Engineering, Faculty of Engineering, Yamagata University, Yonezawa 992, Japan
[2]Central Research Laboratory, Nitto Denko Corporation, 1–1–2 Shimohozumi, Ibaraki 567, Japan

A new photosensitive polyimide precursor based on polyisoimide (PII) and nifedipine (DHP) as a photoreactive compound has been developed. PII was prepared by the ring-opening polyaddition of 4,4'-hexafluoroisopropylidenebis (phthalic anhydride) (6FDA) and 4,4'-hexafluoroisopropylidenebis(p-phenyleneoxy) dianiline (BAPF), followed by treatment with trifluoroacetic anhydride-triethylamine (TEA) in N-methyl-2-pyrrolidone (NMP). The dissolution behavior of the PII film containing 20 wt% of DHP after exposure and post-exposure bake (PEB) has been studied and we found that DHP in the unexposed PII film acts as a dissolution inhibitor in DMAc after PEB at 150 °C. The photogenerated product in the exposed PII film does not affect the dissolution rate. Because of this change in solubility, PII containing DHP functioned as a photosensitive resist. The resist had a sensitivity of 450 mJ/cm^2 and a contrast of 2.5 when postbaked for 10 min at 150 °C and developed with DMAc.

Photosensitive polyimides are currently receiving considerable attention for their potential use in the fabrication of semiconductor devices and multichip modules, since they enable the number of process steps to be reduced by avoiding the use of classical photoresists. In most cases, the photosensitive groups are attached to the pendant carbonyl groups of poly(amic acid)s (polyimide precursor). A typical example is the acrylate ester of poly(amic acid) which after exposure to UV radiation results in a crosslinked polyimide precursor (*1*). After removing the unexposed polymer, the crosslinked photoreactive groups are thermolyzed during curing to leave polyimide.

Positive resists based on a novolak resins with o-diazonaphthoquinone (NQD) are standard materials used in semiconductor manufacturing, where NQD acts as a dissolution inhibitor for aqueous base development of the novolac resin. Upon exposure to light, NQD is converted to indenecarboxylic acid that increases the dissolution rate of the novolac matrix in the regions where exposure has occurred.

Nifedipine [1,4-dihydro-2,6-dimethyl-4-(nitrophenyl)-3,5-pyridinedicarboxylic acid dimethylester] (DHP) is well known not only as a Ca-antagonist but also as a photosensitive compound, and is converted to a corresponding pyridine derivative after exposure to UV light (eq.1) (*2-3*).

0097–6156/94/0579–0242$08.00/0

The photochemical transformation in eq.(1) has been utilized in a photoresist formulation (*4-5*). Recently, Omote et al (*6*). reported that nifedipine acts as the dissolution inhibitor in a poly(amic acid) after post-exposure bake (PEB). This system exhibits good sensitivity and contrast.

In a previous paper (*7*) we studied the preparation and properties of polyisoimide (PII) as a polyimide precursor. This investigation revealed that the solubility of PII in organic solvents is higher than that of corresponding polyimide. The isomerization reaction of isoimide to imide is catalyzed by an acid or a base, and PII is converted easily to polyimide by high temperature thermal treatment without the elimination of volatile compounds.

These findings prompted us to develop a new positive resist by using PII and DHP as the polymer matrix and the photosensitive compound, respectively. This paper describes the preparation and properties of a novel positive photoreactive polyimide precursor consisting of PII and DHP.

$$\text{Nifedipine(DHP)} \xrightarrow{h\upsilon} \text{NDMPy} \quad (1)$$

Experimental

Materials. Cyclohexanone, N,N-dimethylacetamide (DMAc), N-methyl-2-pyrrolidone (NMP), and triethylamine (TEA) were purified by distillation. 4,4′-Hexafluoroisopropylidenebis(p-phenyleneoxy)dianiline (BAPF) was purified by recrystallization from cyclohexane and chloroform. 4,4′-Hexafluoroisopropylidenebis (phthalic anhydride) (6FDA) was obtained from American Hoechst Co. Other reagents and solvents were obtained commercially and used as received. Nifedipine was synthesized according to the reported procedure (*6*).

Polymer (PII) synthesis. A solution of BAPF (2.22 g, 5.0 mmol) in NMP (43.2 mL) was cooled with an ice-water bath. To this solution, 6FDA (2.59 g, 5.0 mmol) was added with constant stirring . The mixture was stirred at room temperature for 4 h. The resulting viscous solution was diluted with NMP (48.2 mL) and TEA (1.4 mL, 10.0 mmol) was added dropwise while continuously stirring. Then, the reaction mixture was cooled with an ice-water bath, and trifluoroacetic anhydride (1.54 mL, 11.0 mmol) was added dropwise with stirring. The mixture was stirred at room temperature for 4 h and poured into 2-propanol (1000 mL). The polymer precipitated was filtered off and dried *in vacuo* at 40 °C. The yield was 4.54 g (98 %). The inherent viscosity of the polymer in DMAc was 0.42 dL/g at a concentration of 0.5 g/dL at 30°C. IR (KBr): ν 1800 cm^{-1} (C=O), 930 cm^{-1} (C-O). Anal. Calcd for (C_{46} H_{22} N_2 O_6 F_{12})$_n$: C, 59.62% ; H, 2.39% ; N, 3.02%. Found : C, 59.44% ; 2.70% ; N, 3.03%.

Measurements.   Infrared spectra were recorded on a Hitachi I-5020 FT-IR spectrophotometer.
Viscosity measurements were carried out by using an Ostwald viscometer at 30 °C. Thermal analyses were performed on a Seiko SSS 5000-TG/DTA 200 instrument with a heating rate of 10 °C/min for thermogravimetric analysis (TGA) and a Seiko SSS 5000 DSC220 using a heating rate of 10 °C/min for differential scanning calorimetry (DSC) under a nitrogen atmosphere. Molecular weights were determined using a gel permeation chromatograph (GPC) calibrated with polystyrene using a JASCO HPLC system equipped with a Shodex KD-80M column at 40 °C in dimethylformamide (DMF).   Film thicknesses were measured on a Dektak 3030 system (Veeco Instruments Inc.)

Dissolution rate. PII was dissolved at 15 wt% in cyclohexanone, to which was added DHP (5-50 wt% of the total solid). Films spin-cast on NaCl plates or silicon wafers were prebaked at 80 °C for 10 min and exposed through a filtered super high pressure mercury lamp SH-200 (Toshiba Lighting & Technology Corporation). Imagewise exposure was carried out in a contact mode. Exposed films were postbaked at 80-220 °C for 10 min and subjected to IR measurement or developed with DMAc at room temperature.

Photosensitivity. Five- μm thick PII films on silicon wafers were exposed to 436 nm wavelength light from the filtered super high mercury lamp. Exposed films were postbaked at 150 °C for 10 min and developed for 30 sec in DMAc at room temperature. The characteristic curve was obtained by plotting a normalizing film thickness against logarithmic exposure energy.

Result and Discussion.

Polyisoimide (PII) Synthesis. The polyisoimide (PII) based on 6FDA and BAPF was selected as the polymer matrix, because PII was expected to display excellent solubility characteristics in organic solvents as a result of ether and hexafluoroisopropyliden linkages in the polymer backbone. In a previous paper (7), we showed that trifluoroacetic anhydride-TEA was a suitable dehydrating agent for the formation of isoimides from amic acids. Thus, PII was prepared by the ring-opening polyaddition of 6FDA and BAPF, followed by treatment with trifluoroacetic anhydride-TEA in NMP (eq.2).
      The polymer structure was confirmed to be the corresponding PII by means of infrared spectroscopy and elemental analysis. The IR spectra exhibited a characteristic absorptions at 1800 cm^{-1} due to the isoimide carbonyl. Imide contents determined by IR spectroscopy were less than 5%. Elemental analysis also supported the formation of expected polymer.
      PII was soluble in a wide range of organic solvents, such as DMAc, DMF, cyclohexanone, dichloromethane, toluene, acetone, and tetrahydrofuran.   A transparent yellow films were cast from a DMAc or cyclohexanone solution of PII. The molecular weight of the polymer having an inherent viscosity of 0.42 dL/g was determined by GPC. $\overline{Mn}$ and $\overline{Mw}$ values were 47,000 and 157,000, respectively, relative to standard polystyrene, and $\overline{Mw}/\overline{Mn}$ was 3.3. The thermal stability of the polymer was examined by TGA. The polymer showed a 10% weight loss at 530 °C in nitrogen. DSC on the polymer powder showed an endotherm at 215 °C and a large exothermic peak at around 270 °C in the first heating process. In the second heating process, these peaks were not observed and new endothermic peak appeared at 245 °C. Furthermore, no weight loss was observed at around 270 °C. Based on these data, the first endotherm at 215 °C and the large exothermic peak at around 270 °C observed in the first heating process are assigned to the glass transition temperature

(Tg) and the thermal isomerization temperature of PII, respectively. The endothermic peak at 245 °C in the second heating process reflects Tg of the final polyimide (PI).

Poly(amic acid)   Polyisoimide (PII)

Polyimide (PI)   (2)

Ar$_1$:   BAPF   Ar$_2$:   6FDA

Lithographic Evaluation. It is very important to clarify the dissolution behavior of exposed and unexposed areas to produce satisfactory resist images. Thus, the PII film (5 μm thick) containing 20 wt% of DHP was exposed (500 mJ/cm^2) to 436 nm UV irradiation. The dissolution rate as calculated from the remaining film thickness after development with DMAc is shown as a function of PEB temperature in Figure 1.

The dissolution rate of the exposed film is obviously faster than that of the unexposed film at PEB temperature below 170 °C. However, raising the PEB temperature above 170 °C reversed the dissolution behavior and the unexposed part became more soluble relative to the exposed part.

In order to further investigate the dissolution behavior described above, PII films containing 20 wt% of DHP and no DHP were spin-coated on silicon wafers. These films were then heated to temperatures ranging from 80 to 150 °C for 10 min. Subsequently, the dissolution rates of the polymer films in DMAc were evaluated. Figure 2 indicates that the dissolution rate of unexposed PII containing DHP decreases markedly at approximately 130 °C compared to that of PII without DHP. However, exposed PII (500 mJ/cm^2) films containing 20 wt% of DHP had a solubility almost identical to that of PII itself. Figure 3 shows the dissolution behavior of PII itself, PII containing DHP baked at 150 °C for 10 min, and PII containing DHP postbaked at 150 °C for 10 min after UV exposure (500 mJ/cm^2), respectively. PII containing DHP after exposure and PEB had a comparable solubility in DMAc to PII itself and dissolved more than 10 times faster than the thermally treated PII containing DHP. These results indicate that thermally treated DHP acts as a dissolution inhibitor in the PII film whereas exposed DHP does not promote the dissolution rate. Similar behavior has been observed for the resist system composed of poly(amic acids) containing DHP. Omote et al (6). interpreted that the dissolution inhibition of poly(amic acids) as promoted by DHP near the PEB conditions results from decreased

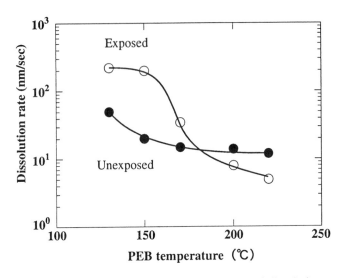

Figure 1 Relationship between PEB temperature and dissolution rate.

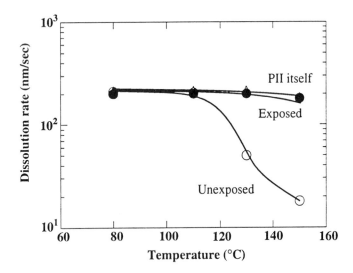

Figure 2 Dissolution rates of PII films with DHP under several
heat treatment temperatures.

hydrophilicity caused by intermolecular hydrogen bonding between DHP =NH and polymer -COOH groups. Furthermore, as one of the reasons for the dissolution promoting effect of DHP after exposure and PEB, the decrease of photochemical product, 4-(2′-nitrosophenyl)-2,3-dicarboxymethyl-3,5-dimethylpyridine (NDMPy), by vaporization has been proposed. To confirm the vaporization of NDMPy, the quantities of photogenerated NDMPy in the films exposed to 500 mJ/cm^2 of 436 nm irradiation and subsequently heated at 150 °C for 10 min were measured by high performance liquid chromatography (HPLC). About 40 % of the NDMPy was actually vaporized out of the film. These findings clearly indicate that post-baking at 150 °C facilitates evaporation of NMDPy.

Based on these results, the dissolution inhibition of DHP in DMAc by simple thermal treatment could be explained in terms of the formation of strong hydrogen bonds between DPH =NH groups and carbonyl groups of PII. The fact that the exposed polymer film had the same dissolution rate as PII itself could be explained as follows; (a) as photogenerated NMDPy does not have any active hydrogens, hydrogen bonds between NMDPy and PII can not be formed and (b) significant amounts of NMDPy evaporate from the film.

On the other hand, the solubility of exposed PII gradually decreased with increasing PEB temperature. Polyisoimide isomerizes easily to the corresponding polyimide by high temperature thermal treatment and the isomerization reaction of isoimide to imide is also catalyzed by an acid or a base. To elucidate the reversed dissolution behavior, the isomerization ratio of PII to polyimide (PI) in the presence of DHP before and after exposure versus PEB temperature was measured. PII films containing 20 wt% of DHP, spin-coated on NaCl plate, were exposed (500 mJ/cm^2) to 436 nm UV irradiation. The conversion of PII to PI was determined by comparing the absorptions of imide (1380 cm^{-1}) with an internal standard peak at 1500 cm^{-1}. The result is represented as a normalized value because PII included approximately 5% of the imide form (Figure 4). Although, there is no remarkable difference on the conversion of PII to PI between the exposed and unexposed films below 170 °C of PEB temperature, at a PEB temperature of 200 °C the exposed film isomerized much faster than the unexposed film. Therefore, the reversed dissolution behavior at higher PEB temperatures may be interpreted with the formation of PI catalyzed by photochemical product, NDMPy, which is a weak base.

Table I shows the results of qualitative solubilities of exposed and unexposed PII films containing DHP after postbaking at 150 °C. Exposed polymer films exhibited excellent solubility toward various solvents as compared to the unexposed polymer.

The effect of the DHP loading on the dissolution rate of PII in DMAc was studied as shown in Figure 5. About 10 wt% of DHP to polymer was enough to achieve a satisfactory dissolution contrast.

After a preliminary optimization study involving DHP loading, postbaking temperature, and developer temperature, we formulated a photosensitive resist consisting of PII and 20 wt% of DHP. The sensitivity curve for a 5 μm thick resist shown in Figure 6 is resulting from consistent with the dissolution behavior study. This indicates that the resist can be imaged at 450 mJ/cm^2 with a contrast of 2.5.

In Figure 7(a) are presented scanning electron micrographs of positive images made from PII containing 20 wt% of DHP by postbaking at 150 °C for 10 min exposure of 450 mJ/cm^2. This resist is capable of resolving 10 μm features. Furthermore, the positive image of PII is converted to the positive image of PI by heating at 280 °C for 1 h without any deformation as indicated in Figure 7(b).

Conclusion

We have prepared a new positive-type photosensitive polyimide precursor by using polyisoimide as a polymer matrix and DHP as a photoreactive compound. The resist

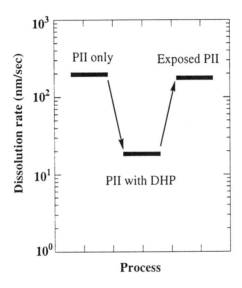

Figure 3 Dissolution contrast for the system of PII and DHP.

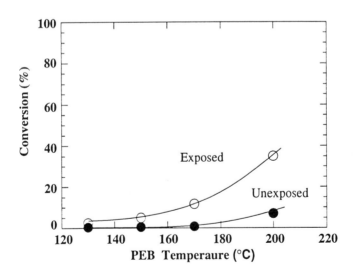

Figure 4 Thermal isomerization of PII in the presence of DHP.

Table I. PEB Solubility Characteristics of Exposed and Unexposed PII Films containing DHP

Solvents	solubility	
	Exposed	Unexposed
Cyclohexanone	+ +	±
Acetone	+ +	+
Methyl ethyl ketone	+ +	+
Isoamyl acetate	+ +	±
Dioxane	+ +	±
1,1,2,2,-Tetrachloroethane	+ +	±
Chloroform	+ +	±
1,2-Dichloroethane	+ +	±
Dimethyl acetamide	+ +	±
Dimethyl formamide	+ +	±
2-Methoxy ethanol	−	−
2-Propanol	−	−
Methanol	−	−
10% TMAH	±	−

+ +: soluble at room temperature , + : soluble by heating ,
± : partially soluble or swelling , − : insoluble

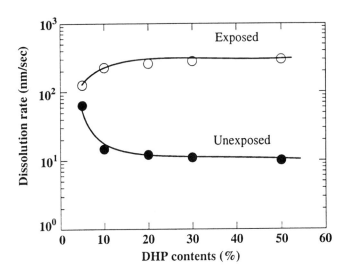

Figure 5   Relationship between DHP content and dissolution rate.

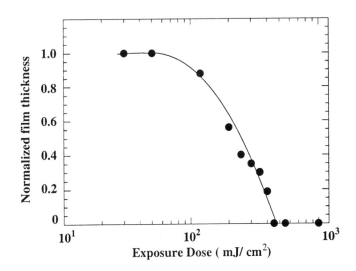

Figure 6 Characteristic exposure curve for the system of PII and DHP.

Figure 7(a) Scanning electron micrographs of Pattern made from PII
containing DHP (after development with DMAc).

Figure 7(b) Scanning electron micrographs of Pattern made from PII
containing DHP (after thermal treatment at 280 °C for 1 h).

is designed such that polyisoimide and DHP provide a soluble polyimide precursor and a photoreactive dissolution inhibitor, respectively. Unlike poly(amic acids), polyisoimides do not eliminate volatile compounds upon conversion to the final polyimide and, thus, represent a much more desirable polyimide precursor.

Acknowledgment

The authors are indebted to Hitoshi Nagasawa and Sadao Kato for their technical assistance and Takeyoshi Takahashi for performing the elemental analyses. We also wish to acknowledge the financial support from the Ministry of Education, Science and Culture of Japan (No.05453140).

Literature Cited

1. Rubner, R.; Ahne, H.; Kühn, E.; Koloddieg, G. *Photogr. Sci. Eng.*, **1979**, *23(5)*, 303
2. Berson, J. A.; Brown, E. *J.Am.Chem.Soc.*, **1955**,*77* , 447
3. Yamaoka, T.; Watanabe, H.; Koseki, K.; Asano, T. *J.Imaging Sci.*, **1990**,*34* , 50
4. Ranz, E. Japan Patent Disclosure, kokai 49-60733
5. Leuschner, R.; Ahne, H.; Marquardt, U.; Nickel, U.; Schmidt, E.; Sebald, M.; Sezi, R. *Microelectronic Enginieering* ., **1993**, *21*, 255
6. Omote, T.; Yamaoka, T. *Polym. Eng. Sci.*, **1992**, *32(21)*, 1634
7. Mochizuki, A.; Teranishi, T.; Ueda, M. *Polym.J.*, **1994**, *26*, 315

RECEIVED September 13, 1994

Chapter 20

Photochemical Behavior of Nifedipine Derivatives and Application to Photosensitive Polyimides

T. Yamaoka, S. Yokoyama, T. Omote, K. Naitoh, and K. Yoshida

Department of Image Science, Faculty of Engineering, Chiba University, Chiba 263, Japan

Photochemical conversion of 4-(2'-nitrophenyl)-1,4-di-hydropyridine derivatives was found to act as a chromophore for photosensitive polyimide precursors giving a dual imaging-mode characteristics. The sensitivities and contrasts for the negative and positive-working poly(amic-acid)'s were evaluated. The photochemical behaviors such as the quantum yields of the photoemission and the conversion, the reaction products and the transient species were studied. The mechanism of photosensitizing poly(amic-acid) was proposed.

Since the invention of photosensitive polyimide precursors based on crosslinking reaction of the bichromate by Kerwin (1) and on radical polymerization of methacryloyl groups by Rubner (2), extensive efforts to use these new types of photopolymers for production of electronic devices have been made (3-5). From the standpoint of application, the photosensitive polyimides have still to be improved for various specifications such as mechanical, rheological and electrical properties in addition to photosensitivity. One of the mechanism to provide photo-functionalized poly(amic acid) is to simply mix photosensitive compounds such as multifunctional acrylate monomers, bisazides or o-naphthoquinone diazide (6). In 1955, Berson et al. reported a spectroscopic study of 2,6-dimethyl-3,5-dicarboxy-alkyl-4-(2'-nitrophenyl)-1,4-dihydropyridine (abbreviated hereafter to o-NMHP) in solution (7). Yamaoka et al. reported tha to-NMHP is a novel photosensitizer to produce a positive-working photoresist in combination with novolak resin (8). Omote et al.reported a positive-working photosensitive poly(amic acid) using o-NMHP as a photosensitizer (9). Leuschner et al. used an analogous compound as a photobase in development of a CARL process based on a chemical amplification mechanism (10). In this study, a series of novel photosensitive

0097–6156/94/0579–0253$08.00/0

polyimide precursors consisting of poly(amic acid) or its derivatives are investigated and it was found that these photosensitive
polymers provide a dual imaging mode, that is, a positive and a
negative working character depending on the process variation after
exposure. The photochemical reactions of o-NMHP and its derivatives.
The nifedipine derivatives used for the spectroscopic and mechanistic studies beside o-NMHP involve 2,6-dimethyl-3,5-dicarboxyalkyl-4-
(4'-nitrophenyl)-1,4-dihydropyridine (p-NMHP), 2,6-dimethyl-3,5-di-
carboxyalkyl-4-(2',4'-dinitrophenyl)-1,4-dihydropyridine (o,p-DNMHP)
The mechanism of the dual-mode patterning will be described.

Poly(amic acid) Sensitized with NMHP and its Derivatives

Poly(amic acid) having the structure shown below is used as a polyimide precursor. NMHP and its derivatives synthesized in the present
study are summarized in Table 1 together with their spectroscopic
data.

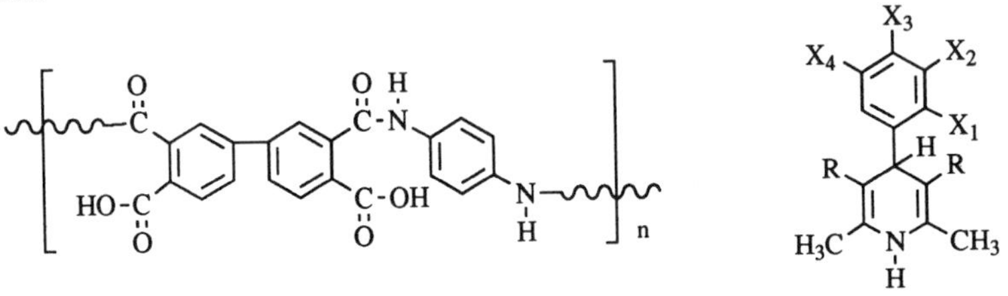

Molecular structures of poly(amic acid) and 1,4-dihydropyridine
derivatives.

Table 1   Spectroscopic Data of Synthesized 1,4-Dihydropyridine
          Derivatives

Derivativs	X_1	X_2	X_3	X_4	R	λ_{max} (nm)	ε_{365} (L/mol·cm)	ε_{436} (L/mol·cm)
	NO_2	H	H	H	$COOCH_3$	354	4500	–
	NO_2	H	H	H	CN	338	3310	–
	NO_2	H	H	H	$COCH_3$	380	4530	755
	NO_2	H	NO_2	H	$COOCH_3$	380	2455	793
	NO_2	H	NO_2	H	CN	375	1820	170
	NO_2	H	NO_2	H	$COCH_3$	400	3585	1890
	NO_2	OCH_3	H	H	$COOCH_3$	348	4400	–
	NO_2	OCH_3	OCH_3	H	$COOCH_3$	335	6600	–
	H	H	NO_2	H	$COOCH_3$	354	4500	–

Figure 1 (solid line) shows the positive-mode sensitivity curve of
poly(amic acid) containing o-NMHP, indicating that the sensitivity
is 170 mJ/cm² at a 365 nm light. The dissolution rate of the photo-

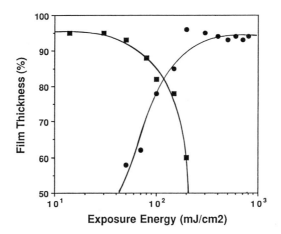

Figure 1. Characteristic curves of poly(amic acid) containing NMHP. ($-\blacksquare-$) : Positive-working mode. ($-\bullet-$) : Negative-working mode.

sensitive poly(amic acid) in an aqueous base developer is governed
by the temperature of the post exposure baking as well as the
exposure energy. Figure 2(a) and 2(b) show the dissolution rates of
the poly(amic acid) films containing o,p-DNMHP or o-NMHP as a function of PEB temperature. In the case of o,p-DNMHP, the dissolution
rate for both exposed and unexposed films is decreased with an
increase in the PEB temperature. On the other hand, the behavior of
the poly(amic acid) film containing o-NMHP against the PEB temperature is quite different from that containing o,p-DNMHP. As can be
seen in Figure 2(b), the dissolution rate of the exposed film exhibits a continuous decrease with increasing PEB temperature 148°C.
However, at about 150°C, it suddenly increases to the dissolution
rate of 140°C. This discrete change of the dissolution rate at 148°C
suggests that some chemical reaction occurs in the poly(amic acid)
film at this temperature. The dependence of the dissolution rate on
the PEB temperature shown in Figure 2(b) suggests that the poly(amic
acid) film containing o-NMHP gives positive or negative-working
characteristics depending on the PEB temperature. If the PEB temperature is lower than 148°C, the dissolution rate of the exposed films
is faster than that of the unexposed, resulting in positive patterns
after development. If the PEB temperature is higher than 150°C, the
unexposed areas may be dissolved faster than the exposed areas to
give negative patterns. The characteristic curve of poly(amic acid)
containing p-NMHP obtained with the PEB temperature of 200°C as
shown in Figure 1 together with that of the positive-working mode.
From the characteristic curve, the sensitivity and contrast value
for the negative-working mode were determined to be 250 mJ/cm^2 and
2.5, respectively. Figure 3 shows a SEM picture of the negative
patterns.

Photochemical Reaction of NMHP Derivatives

The absorption spectra of NMHP derivatives, nitrobenzene (NB) and
dihydropyridine (DHP) are shown in Figures 4(a) and 4(b). In terms
of π-electronic conjugation system, the electronic spectrum of a
NMHP derivative is considered to be almost the sum of the spectra of
NB and DHP. Because, the π-electronic conjugation systems of the NB
moiety and the DHP moiety are separated by the sp^3 hybrid orbitals
of the carbon atom to connect these two moieties and unconjugated.
Actually, the absorption spectrum of p-NMHP is almost a simple sum
of NB and DHP. However, the spectrum shown in Figure 4(a) is not a
simple sum of those for two compounds, suggesting the existence of
some interaction between the two moieties corresponding to DHP and
NB. As an example, the absorption spectrum and its change due to
exposure to light are shown for o-NMHP in Figure 5. The spectrum of
o-NMHP in acetonitrile changes due to exposure to 365 nm light and
the isobestic points at 260 and 310 nm are kept constant during the
exposure, indicating that the photochemical conversion of o-NMHP
proceeds along a single reaction path.
    DHP absorbing the long wavelength light of up to 430 nm shows
almost no sensitivity to the light at 365 nm judging from the fact

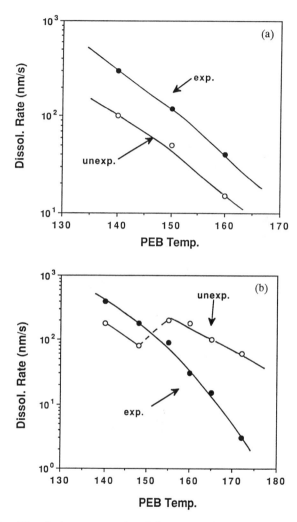

Figure 2a. Dissolution rates of poly(amic acid) containing o,p-DNMHP before and after exposure, as function of PEB temperature. 2b. Dissolution rates of poly(amic acid) containing o-NMHP before and after exposure, as function of PEB temperature.

Figure 3. A SEM picture of a cross section of line and space patterns formed by the negative mode.

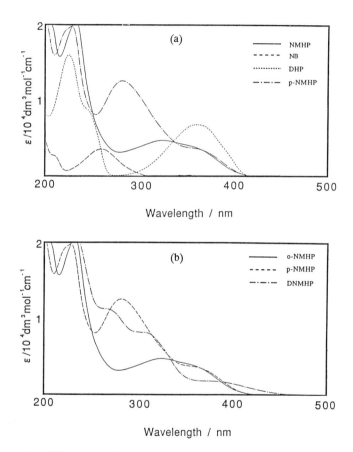

Figure 4a. Electronic absorption spectra of o-NMHP, o,p-DNMHP, DHP, and nitrobenzene. 4b. Electronic absorption spectra of o-NMHP, p-NMHP, and o,p-DNMHP.

that the DHP spectrum in a N_2 bubbled acetonitrile solution is unchanged due to the irradiation of a 365 nm light. However, the absorption intensity decreases by irradiation without N_2 bubbling. If the solution is exposed to the light during the bubbling of O_2 gas, the absorption intensity shows an obvious decrease. The quantum yields of the photochemical conversion of DHP measured by the intensity decrease were 0.002 and 0.005, respectively (Table 2).

Table 2   Quantum Yields of o-NMHP, o,p-DNMHP, and DHP at 365 nm and 436 nm

	o-NMHP	o,p-DNMHP		DHP
Concentration (mol/L)	365	365	436	365
5.0×10^{-4}	0.23	0.16	0.32	0.002 (in air)
4.2×10^{-4}		0.16	0.32	~ 0 (N_2 bubbling)
2.0×10^{-4}	0.23	0.14	0.26	0.005 (O_2 bubbling)
4.2×10^{-5}		0.13	0.23	(1.0×10^{-4} mol/L)
2.0×10^{-5}		0.15	0.17	
1.0×10^{-5}	0.22	-	0.16	
6.0×10^{-5}		0.12	0.15	
4.0×10^{-6}		0.10	0.11	

This fact means that the photochemical conversion of DHP requires oxygen and DHP in the excited state may react with oxygen since oxygen molecules do not absorb the 365 nm light.   The quantum yields of photochemical conversion of o-NMHP, o,p-DNMHP and DHP in aceto-nitrile are summarized in Table 2. Regarding these, it is noteworthy to point out that the quantum yields of o,p-DNMHP depend on exciting light wavelength. The quantum yields of o,p-DNMHP for excitation using 436 nm varies from 0.32 to 0.11, depending on the concentra-tion of the solution, while those values for the excitation using 365 nm light are constant regardless of the solution concentration.

Photochemical behavior of DHP

DHP emmits both fluorescene and phosphorescence with the peaks at 452 and 680 nm, respectively, as is shown in Figure 6, while no emission was observed with o,p-DNMHP and o-NMHP. The physical data of DHP are summarized in Table 3. The quantum yields of the fluorescence (ϕ_f) and the phosphorescence (ϕ_p) in 77 K EPA were determined to be 0.015 and 0.32, respectively. The fluorescence of DHP is quenched by NB and DNB at almost a diffusion controlled rate. The values of free energy change (ΔG) accompanying the electron transfer calculated based on the Rehm-Weller equation (10) with the

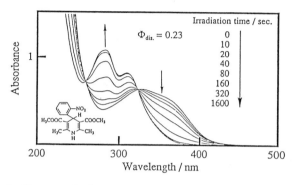

Figure 5. Change of absorption spectrum of o-NMHP due to the exposure to the light. (x 10⁻⁴mol/L in methylalcohol)

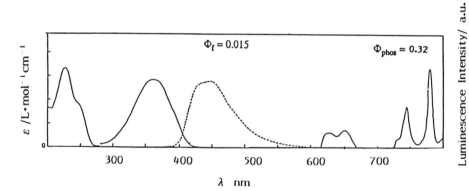

Figure 6. Absorption, fluorescence, and phosphorescence spectra of DHP.

redox potentials for DHP, NB and DNB are summarized in Table 4. These values predict the possibility of the spontaneous electron transfer from the excited singlet state of DHP to NB or DNB. In spite of the effective quenching of DHP fluorescence by NB, the quantum yields for the DHP conversion are not increased even in the presence of NB or DNB. It appears that the photochemical conversion of DHP proceeds independently even in the presence of NB or DNB. In terms of the free energy change(ΔG), the electron transfer from DHP to NB or DNB is possible for the lowest singlet excited state of DHP but not for the lowest triplet excited state. However, the short lifetime of the lowest singlet excited state, the small ϕ_f, and the

Table 3   Physical Data of DHP

λ_{max} (nm)	$\varepsilon(\lambda_{max})$ (L/mol·cm)	$^1E_{00}$ (kcal/mol)	$^3E_{00}$	$E_{ox}{}^a$ (V vs SCE)	Φ_f	Φ_p	$K_{ISC}{}^b$	$\tau_s{}^c$	$\tau_T{}^d$
227	26660	70.6	45.4	1.04	0.015	0.32	6.4	2.3	2.9
363	11390								

^{a}(V vs SCE), ^{b}(x 10^{10}), ^{c}(x 10^{-9}), ^{d}(x 10^{-6}).

Table 4   The Free Energy Changes (ΔG's) Accompanying the Electron Transfer

	ΔG^s (kcal/mol)	ΔG^T (kcal/mol)
DHP / Nitrobenzene	−10.02	15.18
DHP / 1,3-Dinitrobenzene	−13.99	12.09

large k_{ISC} may make the electron transfer path unimportant in this reaction system. These observations suggest the following paths for the DHP reaction.

DHP $\longrightarrow$ ^{1}DHP
^{1}DHP $\longrightarrow$ ^{3}DHP
^{3}DHP $\longrightarrow$ DHP

^{1}DHP + (D)NB $\longrightarrow$ DHP$^+$ + (D)NB $\longrightarrow$ Product
^{3}DHP + O$_2$ $\longrightarrow$ ^{3}DHP $\cdots$ O$_2$ $\longrightarrow$ Product

Transient Spectra of DHP, o-NMHP, and o,p-DNMHP

The physical data of the transient species detected using a laser flash spectroscopy are summarized in Table 5. Transient absorptions for DHP appeared at 420, 460, and 660 nm at 0.1 µs after the excitation, and decayed with the rate constant of $3.4 \times 10^4 s^{-1}$. The transient absorptions can be assigned to T-T absorption of DHP. o-NMHP shows the transient absorptions in the vicinity of 320 and 420 nm during the photochemical reaction (Figure 7). At 0.2 µs after the excitation, the absorptions appear in the vicinity of 420 and 700 nm, and at 1.0 µs, the absorption at 420 nm decreases while the absorption at 700 nm still increases. The kinetic curves of the

Table 5   Transient absorption of DHP, NMHP, DNMHP

	T-T absorption (nm)	Decay rate constant (s^{-1})	2nd transient absorption (s^{-1})
DHP	420, 460, 660	2.9×10^5	......
NMHP	320, 420	4.9×10^5	700 nm[a]
DNMHP	480	7.8×10^6 ($\sim 10^{-3}$ mol/L)	......
		3.5×10^6 (7×10^{-4} mol/L)	......

[a] The raise and decay rate constants are 2.4×10^6 and 2.8×10^5 (s^{-1}), respectively.

absorptions at 420 and 700 nm are shown in Figures 7(b) and 7(c). The decay rate constant of the absorption at 420 nm is determined as $4.9 \times 10^5 s^{-1}$, which may be attributable to the T-T absorption of o-NMHP. The absorption at 700 nm rises with the constant of 2.4×10^6 s^{-1} and decays with the rate constant of $2.8 \times 10^5 s^{-1}$. The rate constant of the rising curve for the absorption at 700 nm is slower than that of the absorption at 420 nm. The discrepancy of these rate constants suggests that these two transient species are produced independently of each other. Judging from the fact that the photochemical reaction of o-NMHP produces only a pyridine derivative the transient species observed in the 700 nm region is assigned to the precursor to the pyridine derivative.

o,p-DNMHP shows a transient absorption at 480 nm, the decay rate of which depends on the concentration of the solution. The decay rate constants of the transient absorption are $3.5 \times 10^6 s^{-1}$ for 10^{-3} mol/L, and 7.8×10^6 for 7×10^{-4} mol/L, respectively, indicating the faster decay rate for the higher concentration of the solution.

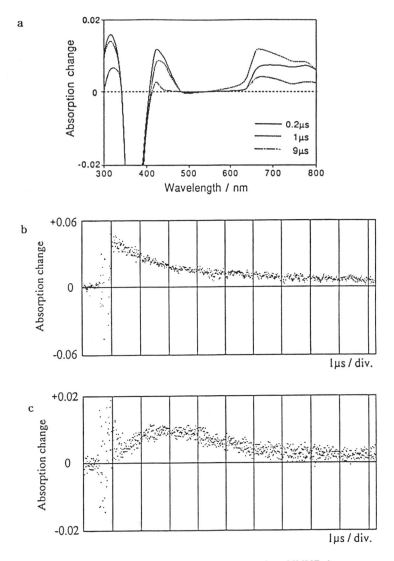

Figure 7a. Time-resolved transient spectra of o-NMHP in an acetonitrile. 7b. A decay curve of the transient species of o-NMHP
measured at 420 nm. 7c. Raise and decay curves of the transient
species of o-NMHP measured at 700 nm.

Reaction Path of o-NMHP and o,p-DNMHP

The apparent reaction paths of o-NMHP and o-,p-DNMHP observed by the experiments, are summarized as the following Scheme.

Although o,p-DNMHP is composed of two molecules, DNB and a DHP derivative from the standpoint of π -electronic conjugation, the absorption spectrum of o,p-DNMHP is different from the simple sum of those two molecules. It may suggest a delocalization of the π - electronic conjugation between both moieties. The absorption spectra in Figure 4 show that DHP has an absorption in the longer wavelength region than those of nitro- or dinitrobenzene. If there is some interaction between the both moieties, the π -electronic system can be expressed with the function Ψ based on the electronic configur- ations shown in Figure 8. The ground state, the lowest and the second lowest excited state are given by Eq.(2), (3) and (4). The wavelength dependence is not observed for o-NMHP because the photo- chemical conversion of o-NMHP is not occured using 436 nm light due to the low absorption coefficient. However the photochemical conver- sion of o-NMHP may proceed by a similar mechanism to that of o, p-DNMHP.

$$o,p\text{-DNMHP} \xrightarrow{365 \text{ nm}} {}^1o,p\text{-DNMHP}^* \xrightarrow{\text{ISC}} {}^3o,p\text{-DNMHP}^* \rightarrow \text{Another transient species.}$$

```
                    \            /  1st order
                     ┌──────────────────────┐
                     │  Pyridine  derivative │
                     └──────────────────────┘
                     /            \  2nd order
                   [X]             [X]

                    +               +
```

$$o,p\text{-DNMHP} \xrightarrow[436 \text{ nm}]{} {}^1o,p\text{-DNMHP}^* \xrightarrow[\text{ISC}]{} {}^3o,p\text{-DNMHP}^*$$

$$\Psi = C_1 \phi (G) + C_2 \phi (LE_{D1}) + C_3 \phi (LE_{D2}) + \cdots \cdots + C_4 \Phi (LE_{N1})$$
$$+ C_5 \phi (LE_{N2}) + \cdots + C_6 \phi (CT_1) + C_7 \phi (CT_2) + C_7 \phi (CT_2) + \cdots$$
$$\cdots \cdot (1)$$

$$\Psi_G \doteqdot C_1 \phi (G) + \cdots \qquad\qquad \cdots (2)$$

$$\Psi_1 \doteqdot C_1 \phi (G) + C_2 \phi (LE_{D1}) + C_3 \phi (LE_{D2}) + \cdots + C_4 \Phi (LE_{N1}) \cdots (3)$$
$$C_2 \gg C_1, \ C_3, \ C_4$$

$$\Psi_2 \doteqdot C_1 \phi (G) + C_2 \phi (LE_{D1}) + C_3 \phi (LE_{D2}) + \cdots + C_4 \Phi (LE_{N1}) \cdots (4)$$
$$C_4 \gg C_1, \ C_2, \ C_3$$

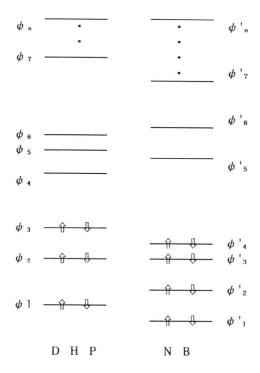

Figure 8. The molecular orbitals and electronic configuration of DHP and NB derivatives.

Where,

$\Phi(G)$

$\Phi(LE_{D1}) = \Phi(\phi_3 \Rightarrow \phi_4)$ $\Phi(LE_{N1}) = \Phi(\phi'_4 \Rightarrow \phi'_5)$

$\Phi(CT_1) = \Phi(\phi_3 \Rightarrow \phi'_5)$

$\Phi(LE_{D2}) = \Phi(\phi_3 \Rightarrow \phi_6)$ $\Phi(LE_{N2}) = \Phi(\phi'_4 \Rightarrow \phi'_6)$

$\Phi(CT_2) = \Phi(\phi_3 \Rightarrow \phi'_6)$

$\Phi(LE_{D3}) = \Phi(\phi_2 \Rightarrow \phi_4)$ $\Phi(LE_{N3}) = \Phi(\phi'_3 \Rightarrow \phi'_5)$

$\Phi(CT_3) = \Phi(\phi'_4 \Rightarrow \phi_4)$

Judging from the spectra of o,p-DNMHP, o-NMHP, DHP, NB, and DNB, the approximated functions for the lowest and the second lowest state will be expressed by Eqs.(3) and (4). o,p-DNMHP excited to the Ψ_1 or Ψ_2 state is mostly contributed by $\Phi(LE_{D1})$ and $\Phi(LE_{N1})$, respectively. In general, o,p-DNMHP excited to the higher state Ψ_2, may be deactivated to the lowest excited state, Ψ_1 in the present case, and react at this state. The excitation wavelength dependence of the o,p-DNMHP quantum yield suggests that it reacts directly at the Ψ_2 state if it is excited using a 365 nm light.

The concentration dependence of the quantum yield is more remarkable for the excitation by 436 nm than that by 365 nm. The second order kinetics of the o,p-DNMHP quantum yield suggests that the conversion at the Ψ_D state is not unimolecular but bimolecular with a mechanism involving the interaction of the excited species on the Ψ_D state with other excited species on the Ψ_N such as an energy transfer or electron transfer. The energy diagram of DHP, NB, DNB, o-NMHP, p-NMHP, and o,p-DNMHP, are illustrated in Figure 9a and 9b. The excited states of o-NMHP seem to contain the mixture of the excited states of DHP and NB. On the other hand, the lowest and the second lowest excited states of p-NMHP have the almost similar values to those of DHP and NB, suggesting that the interaction of the both conjugation systems in p-NMHP may be small to result in no photochemical conversion of p-NMHP. Figure 9b also compares the excited states of o, p-DNMHP with those of DHP and DNB. The excited states of o,p-DNMHP seem to contain the components of the excited states of DHP and DNB. The main component of the lowest excited state is that of DHP judging from the low excitation energy 3.0 eV. The second lowest excited state of o,p-DNMHP may contain the lowest excited state of DNB to some extent suggesting that the excitation to this state may lead to the excitation of o,p-dinitrophenyl moiety . One of the possible reaction mechanism for o-NMHP and o,p-DNMHP based on the observation in the present study is proposed as follows although it is not conclusive.

$$\underset{\Psi_G}{DNMHP(S_0)} \xrightarrow{365 \text{ nm}} \underset{\Psi_2}{DNMHP(S_2)} \longrightarrow P$$

$$\underset{\Psi_1}{DNMHP(S_1)} + \underset{\Psi_G}{DNMHP(S_0)} \longrightarrow DNMHP^+ + DNMHP^-$$

$$\underset{\Psi_G}{DNMHP(S_0)} \xrightarrow{436 \text{ nm}} \underset{\Psi_1}{DNMHP(S_1)}$$

$$DNMHP(S_1) \longrightarrow DNMHP(T_1)$$

$$DNMHP(T_1) + DNMHP(S_0) \longrightarrow P$$

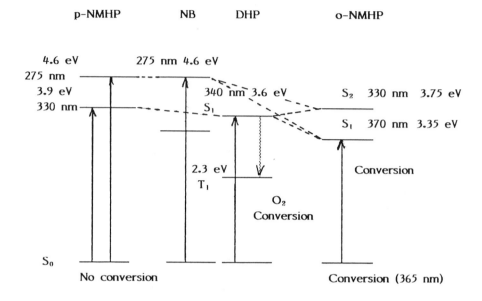

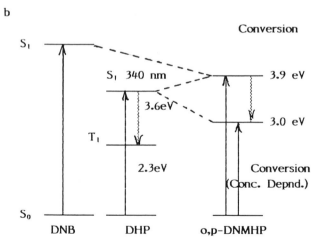

Figure 9(a, b). The energy diagram of DHP, NB, DNB, o–NMHP, p–NMHP, and o,p–DNMHP.

The possibility of electron transfer and T-T energy transfer mechanisms between dinitrobenzene or nitrobenzene and DHP were discussed
to explain the quenching of fluorescence of DHP, and the concentration dependence of the quantum yield of o,p-DNMHP. It is more reasonable to consider that these bimolecular interactions should occur
more efficiently for o,p-DNMHP or o-NMHP which are composed of the
combined molecules of DHP and dinitrobenzene or nitrobenzene.
However, the explanation of the wavelength and the concentration
dependence of the quantum yields for o,p-DNMHP requires a bimolecular interaction. This discrepancy may indicate that the electron
transfer and T-T energy transfer between the two conjugation systems
necessitate the slight contact of the the π-electrons of the two
systems which is inhibited for o,p-DNMHP or o-NMHP due to the steric
hindrance.

Experimental

**Synthesis of 2,6-Dimethyl-3,5-dicarbomethoxy-4-(2'-nitrophenyl)-1,4-
dihydropyridine (o-NMHP).** An ethanolic solution of o-nitrobenz-
aldehyde was cooled in an iced bath, and to this solution a 2.0
molar equivalent methyl acetoacetate was added dropwise and the
solution was stirred for about one hour with stir. Then, a 28 %
aqueous solution containing 1.5 molar equivalent ammonium was
gradually added and reacted at 70-80°C for five hours. As the
reaction proceeded, the product precipitated. The o-NMHP thus
produced was purified with recrystallization. MP, 172°C ; δ_H(400
MHz, CDCl$_3$) 2.34[6H, s, 2,6-CH$_3$], 3.59[6H, s, 3,50COOCH$_3$], 5.66[1H,
s, NH], 5.72[1H, s, 4-PyH], 7.23-7.69 [4H, m, ArH]; ν_{max}/cm^{-1}
3330(NH), 2800-300(CH$_2$, CH$_3$), 1680(C=O), 1530, 1350(NO$_2$).

Other derivatives of NMHP were also synthesized using this
method for NMHP. 2,6-Dimethyl 3,5-dicarbomethoxy -1,4-dihydro-
pyridine (DHP) was synthesized from formaldehyde and methyl acetoacetate and purified through sublimation under reduced pressure
($\sim 10^{-4}$Torr) at 120°C.

Spectroscopy. Electronic and emission spectra were measured using a
Hitachi spectrophotometer U-3000 and a Hitachi fluorescenece
spectrophotometer F-4010. IR spectra were measured with a Horiba
Furie infrared spectrophotometer FT-200. The lifetime of the
fluorescenece was measured with a Horiba time-resolved fluorescenece
spectrophotometer. The absorption spectra of the transient species
and T-T absorption were measured with a nanosecond time-resolved
spectrophotometer NASA-1 manufactured by Hamamatsu Photonics Co..
Third harmonic wave of Continuum Q-switched Nd:YAG laser (355nm, the
halfwidth of 5~7 ns) was used as the excitation light.

**Quantum Efficiency of Photochemical Conversion of DHP and (D)NMHP
Derivatives.** When the photochemical conversion of o-NMHP or o,p-
DNMHP proceeds according to the photochemical process given by

eq.(5) – (7), the rate of the conversion under irradiation of stationary light is described by eq.(8)

$$(D)NMHP \longrightarrow (D)NMHP^* \qquad (5)$$

$$(D)NMHP^* \longrightarrow (D)NMHP \qquad (6)$$

$$(D)NMHP^* \longrightarrow Product \qquad (7)$$

$$\ln \cdot \frac{\exp(2.303\,\varepsilon_\lambda[(D)NMHP_0]\cdot L) - 1}{\exp(2.303\,\varepsilon_\lambda[(D)NMHP_t]\cdot L) - 1} = 2.303\,\varepsilon_\lambda \cdot \frac{k_p}{k_p + k_d} I_0 \cdot t \qquad (8)$$

k_p: the rate of photodissociation
k_d: the rate of deactivation
ε_λ: molar extinction coefficient at λ nm
I_0: light intensity (einstein/cm^{2}s)
t : irradiation time (s)
$[(D)NMHP_0]$: initial concentration of (D)NMHP
$[(D)NMHP_t]$: concentration of (D)NMHP at t time

Here the quantum yield(ϕ_r) is given by eq.(9)

$$\phi_r = \frac{k_p}{k_p + k_d} \qquad (9)$$

Literature Cited
1) Kerwin, R. E.; Goldrick, M. R. Polym. Eng. Sci. 1971, 11, 426.
2) Ahne, H.; Domke, W.; Rubner, R.; Schreyer, M. in "Polymer for High Technology Electronics and Photonics", ACS Symposium Series 346, M. J. Bowden and S. R. Turner, Eds., Amercan Chemical Society, Washington, DC., 1978.
3) Neder, A. E.; Imai, K.; Craig, J. D.; Lazaridis, C. N.; Murray, D.O.;Pottiger, M. T.; Dombchik, S. A.; Lauttenberger, W. J. Conf. Preprints, Photopolymers Principles Progresses and Materials; p.333, 1991, Ellenville, N.Y.
4) Rhode, O.; Riediker, M.; Schaffner, A.; Batemann, J. Solid State Technol. 1986, 29, p.109.
5) Yoda, N.; Hiramoto, H. J. Macromol. Sci.(Chem.), A21, 1984, 1641.
6) Khanna, D. N.; Mueller, W. H. Technical Papers, p.429, 1988, 8th Intn'l Conf. on Photopolym., Ellenville, N. Y. Mid-Hudson Section, Soc. Plastics Engineers Inc..
    Khanna, D. N.; Mueller, W. H. J. Polym. Eng. Sci., 1989, 29, 14, 954.
7) Berson, J.A.; Brown, E. J. Am. Chem. Soc. 1955, 77, 447.
8) Yamaoka, T.; Watanabe, H.; Koseki, K.; Asano, T. J. Imaging Sci. 1990 34, 2050.
9) Omote, T.; Yamaoka, T. J.Photopolym.Sci.Technol. 1991, 4, 379.
10)Rehm D.; Weller, A. Isr. J. Chem. 1970, 8, 259.

RECEIVED September 13, 1994

OPTOELECTRONIC, CONDUCTING, AND PHOTORESPONSIVE POLYMERS

Chapter 21

Waveguiding in High-Temperature-Stable Materials

C. Feger[1], S. Perutz[1,4], R. Reuter[1,5], J. E. McGrath[2], M. Osterfeld[3], and H. Franke[3]

[1]IBM T. J. Watson Research Center, Yorktown Heights, NY 10598
[2]Department of Chemistry, Virginia Polytechnic Institute and State University, Blacksburg, VA 24061
[3]Universität Duisburg, Fachbereich Physik, Duisburg, Germany

High temperature stable polymeric waveguides are needed in several applications. Two such applications are described: optoelectronic connection between chip and module in multichip module (MCM) applications where high process temperatures are experienced during the chip joining process, and optical, environmental sensors which may be exposed to other hostile conditions besides high temperatures in the sensing environment. Requirements, material options and some results will be described.

Until recently lightguiding has been limited to inorganic and low temperature stable organic materials. Inorganic materials still offer the lowest optical losses of any material and therefor glass fibers are the choice for long distance transmission of optical signals at 1310 and 1550 nm. These wavelengths are chosen to coincide with absorption minima of these fibers. Organic fibers have much higher optical losses at these wavelengths (on the order of 200 dB/cm). However they have increasingly been used in short to medium distance applications mostly because they are less brittle than glass. Also they are very easily processed which allows production of cheap switches particularly for telecommunication applications. Furthermore, organic waveguides can act as host for a wide variety of dopants which is important for non-linear optical applications.

One of the limits of organic materials is their relative sensitivity to temperature changes. Thermal stresses already at moderate temperature excursions can lead to local refractive index changes. Worse, the polymer might reach its glass transition and begin to flow, destroying any optically active structures (e. g. reflectors and lenses) produced in them. This is of particular concern in main frame computer opto-electronic applications. Due to the

[4]Current address: Department of Materials Science, Cornell University, Ithaca, NY 14853
[5]Current address: Görreshöhle 5, D–5305 Alfter, Germany

enormous heat production in main frame chips cooling has to be provided from the backside (thermal conduction module). This forces the use of flip chip or C4 technology in which the chip is connected to its substrate with the active side facing to the substrate. While optical fibers in consumer electronics applications can be connected to the top (active) surface of the chip, optical fibers (diameter > 100 μm including cladding) do not fit between substrate and a chip connected by C4s (Fig. 1). Thus optical waveguides for optical multichip modules [1] have to be manufactured on top of the substrate before chip joining. The latter is often done at temperatures exceeding 350 °C. Although the exposure time to this high temperature during one chip join cycle is for a relatively short period of time only, the cumulative exposure time caused by rework operations can be substantial. At such high temperatures most organic waveguides decompose.

Another parameter that is defined by the application in main frame computers is the wavelength at which the signal transmission operates.   In these applications solid state lasers integrated in optical chips are envisioned. The most likely wavelength of transmission would be around 830 nm. Because of the ease of working with visible light most waveguiding studies, however, have been executed using HeNe lasers (632.6 nm).

The advent of organic polymers - among them polyimides - which are stable to temperatures exceeding 400 °C provided the first opportunity to use organic materials in such thermally demanding opto-electronic interconnect applications.   Although the necessary signal transmission distances translate to optical loss requirements of only about 0.5 - 1 dB/cm, the first commercial polyimides were unsuitable for waveguiding applications because of their very high optical losses [2-4].

We and others investigated the source of the optical losses in polyimides and found that they are caused by absorption due to the presence of charge transfer complexes (CTCs) and solvent impurities and by scattering due to ordering of the polyimide chains in crystalline or liquid crystalline domains [5-9] Due to the prevalent use of HeNe lasers for studies of waveguiding properties it was believed that only colorless polyimides will make good waveguides. By utilizing a 830 nm solid state laser we showed that some polyimides with significant optical losses attributed to CTC absorption at 633 nm exhibit relatively low losses at the higher wavelength (Fig. 2) at which (in polyimides) only scattering losses are relevant. This behavior is typical for polyimides which are essentially amorphous but exhibit CTC formation [8]. However, for NLO applications, particular for use as matrix polymer, and for photosensitive polyimides colorless polyimides which allow waveguiding at a wide range of wavelengths are of considerable interest.

One of the major advantages of polyimide chemistry is the large variety of possible molecular structures which can be realized because many dianhydrides and diamines are available. Choosing the right combination of polyimide structure should allow to avoid CTC formation and chain ordering. In particular the tendency to CTC formation is reduced by lowering the acceptor strength of the dianhydride and/or the donor strength of the dianhydride moiety (increased ionization potential). Chain ordering and CTC formation can be hindered sterically by the introduction of bulky side groups [7].

An ideal candidate group to accomplish suppression of chain ordering is the trifluoromethyl (CF3) group. The first CF3 group containing moiety to be found in polyimides was the hexafluoroisopropyliden (6F) group introduced into the dianhydride moiety (such as in DuPont's Pyralin 2566 which is synthesized from hexafluoro - isopropylidene - 2, 2' - bis (phthalic anhydride) (6FDA; Fig. 3) and 4,4'- diaminophenyl ether (oxydianiline,

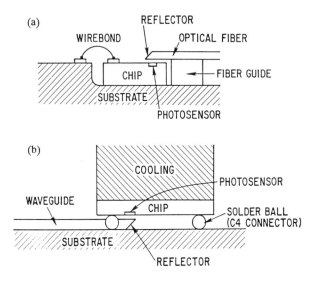

Figure 1. Schematic representation of a) fiber-to-chip connection and b) opto-electronic multichip module

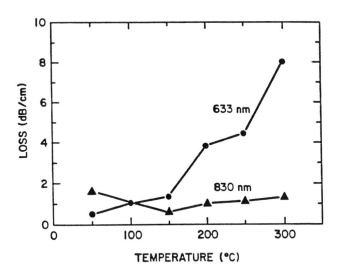

Figure 2. Optical losses in BTDA-6FDA polyimide slab waveguides at 633 and 830 nm. (Reproduced with permission from reference 8. Copyright 1991 SPE.)

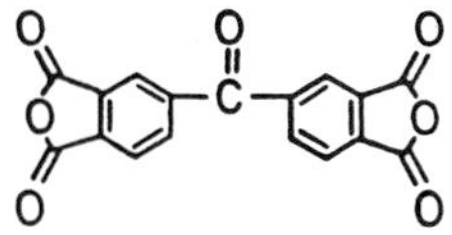

BTDA

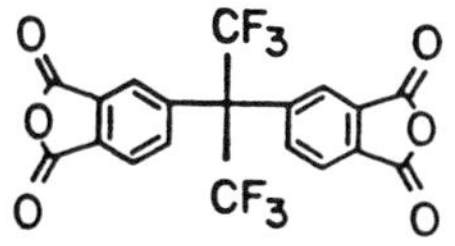

6FDA

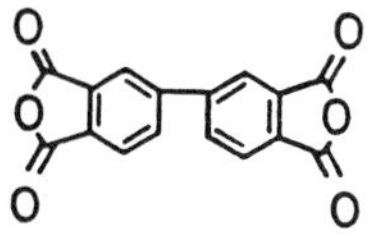

BPDA

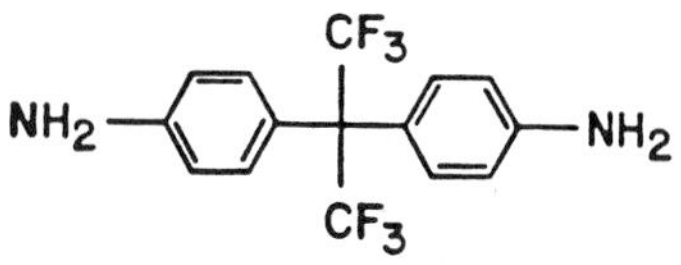

6FDAM

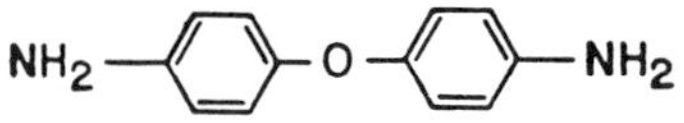

ODA

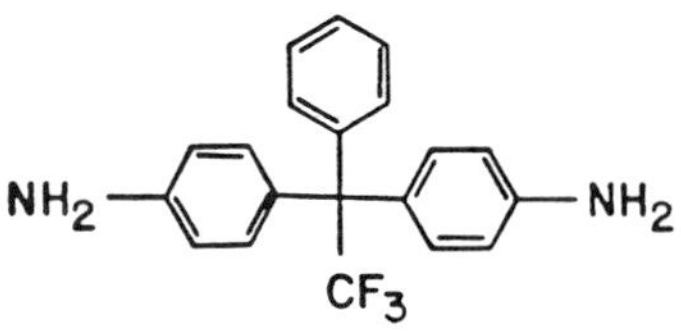

3FDAM

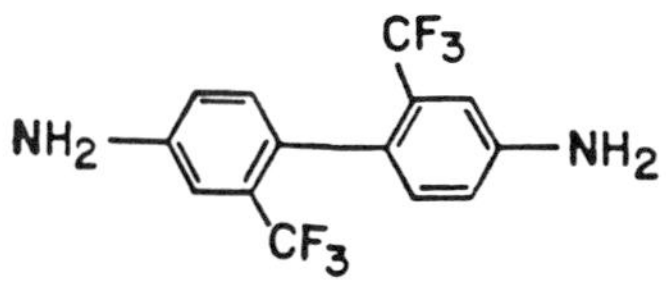

PFMB

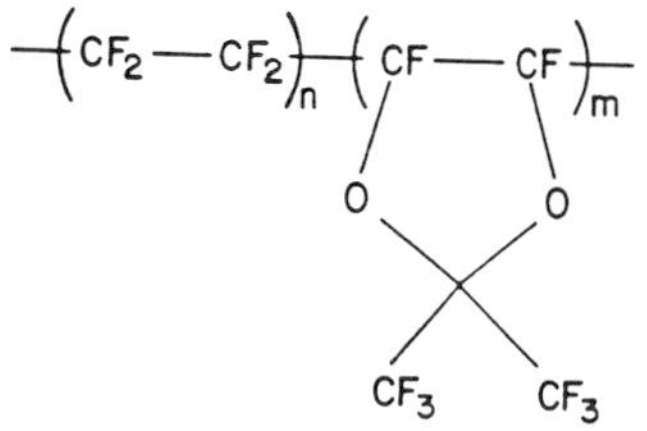

Figure 3. Molecular structure of some diamines and dianhydrides.

ODA; Fig. 3)). Later this group was introduced into the diamine which led to the production of SIXEF-44® (Hoechst - Celanese Corp.) in which both dianhydride and diamine bear a 6F group (6FDA and hexafluoroisopropylidene - 2, 2 - bis(4-aminobenzene), 6FDAm; Fig. 3). From the latter slab waveguides with good, long time thermo-optical stability at 300 °C and optical losses as low as 0.1 dB/cm (depending on the casting solvent) can be produced [5, 7].

Unfortunately, while the optical performance is improved by introduction of the 6F group into the polyimide backbone, the glass transition temperature is lowered. Thus, the maximum temperature to which optical components made from such polyimides can be exposed is limited to about 300 °C. To circumvent this, polyimides with 2,2′ - bis(trifluoromethyl) 4,4′ - diamino biphenyl (PFMB or TFMB; Fig. 3) [10-13]. were synthesized.

PFMB Based Polyimides

The first polyimides containing PFMB diamines were synthesized by Harris et al. [12] and by Matsuura et al. [14, 15]. Some of these polyimides are intriguing because of their lyotropic liquid crystalline behavior (in BPDA based polyimides), their low dielectric constants and high thermal stability. Because of the high chain rigidity and consequently significant chain ordering behavior, high scattering losses are expected for these polyimides. However, the absence of absorption in the visible promises good waveguiding properties if ordering can be suppressed. Following our previous studies, polyimides with the highly kinked 6F unit should hinder chain ordering. Indeed, scattering losses of 6FDA-PFMB are as low as in 6FDA-6FDAm as seen in the optical loss behavior of the former at 830 nm (Fig. 4). Clearly, 6FDA-PFMB does not exhibit liquid crystal formation or any significant amount of other chain ordering. Comparison of the loss value recorded at 633 nm (2.5 dB/cm) to the one recorded at 830 nm (1.1 dB/cm; both for films cured at 300 °C for 30 min.) shows still a small amount of absorption in the visible. Fig. 4 indicates further that the thermo-optical stability of 6FDA-PFMB is higher than 6FDA-6FDAm; also the latter does not loose much weight even at 400 °C (about 0.1 wt.-%/hour), it turns brown at about 330 °C and exhibits losses above 10 dB/cm when cured for 30 min. at 350 °C. The still relatively low loss of 1.85 dB/cm in 6FDA-PFMB after 30 min. at 350 °C indicates that these films might survive short term exposure to high temperature chip join conditions. Compared to 6FDA-6FDAm polyimide 6FDA-PFMB also shows a higher glass transition (356 °C compared to 310 °C for 6FDA-6FDAm). Less intriguing is the processing behavior of PFMB based polyimides. The high rigidity provided by the PFMB fragment imparts a marked process dependence on properties such as birefringence and CTE (C. Feger, unpublished results). This is seen in Fig. 5, which shows a comparison of birefringence vs film thickness data. Compared to 6FDA-6FDAm and 6FDA-ODA, 6FDA-PFMB shows higher birefringence indicating marked anisotropy, which has been shown to cause process dependent properties (C. Feger, S. M. Perutz and M. F. Rubner, in preparation).

Triphenylphosphine Oxide Containing Polymers

Another recent entry in the ever growing list of diamines for polyimides is a group of diamines containing triphenylphosphine oxide (TPPO) units (Fig. 3) [16, 17]. They are similar in structure to the also attractive 3F diamine (1, 1 -

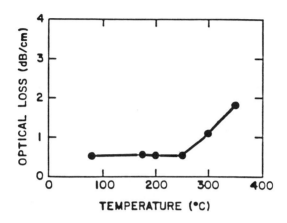

Figure 4. Optical loss of 6FDA-PFMB polyimide slab waveguides at 830 nm vs maximum cure temperature.

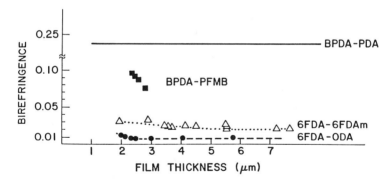

Figure 5. Comparison of birefringence vs film thickness for three polyimides.

bis (4-aminophenyl) - 1 phenyl - 2,2,2 trifluoroethane; Fig. 3) pioneered by Alston [18,19]. While 3F containing polyimides exhibit similar glass transitions compared with their 6F analogs the T_gs of the analog TPPO containing polyimides are higher throughout. Because the TPPO unit provides again a kink in the polymer chain polyimides with this moiety in the backbone should be isotropic. However, at this point TPPO containing polyimides were not available to us. Instead we analyzed the optical properties of a TPPO containing polyaryl ether (Fig. 6.a) with an absorption edge in the UV/vis transmittance spectrum at the low wavelength of 360 nm (Fig. 6.b). The low absorption above 400 nm promises good waveguiding properties over a wide range of wavelengths. Unfortunately the T_g of the investigated TPPO containing polyaryl ether is only 220 °C (DSC, 5 °C/min heating rate). On the other hand this polymer is soluble in aprotic polar solvents such as NMP. Thus, processing is reduced to casting and drying. The expected low optical losses together with the high glass transitions make this material an attractive choice for passive waveguides or as matrix for NLO molecules.

High Temperature Stable Polymer Waveguides in Optical Sensors

Sensitivity to the environment is usually one of the shortcomings of organic fibers which can be avoided by using an insensitive cladding material. However, the interaction of the environment with a polymer can be used to advantage in environmental sensors [20, 21]. Environment induced changes in waveguiding properties such as a shift in the incoupling angle of a given mode are at the core of the sensing principle. One possible configuration (Fig. 7.) consists of light source, signal analyzer, and a 90° prism with a waveguide coated on one of the faces. To increase the sensitivity of the set-up an attenuated total reflection (ATR) coating of Ag is placed between prism and waveguide. The sensor works by locking in on one of the modes of the waveguide. The mode shifts with exposure of the waveguide to an environmental influence such as a solvent vapor. For instance, the central line of the fourth TM mode in Teflon® AF 1600 (DuPont) shifts by 0.515° which is a very large effect considering that the angular resolution is 0.0001°. Besides recording the mode shift after sufficient exposure times between a few seconds and several minutes the kinetics of the vapor sorption and desorption can be studied as well. Sorption and desorption curves are characterized by differing time constants which however seem to be characteristic for a given vapor/waveguide pair.

The described method allows quantitative analysis because the mode shifts are linear with concentration (Fig. 8) at least in all cases studied so far. Not all vapors interact with a waveguide of a given structure which makes it possible to analyze vapors in the presence of other volatiles. For instance, water does not affect the waveguide properties of Teflon® AF. This allows the analysis of hygroscopic solvent vapors under ambient conditions.

While the inertness of Teflon® AF waveguides towards water is not unexpected, it was not expected to find selective behavior of these waveguides towards linear and branched hydrocarbons which play a significant role in the oil refinery industry. Fig. 9a shows the phase shift of the fourth mode in a Teflon® AF waveguide exposed to 51 % n-heptane in N_2 vs time. Also shown (Fig. 9b) is the phase shift vs time for the same waveguide and n-heptane in a mixture with iso-octane vapor. In the presence of iso-octane the time constant of n-heptane sorption is changed but the same final mode shift is reached. iso-Octane alone does not affect the mode angle. The slower up-take

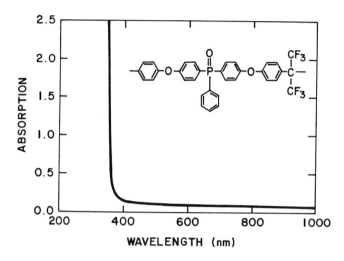

Figure 6. Structure of a TPPO containing polyaryl ether and its UV/vis absorption spectrum.

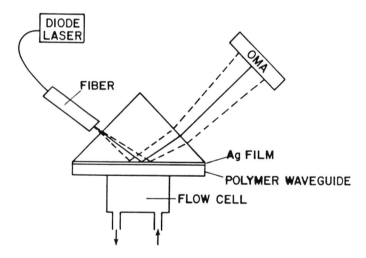

Figure 7. Experimental configuration for the measurement of mode shifts in waveguides. OMA: optical multichannel analyzer

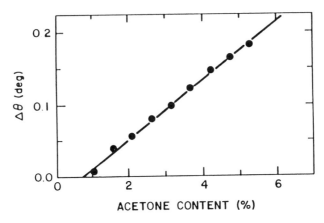

Figure 8. The mode shift $\Delta \Theta$ versus acetone concentration in dry nitrogen.

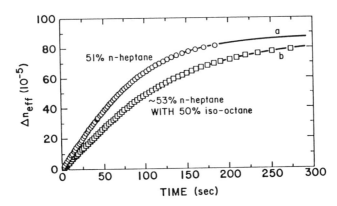

Figure 9. The mode shift $\Delta \Theta$ versus time for a) 51% n-heptane in nitrogen and b) a mixture of n-heptane (53%) and iso-octane (50%) in nitrogen.

might be due to surface sites occupied by iso-octane. However, the latter seems not to be able to penetrate into the Teflon® AF waveguide. In summary the application of waveguides to detect vapors quantitatively using metal film enhanced leaky mode spectroscopy is versatile and promising.

Conclusions

Ever new diamines available for polyimide synthesis make it very likely that polyimides with the right combination of properties required for integrated waveguides in opto-electronic multichip modules will be ready when needed. Although the current economic climate has relaxed the schedule for such a need, other applications for organic waveguide materials, for instance in optical sensors, are emerging.

Experimental

The materials investigated were obtained from commercial sources (SIXEF®-44 from American Hoechst, Teflon® AF 1600 from DuPont) or synthesized from dianhydrides and diamines in γ-butyrolactone (GBL). PFMB was obtained from Marshalton Labs. The synthesis of the TPPO containing polyaryl ether has been described elsewhere. [17].

Planar waveguides were produced by spin-casting films on glass substrates with thickness of about 5 μm. Thick films (thickness > 50 μm) for absorption spectra measurements were prepared by doctor-blading. Curing was performed by heating under nitrogen in an oven. The measurement procedure using an optical multichannel analyzer has been described elsewhere [8].

UV/visible absorption spectra were recorded with a Perkin Elmer spectrophotometer.

Acknowledgments

We are indebted to American Hoechst for providing samples of SIXEF®.

Literature Cited

1. *Multichip Module Technologies and Alternatives: The Basics*; Doane, D. A.; Franzon, P. D., Eds.; Van Nostrand Reinhold: New York, NY, 1993.
2. Russell, T. P.; Gugger, H.; Swalen, J. D.; *J. Polymer Sci., Polym. Phys. Ed.* **1983**, *21*, 1745.
3. Franke, H; Crow, J. D. *Integr. Optical Circ. Eng., Proc. SPIE* **1986**, *651*, 102.
4. Franke, H.; Knabke, G.; Reuter,R. *SPIE Proc.* **1986**, *682*, 191.
5. Reuter, R.; Franke, H.; Feger, C. *Appl. Optics 1988*, *27*, 4565.
6. St. Clair,A. K.; St. Clair, T. L.; Slemp, W. S. In *Recent Advances In Polyimide Science and Technology*; Weber, W. D.; Gupta, M. R., Eds.; Mid Hudson Sect. of the Soc. Plast. Eng.: Poughkeepsie, NY, 1987; p. 16.
7. Feger, C.; Reuter, R.; Franke, H. In *Polymers in Information Storage Technology*; Mittal, K. L., Ed.; Plenum Press: New York, NY, 1989; pp 227-233.
8. Feger, C.; Perutz, S.; Reuter, R. *SPE Tech. Pap.* **1991**, *37*, 1594.

9. Feger, C. *Polym. Prepr.* **1991**, *32*, 76.
10. Hall, D. M.; Harris, M. M. *Proc. Chem. Soc.* **1959**, *1959*, 396.
11. Rogers, H. G.; Gaudiana, R. A.; Hollinsed, W. C.; Kalyanaraman, P. S.; Manello, J. S.; McGowan, C.; Minns, R. A.; Sahatjian, R. *Macromolecules* **1985**, *18*, 1058.
12. Harris, F. W.; Hsu, S. L-C., Tso, C. C. *Abstr. of Papers, Part 1, 1989*, 1989 Intl. Chem. Congr. Pacif. Basin Soc.: Honolulu, HI, pp 123.
13. Matsuura, T.; Hasuda, Y.; Nishi, S.; Yamada, N. *Macromolecules* **1991**, *24*, 5001.
14. Matsuura, T.; Nishi, S.; Ishizawa, M.; Yamada, Y., Hasuda, Y. *Abstr. of Papers, Part 1, 1989*, 1989 Intl. Chem. Congr. Pacif. Basin Soc.: Honolulu, HI, pp 87.
15. Matsuura, T.; Ishizawa, M.; Hasuda, Y.; Nishi, S. *Polymer Prepr. Japan* **1989**, *38*, 434.
16. Hirose, S.; Nakamura, K.; Hatakeyma, T.; Hatakeyama, H. *Sen-I Gakkaishi* **1988**, *44*, 563.
17. Grubbs, H. J.; Smith, C. D.; McGrath, J. E. *Polym. Mater. Sci. Eng.* **1991**, *65*, 111.
18. Alston, B. W., Gratz, R.F. *US Patent* 4,885,116, 1989.
19. Rogers, M. E.; Grubbs, H.; Brennan, A.; Rodrigues, D.; Wilkes, G. L.; McGrath, J. E. In *Advances in Polyimides Science and Technology*; Feger, C.; Khojasteh, M. M.; Htoo, M., Eds.; Technomic Publ.: Lancaster, PA, 1993; pp 33.
20. Galipeau, W.; Vetelino, J. F., Lee, R.; Feger, C. *Sens. Actuators B, Chem.* **1991**, *B5(1-4)*, 59.
21. Osterfeld, M.; Franke, H.; Feger, C. *Appl. Phys. Lett.* **1993**, *62*, 2310.

RECEIVED October 21, 1994

Chapter 22

Rodlike Fluorinated Polyimide as an In-Plane Birefringent Optical Material

Shinji Ando[1], Takashi Sawada[1], and Yasuyuki Inoue[2]

[1]NTT Interdisciplinary Research Laboratories, Midori-cho,
Musashino-shi, Tokyo 180, Japan
[2]NTT Optoelectronics Laboratories, Tokai, Ibaragi 319–11, Japan

Fluorinated polyimide that has a rod-like structure is investigated as a novel in-plane birefringent optical material whose birefringence and/or retardation can be precisely controlled. This material has good flexibility, high transparency, thickness controllability, and high thermal stability. Three methods for generating in-plane birefringence in polyimide films are presented. The in-plane birefringence can be controlled between 0.02 and 0.17, and the optical retardation is controlled to within 1%. A thin, flexible optical half-waveplate at 1.55 μm, only 14.5 μm thick, which is 6.3 times as thin as a zeroth-order quartz waveplate, was prepared. Its retardation was retained after annealing at 350°C for 1 hour.

The most commonly used birefringent materials are calcite, the trigonal crystal of calcium carbonate, and quartz. They are used for a variety of optical polarization components such as waveplates, polarizers, and beam splitters. With the development of planar lightwave circuits (PLC), opto-electronic integrated circuits (OEIC), and multichip interconnections for optical communication, these polarization components will be reduced in size and incorporated into such circuits and modules. For example, a small quartz waveplate is inserted into a silica-based waveguide as a TE/TM polarization mode converter (1). Such inorganic crystals have high optical transparency at the wavelengths of optical communication, 1.0-1.7 μm, and have good thermal and environmental stability. However, their birefringence cannot be changed and they are difficult to make into thin plates or small components. The thickness of a zeroth-order half-waveplate of calcite at 1.55 μm is about 5 μm which is almost impossible to grind and polish because of its fragility. On the other hand, the low birefringence of quartz makes a half waveplate rather thick (92 μm). This results in considerable excess loss when the waveplate is inserted into a waveguide (1).

Polymeric materials such as polycarbonates and poly(vinyl alcohol) are also known as birefringent optical materials. They are used as retardation plates for liquid crystal displays, with their birefringence and thickness being controlled by uniaxial drawing or rolling of the films. However, it is difficult to generate large in-plane birefringence because of their small polarizability anisotropy and flexible molecular structure. The in-plane birefringence of polymeric retardation plates is only as large as that of quartz waveplates. In addition, the current manufacturing process for OEICs and multichip modules includes soldering at 260°C and short-term processes at temperatures of up to 400°C. On the other hand, the PLCs for optical communication require very high reliability and durability. The above-mentioned polymeric materials do not have such thermal and environmental stability. Furthermore, their optical losses at the wavelengths of optical communication are much higher than in the visible region (0.4-0.8 μm) because their two or three types of C-H bonds which harmonically absorb infrared radiation give broad and strong absorption peaks at these wavelengths (2-4). There has been a strong demand for new optical materials having much larger birefringence than quartz while having good processability, good tractability, and high thermal and environmental stability.

The transparency and optical anisotropy of polyimides have been investigated because their excellent thermal, chemical and mechanical stability could make them good waveguide materials (4-6). Optimally cured fluorinated polyimides can decrease optical losses to below 0.1 dB/cm in the visible region, and these losses are stable at temperatures up to 300°C (6). We have recently fabricated heat-resistant multi-mode (ridge-type) and single-mode (buried-type) optical waveguides using fluorinated polyimides with the propagation loss of about 0.3 dB/cm at 1.3 μm (7,8). Polyimides are known as optically anisotropic materials, in which the birefringence of polyimide films is, in general, defined as the difference of refractive indices of the two polarizations, parallel (n_{TE}) and perpendicular (n_{TM}) to the film plane. A very large birefringence between n_{TE} and n_{TM} of 0.24 at 0.633 μm was observed for poly(p-phenylene biphenyltetracarboximide) (9). Polyimide films prepared on isotropic substrates, such as glass plates or silicon wafers, have no refractive index anisotropy in the film plane (10). However, uniaxial drawing of Kapton-type poly(amic acid) films can give highly oriented polyimide films with large in-plane birefringence (11). A maximum elongation of 83% was obtained for the poly(amic acid) film with 40wt% solvent. Refractive index ellipsoids of polyimide prepared on isotropic substrates and uniaxially drawn polyimide are schematically shown in Figure 1. The in-plane birefringence of polyimide films (Δn) is defined as the difference between the largest (n_{TE1}) and the smallest in-plane refractive index (n_{TE2}). The optical axis of n_{TE1} coincides with the drawing direction. The Δn of 0.18 at 0.633 μm obtained for the Kapton-type polyimide is larger than that of calcite (11). Several methods of drawing poly(amic acid) films uniaxially have been developed in order to improve the tensile mechanical properties of polyimides (12,13), but no methods of controlling the in-plane birefringence or the fabrication of polarization components from polyimide have been reported yet.

In this study, we investigate fluorinated polyimide that has a rod-like structure as a novel in-plane birefringent optical material whose birefringence and retardation (birefringence x thickness) can be precisely controlled. The fabrication of a thin, flexible half-waveplate is described and its thermal and optical properties are presented. We have already reported fluorinated polyimides using 2,2'-bis(trifluoromethyl)-4,4'-diaminobiphenyl (TFDB) as a diamine (14-17). These polyimides exhibit high transparency in the visible-near-infrared region, low water absorption, and low refractive indices. These properties are desirable for materials for components used in optical circuits and modules, and are superior to those of conventional nonfluorinated polyimides, such as Kapton. Nonfluorinated polyimides have high water absorption of about 2-3 wt%, which causes optical loss at the wavelengths of optical communication.

EXPERIMENTAL

Materials. A fluorinated polyimide, PMDA/TFDB (Figure 2), synthesized from pyromellitic dianhydride and 2,2'-bis(trifluoromethyl)-4,4'-diaminobiphenyl is used in this study. The preparation of poly(amic acid), the precursor of the polyimide, has been described elsewhere (*14*). Films of poly(amic acid) were prepared by spin-coating N,N-dimethylacetamide solution (15 wt%, 450 poise) onto a 4-inch silicon wafer and drying then in nitrogen at 70°C for 1 h. The solvent content of the films determined from the weight difference before and after curing was 23.8 wt%. Small signals that can be assigned to the polyimide structure were observed in the ^{13}C NMR spectrum of the poly(amic acid) films dissolved in dimethylsulfoxide-d$_6$, even though the polyimide content was less than 5%.

Measurements. The retardation of polyimide films was directly measured by the parallel-Nicole rotation method at 1.55 μm, the wavelength used for long-distance optical communications. An Advantest TQ8143 laser diode was used as light source, and a Newport Model-835 as optical power-meter. The polarizers were Glan-Thompson prisms from Sigma-Koki corporation. Thickness (d) was measured from the interference fringe observed in the near-infrared absorption spectra. The retardation and thickness were measured at the center of the films. In-plane birefringence (Δn) was calculated by dividing retardation by thickness. The largest and the smallest in-plane refractive index (n_{TE1} and n_{TE2}), and the out-of-plane refractive index (n_{TM}) at 1.523 μm were measured by sample rotation using a Metricon PC-2000 refractometer. The Δn and the difference between n_{TE1} and n_{TE2} agree within the experimental error (±0.0005) Thermomechanical analysis (TMA) was conducted using a Sinku-Riko TM-7000 analyzer, in which specimen dimensions were 5 mm wide and 15 mm long. Average thicknesses of poly(amic acid) and polyimide films were 23 μm and 16 μm, respectively.

RESULTS AND DISCUSSION

Methods for Generating In-plane Birefringence. In order to control the in-plane birefringence (Δn) of fluorinated polyimide, we should examine the molecular structure, drawing methods, and curing conditions, because the Δn is determined by the polarizability anisotropy of the repeat unit and the degree of molecular orientation. The PMDA/TFDB polyimide used in this study has a rod-like structure accompanied by a large polarizability anisotropy. This seems advantageous for generating a large in-plane birefringence by uniaxial drawing. This polyimide has a small thermal expansion coefficient of -5 x 10^{-6}°C^{-1}, which reveals the rigidity of the molecular structure.

Use of Anisotropic Substrates (Method 1). Figure 3 shows three methods of

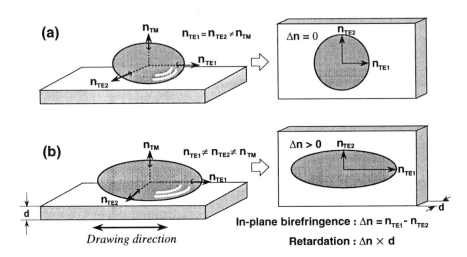

Figure 1. Schematic representation of refractive index ellipsoids of (a) polyimide prepared on isotropic substrates and (b) uniaxially drawn polyimide.

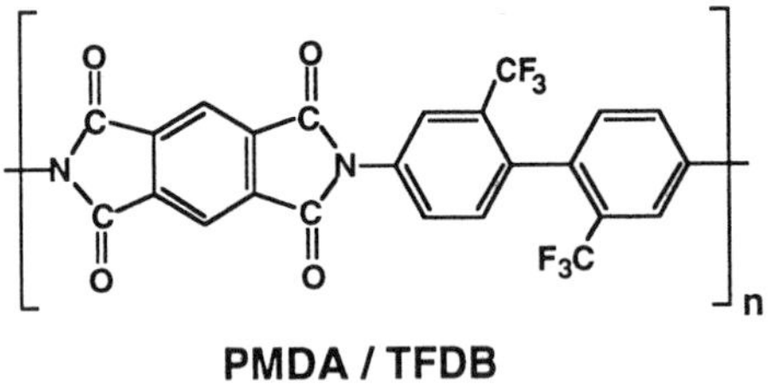

PMDA / TFDB

Figure 2. Rod-like fluorinated polyimide for uniaxial drawing.

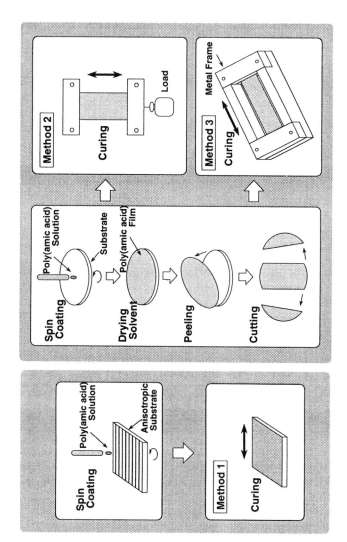

Figure 3. Methods of preparing in-plane birefringent polyimide films.

preparing uniaxially drawn polyimide films. Method 1 and Method 3 are newly devised for this study. In Method 1, the thermal expansion anisotropy of the substrates is used as a drawing force. Poly(amic acid) solution was spin-coated onto one isotropic (4-inch silicon wafer) and three anisotropic substrates (3-inch wafers of lithium niobate (LiNbO$_3$) and lithium tantalate (LiTaO$_3$) and a quartz block (15x15x2 mm)). The c-axes of these crystal substrates were in the planes. After the solvent was removed by drying, the poly(amic acid) films were heated to 350°C at 4°C/min in nitrogen, kept at 350°C for 1 h, and cooled to room temperature. The anisotropy of thermal expansion coefficients ($\Delta\alpha$) of substrates and the Δn of the peeled polyimide films are listed in Table I. The polyimide film cured on a silicon wafer has no birefringence in the film plane. In contrast, anisotropic substrates give rise to in-plane birefringent polyimide films. A larger $\Delta\alpha$ seems to generate a larger Δn from the comparison of LiNbO$_3$ and LiTaO$_3$ even though no simple relationship is observed between Δn and $\Delta\alpha$. From the refractive index measurements, the optical axis of the films with larger refractive indices (n$_{TE1}$) is along the crystal axis with higher thermal expansion coefficient. This indicates that the polyimide molecules orient along the direction of the larger thermal expansion. The Δn of the polyimides, however, are 6-10 times smaller than that for quartz, which are too small to fabricate polarization components. It seems that the substrate $\Delta\alpha$ is insufficient to effectively draw the polyimide chains during curing.

Drawing with a Load during Curing (Method 2). Poly(amic acid) films were uniaxially drawn during curing with constant loads in Method 2. This experiment was conducted using a thermal mechanical analyzer (TMA). The effects of final curing temperature (Tf), heating rate, and load on the in-plane birefringence (Δn) were examined. As shown in Figures 4-6, the Δn are linearly proportional to Tf, heating rate, and load. These relationships are useful for controlling the Δn. In addition, this method gives a larger Δn, which can be controlled between 0.02 and 0.17 by changing the curing and loading conditions.

Varying the final curing temperature is, in particular, effective in generating a large Δn; however, the Δn observed at Tf = 350°C and 440°C are slightly deviated from linearity. The largest Δn of 0.17 is slightly smaller than that observed by Nakagawa (*11*), however the Δn does not saturate even after curing at Tf = 440°C. In contrast, the variation of the heating rate does not cause a considerable change in Δn. The variation of the load showed the best controllability of Δn among the three parameters. Figure 7 shows the temperature profile and TMA curve measured for the films prepared under the standard condition of Method 2 (Tf:350°C, heating rate:4°C/min, load:20g). The poly(amic acid) film begins to shrink at 120°C and then to elongate at 180°C. ^{13}C NMR study on the imidization process of 6FDA/TFDB polyimide synthesized from 2,2-bis(3,4,-dicarboxyphenyl)hexafluoropropane dianhydride and 2,2'-bis(trifluoro-methyl)-4,4'-diaminobiphenyl which has a more flexible dianhydride structure and better solubility in polar solvents shows that imidization reaction becomes significant at 120°C and is almost complete at 180°C (*18*). After that, the film elongates at the constant rate as the temperature increased, but at a slightly higher rate above 300°C. This film elongation causes the polyimide molecules to orient along the loading direction and, hence, to generate in-plane birefringence. The magnitude of Δn in Method 2 can be explained in terms of the elongation after the completion of imidization. As shown in Figure 8, Δn shows a linear relationship with the normalized

Table I. Anisotropy of Thermal Expansion Coefficients of Substrates and
In-plane Birefringence of Polyimides Prepared by Method 1

Substrate	α_c	α_a	$\Delta\alpha$	Δn
		$(\times 10^{-5})$		
Silicon	0.42	0.42	0	0.0000
LiNbO$_3$	0.75	1.54	0.79	0.0009
LiTaO$_3$	0.41	1.61	1.20	0.0011
Quartz	0.80	0.01	0.79	0.0015

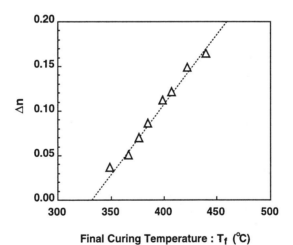

Figure 4. Final curing temperature versus in-plane birefringence of polyimide
films prepared by Method 2 (heating rate:4°C/min,load:20g).

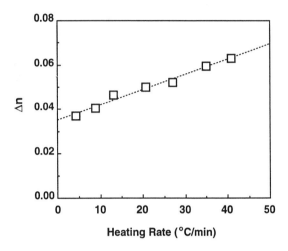

Figure 5. Heating rate versus in-plane birefringence of polyimide films prepared by Method 2 (Tf:350°C, load:20g).

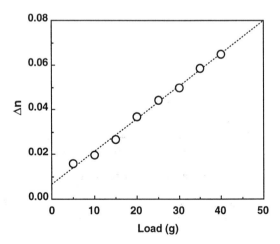

Figure 6. Load versus in-plane birefringence of polyimide films prepared by Method 2 (Tf:350°C, heating rate:4°C/min).

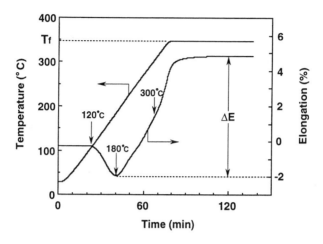

Figure 7. Temperature profile and elongation of poly(amic acid) film prepared under the standard condition of Method 2.

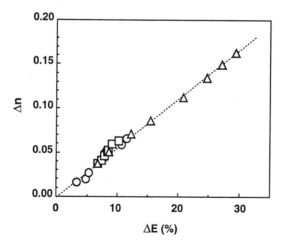

Figure 8. Normalized elongation in TMA curves, ΔE, versus in-plane birefringence of polyimide films prepared by Method 2. Symbols correspond to those in Figures 4-6.

elongation (ΔE) between the most shrunken state at 180°C and the most elongated state at Tf. The larger ΔE, the larger Δn. This relationship is also useful for controlling the Δn.

Figures 9-11 show TMA curves of polyimides prepared under the thermal and loading conditions described above. Figure 9 show that curing above 350°C is much more effective in drawing films than curing below that temperature. This polyimide does not exhibit any apparent glass transition in TMA measurement but molecular re-orientation is appreciably promoted above 350°C. The linear relationship between Δn and Tf shown in Figure 4 is attributed to the linearity of this TMA curve between 370°C and 420°C. Films cured above 420°C (Tf:422°C,440°C) showed very large Δn; however, optical defects with diameter of less than 20 μm were observed under a polarizing microscope throughout the films. This may be ascribed to localized crystallization. As seen in Figure 10, a high heating rate gives a larger elongation at Tf and a small shrinkage at about 180°C. To obtain highly elongated polyimide film, the magnitude of shrinkage occurring between 120°C and 180°C should be small. This suggests that a reaction which inhibits the orientation of polyimide molecules occurs during the imidization process. The ^{13}C NMR study described above showed that depolymerization at amide-groups of poly(amic acid) occurs at around 120°C, and the molecular weight decreases considerably. This low-molecular-weight polymer is converted into a high-molecular weight polyimide by subsequent curing, but this depolymerization process should inhibit the uniaxial orientation of the polymer chain. This is why the fast imidization reaction under high heating rates produce large in-plane birefringence. Figure 11 indicates that a heavy load gives a large elongation at Tf and a small shrinkage at about 180°C. This relationship agrees well with that of Figure 10.

All the films prepared by the above methods were pale yellow, tough, and flexible. However, the polyimide film prepared at a heating rate of 2°C/min was cut during the heating at 190°C. At 190°C, the film was at the most shrunken state with no ΔE. This film cooled to room temperature was brittle and had no in-plane birefringence. This result coincides with the linear relationship between ΔE and Δn (Figure 8). The onset of depolymerization at about 120°C should considerably decrease the tensile strength of the film.

Fix in a Metal Frame during Curing (Method 3). In Method 3, the poly(amic acid) film was peeled from the substrate, two of its sides were cut off, and it was fixed in a brass frame by the other sides. The film was then heated to 350°C at 4°C/min. As the temperature increased, polymer chains seemed to orient along the fixed direction as the result of the tensile stress arising from the shrinkage of the polymer film and evaporation of the solvent. It is worthwhile to note that the polyimide film elongated about 6 % after the curing compared to the poly(amic acid) film along the fixed direction. This indicates that polyimide film spontaneously elongates even after the drawing force has disappeared. The values of Δn, n_{TE1}, n_{TE2}, and n_{TM} for the film 55 mm long and 60 mm wide were 0.053, 1.638, 1.585, and 1.484 respectively. The n_{TE1} and n_{TE2} were observed to be parallel and perpendicular to the fixed direction, respectively. Comparing with the n_{TE} (1.612) and n_{TM} (1.484) of the film prepared on a silicon wafer, the n_{TM} does not show any difference between in-plane birefringent and non-birefringent films. Uniaxial drawing of polyimide films only causes optical anisotropy in the film plane. Considering the thermal expansion of the brass frame, the

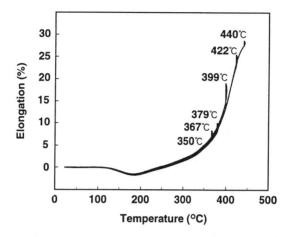

Figure 9. TMA curves with different final curing temperatures in Method 2 (heating rate:4°C/min, load:20g).

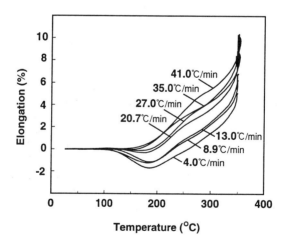

Figure 10. TMA curves with different heating rates in Method 2 (Tf:350°C, load:20g).

poly(amic acid) film is prevented from shrinking and is forced to elongate about 0.1% between 120°C and 180°C. This condition is similar to the TMA behavior that was measured for the film cured at 27.0°C/min in Method 2. As seen in Figure 10, the length of this film is constant below 180°C, and it starts to elongate above that temperature. The Δn and ΔE were 0.052 and 7%, respectively which coincide with the Δn (0.053) and ΔE (6%) obtained for Method 3.

Fabrication of a Half-Waveplate at 1.55 μm. The Δn of 0.053 obtained in Method 3 is large enough to fabricate a thin half-waveplate at the wavelengths of optical communication. Since retardation is the product of the thickness and in-plane birefringence, precise control of retardation requires control of film thickness. It is well-known that the thickness of polyimide film can be controlled by changing the spinning speed of the poly(amic acid) solutions. As shown in Figure 12, the retardation of the polyimide films is linearly proportional to the spinning speed and the thickness. This indicates that Δn of polyimides prepared by Method 3 show a constant Δn of 0.053 for different thicknesses. From this figure, the spinning speed for producing a half wavelength retardation at 1.55 μm should be 570 rpm. 1.55 μm is the wavelength used in current long-distance optical communication systems. The retardation of a film thus obtained was 0.772, which indicates that the retardation of polyimide films can be controlled with an error of less than 1%. In addition, this half-waveplate was only 14.5 μm thick, which is 6.3 times as thin as a quartz waveplate. This will considerably decrease excess loss caused by inserting a waveplate into a waveguide. An attempt to use this polyimide waveplate as a TE/TM polarization mode converter in silica-based planer lightwave circuits is under way (19). Polyimide waveplates are tough and flexible and have superior processability and tractability to inorganic crystals. Polyimides are easily cut by scissors and do not wrinkle even after being folded double.

Thermal Stability of Retardation. Retardation of waveplates must have sufficient thermal stability, in order to be compatible with high performance optical circuits and modules. We examined the effect of thermal treatment by annealing the polyimide waveplates at 250°C, 300°C, 350°C, 380°C, and 400°C for 1 hour. As shown in Figure 13, the retardation does not change below 350°C, the final curing temperature of the polyimide waveplates. Above 350°C, the retardation increases with annealing, suggesting that self-orientation of polyimide molecules takes place. This agrees well with the spontaneous elongation observed for Method 3. A similar phenomenon has been observed by Cheng et al. (20)

CONCLUSIONS

Fluorinated polyimide that has a rod-like structure (PMDA/TFDB) can be used as an in-plane birefringent optical material with good flexibility, thickness controllability, and large birefringence. In order to generate and control the in-plane birefringence (Δn) and/or retardation of the polyimide films, three methods of uniaxial drawing were investigated. Polyimide films cured on anisotropic substrates show very small Δn. The Δn of the polyimides cured with constant loads were linearly proportional to the final curing temperature, the heating rate, and the load. The polyimides fixed in a metal

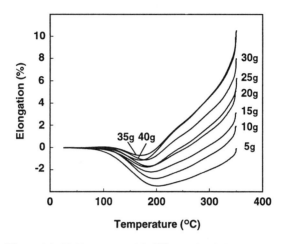

Figure 11. TMA curves with different loads in Method 2
(Tf:350°C, heating rate:4°C/min)

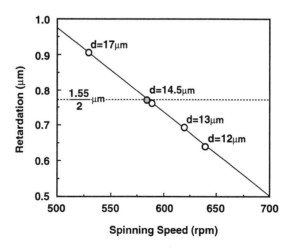

Figure 12. Retardation and thickness of polyimide films versus speed of spin-coating in Method 3.

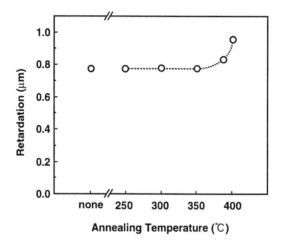

Figure 13. Retardation of a polyimide half-waveplate after annealing for 1h at various temperatures.

frame had the same Δn for different thicknesses. As a result, the Δn of this fluorinated polyimide can be controlled between 0.02 and 0.17, and the retardation can be controlled to within 1%. The retardation of these polyimide films is retained after annealing at 350°C for 1 hour.

REFERENCES

1. Takahashi, H.; Hibino, Y.; Nishi, I. *Opt. Lett.* **1992**, *17*, 499
2. Kaino, T.; Fujiki, M.; Nara, S. *J. Appl. Phys.* **1981**, *52*, 7061
3. Groh, W. *Makromol. Chem.* **1988**, *189*, 2861
4. Franke, H; Crow, J. D. *Proc. SPIE* **1986**, *651*, 102
5. Sullivan, C.T. *Proc. SPIE* **1988**, *994*, 92
6. Reuter, R.; Franke, H.; Feger, C. *Appl. Opt.* **1988**, *27*, 4565
7. Matsuura, T.; Ando, S.; Sasaki, S.; Yamamoto, F. *Elect. Lett.* **1993**, *29*, 269
8. Matsuura, T.; Ando, S.; Matsui, S.; Hirata, H.; Sasaki, S.; Yamamoto,F. *Technical digest of OSA/ACS Topical Meeting (Polymeric Thin Films for Photonic Applications)* Toronto, **1993**, 264
9. Herminghaus, S.; Boese, D.; Yoon, D. Y.; Smith, B. A. *Appl. Phys. Lett.*, **1991**, *59*, 1043
10. Russel, T. P.; Gugger, H.; Swalen, J. D. *J. Polym. Sci.* **1983**, *21*, 1745
11. Nakagawa, K, *J. Appl. Polym. Sci.* **1990**, *41*, 2049
12. Yokota, R.; Horiuchi, R.; Kochi, M.; Soma, H.; Mita, I. *J. Polym. Sci. Polym. Lett.* **1988**, *26*, 215
13. Kochi, M.; Yokota, R.; Iizuka, T.; Mita, I. *J. Polym. Sci. Polym. Phys.* **1990**, *28*, 2463
14. Matsuura, T.; Hasuda, Y.; Nishi, S.; Yamada, Y. *Macromolecules* **1991**, *24*, 5001
15. Matsuura, T.; Ishizawa, M.; Hasuda, Y.; Nishi, S. *Macromolecules* **1992**, *25*, 3540
16. Matsuura, T.; Yamada, N.; Nishi, S.; Hasuda, N. *Macromolecules*, **1993**, *26*, 419
17. Matsuura, T.; Ando, S.; Sasaki, S.; Yamamoto, F. *Macromolecules (in press)*
18. Ando, S.; Matsuura, T.; Nishi, S. *Polymer* **1992**, *33*, 2934
19. Inoue, Y.; Ohmori, Y.; Kawachi, M.; Ando, S.; Sawada, T.; Takahashi, H. *Proc. Integrated Photonics Research* **1994**, 64
20. Cheng, S.; Arnord, F.; Zhang, A.; Hsu, F.; Harris, F., *Macromolecules* **1991**, *24*, 5856

RECEIVED September 13, 1994

Chapter 23

Preparation of Polyphenylene and Copolymers for Microelectronics Applications

N. A. Johnen, H. K. Kim[1], and C. K. Ober

Department of Materials Science and Engineering, Cornell University, Ithaca, NY 14853–1501

High performance polymers with superior physical characteristics are needed for a variety of anticipated applications in both micro- and optoelectronics. Efforts to design such materials involving the copolymerization of 1,3-cyclohexadiene-5,6-bisacetate (DAC) with 4-vinylbiphenyl (VBP) and N-vinylcarbazole (VIC) are described. Polymerization of DAC in the presence of VBP yielded a copolymer while VIC did not copolymerize efficiently with DAC. In addition to the investigation of the chemical synthesis of the copolymers, we report on the physical properties of the spin-cast polymer films. All investigated homopolymers and the DAC/VBP copolymer showed good thermal stability and optical properties equal to or better than poly(phenylene).

Polymeric materials are becoming increasingly important for a variety of applications with both opto- and microelectronics. Their use in microelectronic devices results in efficient transmission of information signals, high space density, and excellent signal bandwidth.

In order to be a candidate for either opto- or microelectronic devices, a polymer has to fulfill a series of requirements including low dielectric constant, ease of processing (preferably using microlithographic techniques), thermal and environmental stability, and low thermal expansion coefficient. For polymers environmental stability means mainly that the polymer is resistant to the effects of moisture. However,

---

[1]Current address: Photonic Switching Section, Electronics and Telecommunications Research Institute, P.O. Box 8, Daeduck Science Town, Taejeon, Korea 305–606

the main requirement for a waveguide material is the efficient and loss-free transmission of light.   In addition, the potential electrical signal speed in such a medium is directly related to the dielectric constant (ε_r) and the refractive index (RI).   These correlations are shown in equation 1.

Refractive index (RI) : $RI = \sqrt{\varepsilon_r}$

Signal Delay Time (TD) : $TD = L \bullet \sqrt{\varepsilon_r / c}$ (1)

In general, the signal propagation speed in a material increases if the dielectric constant (and as a result the RI) decreases.   Therefore both RI and ε_r are important parameters for a waveguide material.

Preparation of Poly(phenylene) from a Soluble Precursor Polymer.
Recently, a new precursor route to the preparation of poly(phenylene) was developed that enabled the preparation of spin-coated films as shown by Ballard et al. (1).   The general scheme for the synthesis of poly(phenylene) that we have adopted is shown in equations 2-4.   1,3-Dihydroxycyclohexadiene was converted to DAC (1,3-cyclohexadiene-5,6-bisacetate).

$$(2)$$

1,3-Dihydroxycyclo-   1,3-cyclohexadiene-5,6-
      hexadiene   bisacetate

DAC can then be polymerized to PDAC (poly(1,3-cyclohexadiene-5,6-bisacetate)) using 2,2'-azobis(isobutyronitrile) (AIBN) as initiator.

$$(3)$$

DAC   PDAC

The soluble polymeric precursors can be transformed in several ways to poly(phenylene).

PDAC   Poly(phenylene)

These conjugated, fully aromatic polymers possess a thermal stability of up to 450°C, which is only matched by a few polymers such as poly(imide) (2-3). The high temperature stability of poly(imides) is paired, however, with a lower environmental stability caused by the imide bond which is readily hydrolyzed.

Alternative Syntheses of Poly(phenylene). Poly(phenylene) has also been prepared by several different approaches. The direct polymerization of benzene in the presence of either aluminum chloride (4) or aluminum chloride and copper(I)chloride (5) as well as the electrochemical polymerization of benzene derivatives (6) have been used to produce poly(phenylene). Kaeriyama et al. (7) prepared poly(phenylene) by coupling methyl-2,5-dichlorobenzoate in the presence of Nickel(0) whereas Tour et al. coupled 1,4-dibromobenzene with buthyl lithium (8). Wegner et al. (9) prepared poly(phenylene) by reacting aryldiboronic ester with para-substituted bromobenzene while Novak et al. (10) did so by reacting the ethyleneglycol diester of 4,4'-diphenylenebis(boronic acid) with 4,4'-dibromobiphenyl. The formation of high molecular weight polymer was claimed in the latter case. In both cases the polymers were substituted to give them solubility, an undesirable property for our purposes. The preparation of unsubstituted, high molecular weight poly(phenylene) by coupling reactions is not possible because insolubility causes a precipitation of the resulting poly(phenylene). This behavior was shown earlier by Yamamoto et al. (11) who prepared poly(phenylene) by coupling 1,4-dibromobenzene with magnesium. Grubbs et al. has used a similar monomer to the one we are employing to prepare linear poly-(phenylene) with a very high para-content (12). Assuming that direct transformation to poly(phenylene) could be achieved in this case, the trimethylsilyl group is 75% heavier than the acetate group, which would cause an even greater weight loss on transformation. In addition, with the high para-content, substantial crystallinity would be expected, which would also be undesirable for use of this polymer as a waveguide.

Advantages of the Precursor Polymer, PDAC. The route for the preparation of poly(phenylene) via the precursor polymer PDAC (equation 4 above), has several advantages compared with the direct synthesis of poly(phenylene), including high molecular weight and compatibility with existing microelectronic fabrication technology. This compatibility is based on the fact that PDAC is soluble and therefore spin-coatable, whereas the resulting poly(phenylene) is nearly insoluble in any solvent. The resulting precursor polymer also has a high content of 1,2-substitution which results in a rather disordered polymer chain. Crystallinity which might tend to scatter light in the waveguide is therefore absent.

As shown in equation 5 the precursor can be thermally transformed or "cured" into poly(phenylene) at temperatures greater than 250°C.

$$\xrightarrow[\text{-2 CH}_3\text{COOH}]{\text{T>250°C}} \qquad (5)$$

Besides heat treatment, it is also possible to transform the precursor into poly(phenylene) through exposure to acid (see equation 6). In our laboratory we have used both sulfonic acids and photoacid generators (PAG), including the triphenylsulfonium salts popularized by Crivello (*13*). The use of a PAG makes lithographic patterning possible (*14-15*) and we are undertaking this area of research with the hope of fabricating waveguides and patterned, low dielectric thin film structures.

$$\xrightarrow[\text{-2 CH}_3\text{COOH}]{\substack{\text{H}^+ \text{ or} \\ \text{PAG and h*v}}} \qquad (6)$$

Theoretical Improvement of Physical Properties of Poly(phenylene). In order to lower the dielectric constant of poly(phenylene) and at the same time adjust the refractive index to more closely match industry standard silica optical fibers, we have explored methods for breaking up the conjugation length of the condensed aromatic rings. We reasoned that such breakage should decrease both the RI and the dielectric constant. To introduce a "break", we have previously explored several styrene-based and maleimide-based monomers. Both copolymers with DAC and the thermally converted copolymers had a lower refractive index (RI= 1.64 - 1.68) than the pure poly(phenylene), which has a RI of 1.68 - 1.72 (16).

This advantage over poly(phenylene) was offset with several disadvantages. Because of the low thermal stability of styrene it was not possible to convert the precursor copolymer of DAC/styrene without decomposition (16). The transformation of the precursor DAC/styrene copolymer into poly(phenylene) was therefore performed with the aid of acid, which was generated by irradiation in the presence of triphenylsulfonium hexafluoroarsenate. These conditions are necessary because thermal transformation of the cyclohexylene units into phenylene units occurred at temperatures above 280°C. When these temperatures were applied to the styrene/DAC copolymer the transparent, uncured polymer film became black and brittle.

One of the attractive features of poly(phenylene) is that it has a moisture uptake of only 0.9 wt% as measured using Forward Recoil Spectrometry. This uptake is due to physical adsorption of water with the polymer and is not due to a chemical reaction. For a poly(imide), such as PMDA-ODA, under similar conditions, water uptake of over 3.0 wt-% has been measured. This uptake can be coupled with a change in the structure of the poly(imide). Specifically the poly(imide) can react with water to form poly(amic acid). Besides this structural change, physical properties such as dielectric constant and therefore the RI can also be affected. A change in the chemical structure can be combined with a change in optical absorption and RI characteristics (coupled with loss of light transmission). For these reasons, further consideration was given to vinyl monomers, since the absence of hydrolyzable groups can render the copolymers more environmentally stable.

As a result of the reported high thermal stability of homopolymers of N-vinylcarbazole (VIC) and 4-vinylbiphenyl (VBP) (18-19), their monomers were interesting candidates for copolymerization with DAC. These copolymers were expected to have several advantages when compared to pure poly(phenylene) based on their possible use as waveguide materials. As mentioned earlier the introduction of the vinyl monomers should reduce the conjugation length, which should in turn result in a lowering of the dielectric constant and the RI of the polymers.

A thermally stable comonomer which does not undergo a conversion reaction also offers another advantage over poly(phenylene) homopolymer. For the complete transformation of PDAC into poly(phenylene) the weight loss is 60.2%, due to the loss of two units of acetic acid per ring. With the replacement of each cyclohexene diacetate unit by a non-degradable comonomer the weight loss and the sample shrinkage of the polymer is reduced.

Experimental

Analytical Methods. Gel Permeation Chromatography (GPC) measurements were performed on a Waters chromatograph system with a 712 WISP injector. Monodisperse polystyrene standards were used for calibration. The detection was performed at UV wavelengths with a Waters 410 Differential Detector. Mass loss determination using thermogravimetric analysis (TGA) was measured with a DuPont thermogravimetric analyzer. Heating rates of 10°C/min were used. For the IR-spectra sodium chloride wafers were coated with a thin film of the polymer and measured using a Galaxy 2020 FTIR spectrophotometer. The absorption spectra of the polymers films, which were spun on quartz wafers, were recorded in the near IR range with a Varian Cary 5 spectrophotometer and in the ultraviolet/visible range with a Perkin Elmer Lambda 4 spectrophotometer. ^{1}H-NMR spectra were measured in deuterochloroform and recorded with a Varian XL200 instrument. Ellipsometry was performed with a Rudolf Research/Auto E L instrument with samples coated on silicon wafers.

For spin casting we prepared a 20% polymer solution in diglyme (2-methoxyethyl ether).

Materials used. 1,3-Cyclohexadiene-5,6-diol (Sigma), diglyme (Aldrich), and deuterochloroform (Aldrich) were used as received. Dichloromethane (Aldrich) was extracted with concentrated sulfuric acid, sodium bicarbonate solution, water, and saturated sodium chloride solution. Dichloromethane was then predried with magnesium sulfate, distilled from calcium hydride, and stored in a brown glass bottle. AIBN, 4-vinylbiphenyl (Aldrich), and N-vinylcarbazole (Aldrich) were recrystallized from ethanol (Aldrich) and dried overnight in vacuum at room temperature. Pyridine (Aldrich) was dried for one night over magnesium sulfate.

Synthesis of 1,3-cyclohexadiene-5,6-bisacetate (DAC). 1,3-Cyclohexadiene-5,6-diol (15 g = 133.8 mmol) was placed in 230 mL of dichloromethane and 70 mL of pyridine in a three neck flask. The flask was equipped with a stirring bar, a condenser, and a dropping funnel with a drying tube (calcium chloride). Acetyl chloride (25 mL = 352 mmol; Aldrich) was placed with 70 mL of dichloromethane in the dropping funnel. After all of the acid chloride solution had been added, the mixture was allowed to stir for another 2 days. Then 40 mL of water were added and the mixture was poured into 10% hydrochloric acid solution. After separation of the organic phase, the aqueous layer was extracted with dichloromethane. The combined organic phases were washed with water, sodium carbonate solution, water, and saturated sodium chloride solution before drying over magnesium sulfate. After evaporation of the solvent, a dark orange oil (26.52 g) resulted. The purification of the crude product was performed by column chromatography (100 g silica gel mesh 60). We used a mixture of hexane/diethyl ether (ratio : 60/40) as eluent. The equal fractions were combined and dried over magnesium sulfate. After evaporation of the solvent the resulting product was a yellow oil (23.47 g = 89.4%) that was fractionally distilled (4 fractions) under vacuum. All fractions were colorless and transparent after distillation. Using ^{1}H-NMR spectroscopy the first two fractions were identified as a mixture of phenylacetate and DAC and fraction three and four (yield of fraction three and four together : 8.9 g = 33.9%) were 100% DAC. Fraction one and two changed their color after several days to yellow whereas fraction three and four remained colorless. The formation of phenyl acetate occured during distillation, due to decarboxylation of one acetate group forming acetic acid, see equation 7.

(7)

Polymerization of DAC. The polymerization of DAC was performed both in bulk and in solution. In the latter case we used dichloromethane as solvent. The effect of the solvent was a decrease of the degree of polymerization without an increase in yield. To polymerize DAC with a small amount of AIBN (1 wt% of AIBN in DAC) we froze and degassed the mixture for 6-7 times to avoid the presence of oxygen during polymerization. The glass tube was then closed tight with a threaded Teflon stopcock and placed into an oil bath at 60-70°C. After 2 days the polymer was dissolved in dichloromethane and the polymerization was precipitated into hexane which contained a small amount of methanol. After drying for one day in vacuum at room temperature the yield was calculated (see Table I).

Polymerization of VBP. The polymerization of VBP (mp = 119-121°C) was performed in solution using dichloromethane as solvent. The glass tube used was closed tightly with a threaded Teflon stopcock after adding VBP, dichloromethane, and AIBN (AIBN : VBP = 1 : 100). After freezing and degassing (6-7 times) the tube was placed into an oil bath at 60-70°C. After 24 h the polymer block was dissolved in dichloromethane and precipitated into methanol. The yield was calculated after drying for one day in vacuum at room temperature.

Polymerization of VIC. The polymerization of VIC (mp = 64-66°C) with AIBN (1 wt% of AIBN related to VIC) in solution was performed as described for the polymerization of VBP.

Copolymerization of DAC with VBP. The copolymerization of DAC with VBP was performed with 1 wt% of AIBN both in bulk (possible if content of VBP is 20 wt% and lower) and in solution (dichloromethane). The content of VBP was chosen as 5, 10, 15 and 20 wt% of the total monomer weight. After freezing and degassing (6-7 times) the glass tube was closed tight with a threaded Teflon stopcock. The workup was identical with the one of PDAC.

Copolymerization of DAC with VIC. The copolymerization of DAC with VIC (content of VIC : 5, 10, 15 and 20 wt%) was performed as described in the copolymerization of DAC with VBP.

Results and Discussion

In order to monitor the transformation of poly(phenylene) precursor (PDAC) into poly(phenylene), IR and UV/VIS spectroscopy were used. For analysis, it was necessary to prepare a thin polymer film which was spun from a 15-20 wt% polymer solution in diglyme after filtration through a 0.5 micron membrane filter. The thickness of the precursor polymer film, which was adjusted by concentration and spin speed (1200-9000 rpm), was measured to be between 1.2 and 0.4 μm. The transformation of precursor PDAC into poly(phenylene) was carried out by heat treatment at temperatures between 300°C and 350°C and monitored by IR-spectroscopy. The advantage of using IR-spectroscopy for our purpose is that it is an easy way to detect the transformation of PDAC into poly(phenylene).

The IR-spectra of PDAC and the poly(phenylene) are shown in Figure 1. The presence of PDAC can be identified by its absorption bands at 1744 cm^{-1} and at 1370 cm^{-1}. The peak at 1744 cm^{-1} is caused by the vibration of the carbonyl group (-COOCH$_3$) present on the cyclohexene repeating units of PDAC and the latter is due to the vibration of the carboxylate unit (-COO-). Both characteristic peaks for poly(phenylene) are located at 1485 cm^{-1} and at 812 cm^{-1}. The first is due to the benzene ring, which is formed after splitting off acetate groups as acetic acid (see equation 5). With each loss of a molecule of acetic acid a double bond is formed. If the second acetate group of the ring system is volatilized, the substituted phenylene ring thus formed is located in the main chain of the polymer.

The incorporation of the cyclohexadiene ring of DAC in the polymer chain, and therefore the symmetry of poly(phenylene), is dependent on the number of participating double bonds. If only one of

Table I. Molecular Weights and Polydispersity for Homopolymers Prepared for this Study

Monomer [a]	Solvent [b]	Yield after 2 days [%]	M_n	M_w	P.D.
DAC (c)	none	46	20.6 K	34.4 K	1.68
DAC (c)	Dichloromethane (u)	10	3.7 K	5.4 K	1.70
DAC (c)	Dichloromethane (p)	30	4.7 K	12.9 K	2.77
DAC (d)	none	70	48.6 K	82.3 K	1.70
VBP (r)	Dichloromethane (p)	91.5	16.3 K	82.9 K	5.08
VIC (r)	Dichloromethane (p)	71	72.5 K	170 K	2.34

[a] Monomers were purified by column chromatography (c), distillation (d), or recrystallization (r).
[b] Dichloromethane was used either unpurified (u) or purified (p).

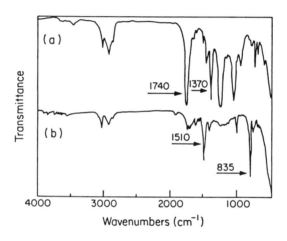

Figure 1. IR-spectra of PDAC recorded before (a) and after (b) heat treatment (T=300°C for 6h).

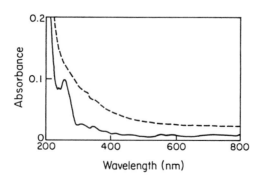

Figure 2. Absorption spectra of PDAC recorded before (———) and after (- - -) dipping in methanesulfonic acid.

the two double bonds of DAC participate in the polymerization 1,2-poly-(5,6-bisacetoxy)cyclohexa-3-ene (1,2-PDAC) is formed. 1,2-PDAC is the precursor for poly(*ortho*-phenylene) (see equation 8).

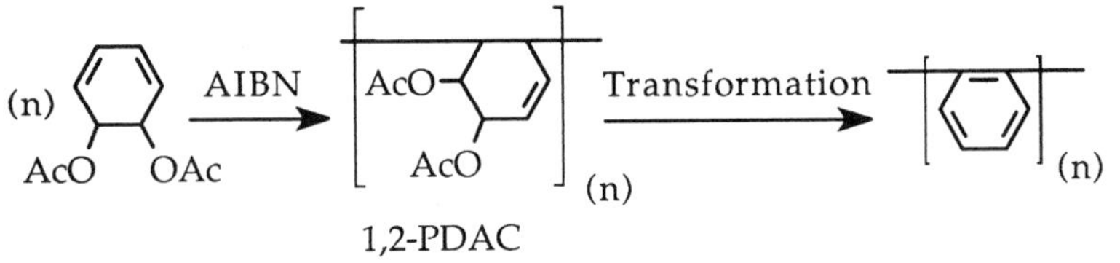

1,2-PDAC

$$(8)$$

With the participation of both double bonds of DAC, 1,4-disubstituted poly(phenylene) is formed after its transformation from 1,4-poly(5,6-bisacetoxy)cyclohexa-2-ene (1,4-PDAC), see equation 9.

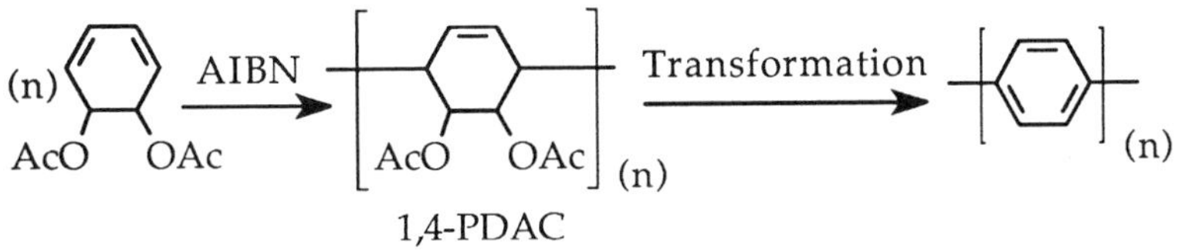

1,4-PDAC

$$(9)$$

Investigations with IR Spectroscopy and GPC. The polymerization mechanism can also be characterized by IR spectroscopy. The IR-spectra in Figure 1 has a strong band at 812 cm^{-1} and a weak band at 760/700 cm^{-1}. The former band is due to the *para*-substituted phenylene and the latter one is caused by an *ortho*-substituted phenylene ring. Therefore it is possible to make a qualitative estimation of the amount of *para*-substituted benzene rings in the poly(phenylene) just by measuring the IR-spectra. In the case shown, the amount of 1,2-substituted poly(phenylene) is about 15% in agreement with earlier ^{1}H-NMR studies (1). UV/VIS spectroscopy was also used to monitor the conversion of PDAC into poly(phenylene), see Figure 2.

The molecular weights of all polymers (homopolymers and copolymers) were investigated with GPC (see Table I). Table I clearly indicates that the highest molecular weights were generally reached with bulk polymerization. Unexpectedly we also obtained the smallest distribution of molecular weight when polymerization was performed in bulk, which is not generally true in radical polymerizations.

Physical properties of the homopolymers. In order to determine the thermal stability of all homopolymers, TGA plots were recorded. The TGA data indicates that PDAC shows a significant decrease in weight at temperatures higher than 280°C, whereas poly(vinylbiphenyl) (PVBP) and poly(vinylcarbazole) (PVIC) show a weight loss at ca. 420°C and 460°C respectively (see Figure 3). The lower nominal thermal stability of

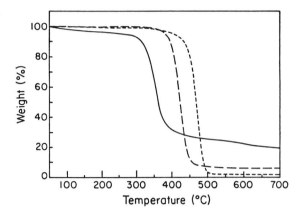

Figure 3. TGA data of PDAC (——), PVBP (— —), and PVIC (- - -).
Heating rate : 10°C/min, nitrogen flow rate : 40 ml/min.

PDAC is due to the transformation of PDAC into poly(phenylene). This transformation is accomplished by the loss of two molecules of acetic acid which creates a theoretical loss of 60.2%. In the TGA-curve of PDAC the loss starting at temperatures higher than 280°C levels off at about 35%. The vinyl polymers degraded at higher temperatures and the weight loss was nearly complete.

The refractive indices of the homopolymers prepared for this study and of some other homopolymers are shown in Table II. Measurement of RI is instructive because of its relationship to the dielectric constant as outlined before in equation (1). The monomers also have relatively simple structures and possess no near-IR absorbing hydroxyl functions, for example, and so would be expected to be relatively near-IR transparent. Near-IR spectra of the homopolymer films and of converted PDAC films are shown in Figure 4a - 4c.

Homopolymers of DAC, VBP and VIC possess a higher thermal stability than poly(styrene). The high thermal stability and the lower RI of PVBP and PVIC as well as near-IR transparency compared to poly(phenylene) suggested that their monomers would be suitable for copolymerization with DAC.

Copolymerization of DAC with VBP. The content of vinyl monomer was measured by both UV/VIS and ^{1}H-NMR-spectroscopy. Besides the detection of the transformation from PDAC into poly(phenylene) described above, UV/VIS spectroscopy allows us to qualitatively detect whether the copolymer includes the desired vinyl monomer (for the UV/VIS spectrum of VBP see Figure 5b). Absorption spectra of PDAC and of the DAC/VBP copolymers are shown in Figure 5. The copolymers (see Figure 5c and 5d) have a strong absorption band at approximately 280 nm where the PDAC homopolymer (see Figure 5a) is nearly transparent. With the incorporation of a higher content of VBP into the copolymer, the absorption at 280 nm increases. The UV/VIS spectrum of PVBP homopolymer has a very strong absorption at this wavelength.

Table II. Refractive Index and Dielectric Constant of Selected Polymers

Polymer	RI	Dielectric constant
PDAC	1.49-1.51	-
Poly(phenylene)	1.68-1.72	3.6
PVBP	1.62-1.64	-
PVIC	1.69	-
Poly(imide) (20), (PMDA-ODA)	-	3.2
(6FCDA-TFMB)[a]	-	2.4
Epoxy	1.55-1.60	-
Poly(methyl methacrylate)	1.49	2.6

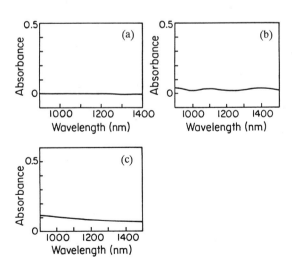

a

(6FCDA-TFMB)

Figure 4. Absorption spectra PVBP (a), PVIC (b), and poly(phenylene) (c).

We also investigated the copolymer of DAC and VBP with ^{1}H-NMR-spectroscopy. Because the aromatic signals of PVBP (see Figure 6b; δ= 7.0-7.6 ppm, multiplet, 7 H) and the 4 vinylic signals of PDAC (see Figure 6a; δ= 4.8-6.1 ppm, multiplet, 4 H) are totally separated, the amount of incorporated VBP could be calculated. The content of VBP in the copolymer (see Figure 6c) was approximately twice as high as in the monomer feed indicating that VBP was used up much faster than DAC*. After determining that DAC forms a copolymer in the presence of VBP we established whether the content of vinyl monomer influenced the aromatization of the cyclohexene bisacetate units. This was measured by both IR (see Figure 7) and UV/VIS spectroscopy (see Figure 5c and 5d). Figure 7 indicates that heat treatment will successfully cause the aromatization of cyclohexene groups in the precursor copolymer. The IR-spectra of the unannealed and the annealed copolymer of DAC and VBP are shown in Figure 7a and 7b respectively. The peaks at 1744 cm^{-1} and at 1370 cm^{-1} were observed for the DAC/VBP copolymer whereas the spectrum for the poly(phenylene) containing copolymer shows characteristic peaks at 1483 cm^{-1} and at 810 cm^{-1}.

Even though the VBP-unit contains two phenylene rings with one 1,4-disubstituted benzene ring, the absorptions at 1483 cm^{-1} and at 810 cm^{-1} are weak because of the low content of VBP in the copolymer. The higher content of the acetate and the carboxylic groups are stronger features due to the PDAC unit. These groups can only be detected in the unannealed copolymer.

The UV/VIS spectrum of the DAC/VBP copolymers shows a lower absorption above 300 nm than the PDAC homopolymer which was treated with methanesulfonic acid. The different content of DAC in the copolymer is responsible for the difference in absorption after acid treatment (see Figure 5c and 5d). The control experiment with PVBP shows that acid treatment did not influence the absorption spectrum, so that the observed increase in absorption of the copolymer is caused exclusivly by aromatization (see Figure 5b).

Copolymerization of DAC with VIC. For the polymerization of DAC in the presence of VIC, we found that both monomers were incorporated into a copolymer. The VIC incorporated within this copolymer was detected by ^{1}H-NMR spectroscopy. In contrast to the vinyl monomer content within the DAC/VBP copolymer, the content of the vinyl monomer in the DAC/VIC copolymer is drastically reduced. This

---

*For a content of 19.8% VBP in the monomer mixture we got a ratio of aromatic protons at δ= 6.5-7.8 ppm to aliphatic protons at δ= 4.3-6.0 ppm as 75 : 58. This indicates a content of 36.5% VBP in the copolymer. The extent of reaction for this particular polymerization was 41.8%. For a monomer mixture with 15.2% VBP and an extent of reaction of 69.4% we found 25.4% VBP in the polymer.

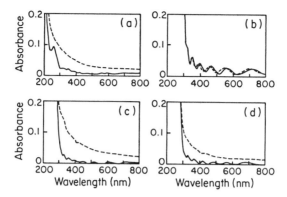

Figure 5. Absorption spectra of PDAC (a), PVBP (b), and DAC/VBP copolymers with copolymer compositions of (90% DAC/10% VBP) (c) and (70% DAC/30% VBP) (d). Spectrum recorded before (———) and after (- - -) dipping in methanesulfonic acid.

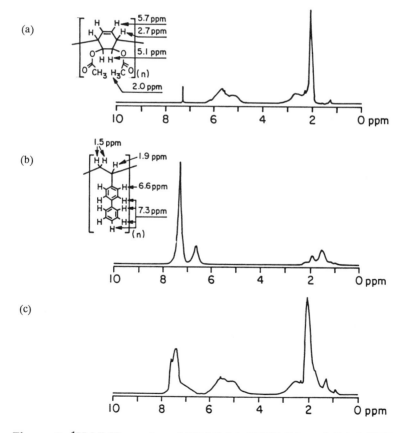

Figure 6. ^{1}H-NMR spectra of PDAC (a), PVBP (b), and DAC/VBP copolymer with a composition of (60% DAC/40% VBP) (c).

comparison is based on 4 different copolymers that were prepared from identical ratios of DAC to vinyl monomer. As can be seen in Table III the monomer ratio of DAC/VIC is directly proportional to the content of VIC in the copolymer. Neither the polydispersity (P.D.) nor the average molecular weights increase with increasing content of VIC.

Table III. Molecular Weight Averages and Polydispersity for PVIC and DAC/VIC Copolymers of Different Monomer Feed Ratio

| Monomer in monomer feed | | Yield[a] | M_n | M_w | P.D. | VIC content in copolymer |
DAC [%]	VIC [%]	[%]				[%]
0	100	70.5	75 500	169 600	2.3	100
80.1	19.9	51.3	33 900	57 700	1.7	6
85.2	14.8	52.1	38 000	61 800	1.6	3
89.6	10.4	53.3	34 100	58 400	1.7	1.7
94.8	5.2	48	35 700	59 200	1.7	0.8

[a]Initiation with AIBN (t= 44 h; T= 65°C; (wt(Monomer)/wt(AIBN)= 100/1).

Physical Properties of Copolymers. In order to check the thermal stability of the copolymers TGA graphs were recorded (see Figure 8). The DAC/VBP and DAC/PVIC copolymers show the same large decrease above 280°C as PDAC does itself. The weight loss of the DAC/VBP copolymer starts later and stops at 370°C. Further heating causes decomposition which stops at a level that is lower than that of pure PDAC. This indicates that some degradation does occur. The difference in the magnitude of the steady state value is due to the content of VBP in the copolymer which is 40% for the DAC/VBP copolymer shown in Figure 8. The decomposition is therefore directly related to the lower thermal stability of VBP as can be seen from Figure 3. The TGA curve of the DAC/VBP copolymer (see Figure 8) as well as the one of PVBP (see Figure 3) shows that the residue weight is about 7%, for both polymers. In contrast the residue weight of polyphenylene is about 35% (see Figure 3). The thermal behavior of the "DAC/VIC copolymer" is nearly identical with that of pure PDAC, because VIC is only introduced to a very small extent. The thermal stability of the copolymer is therefore largely a function of PDAC.

The RI values of the copolymers and PDAC, as measured from thin films using ellipsometry, are shown in Table IV for both precursors and thermally treated polymers.

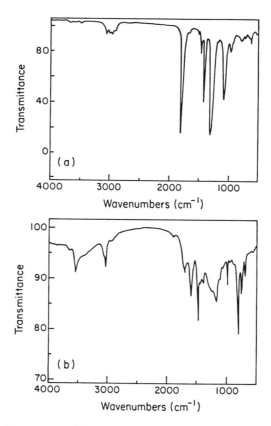

Figure 7. IR-spectra of the DAC/VBP copolymer with a copolymer composition of (90% DAC/10% VBP); recorded before (a) and after (b) heat treatment (T=300°C for 6h).

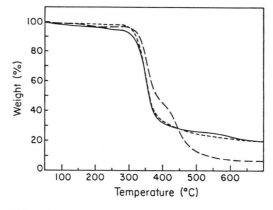

Figure 8. TGA data of PDAC (———) and DAC/VBP copolymers with compositions of (60% DAC/40% VBP) (— —) and (70% DAC/30% VBP) (- - -).

Table IV. RI of DAC/VBP Copolymers, Poly(phenylene), and PVBP

Polymer	Monomer [a] DAC	VBP	RI before transformation	RI after
PDAC	100 (100)	0 (0)	1.49	-
Poly(phenylene)	100 (100)	0 (0)	-	1.68-1.72
P(DAC/VBP)	95 (90)	5 (10)	1.47-1.52	1.84-1.85
P(DAC/VBP)	90 (80)	10 (20)	1.52-1.54	-
P(DAC/VBP)	85 (70)	15 (30)	1.50-1.56	1.9-2.0
P(DAC/VBP)	80 (60)	20 (40)	1.53-1.57	-
PVBP	0 (0)	100 (100)	1.62-1.64	-

[a]Monomer in monomer feed (copolymer composition)

Further research will involve incorporation of different comonomers and studies of the effect of annealing conditions. The requirements for the new comonomers will be higher thermal stability and continuing reduction in the conjugation length of the aromatized copolymer.

Acknowledgment

The authors would like to thank the Air Force Office of Sponsored Research for partial support of this work. Support by the Industry-Cornell Alliance For Electronic Packaging is also appreciated. Finally, the use of the facilities of both the Cornell Materials Science Center and the National Nanofabrication Facility at Cornell is acknowledged.

Literature Cited

(1). Ballard, D. G. H.; Courtis, A.; Shirley, I. M.; Taylor, S. C. *Macromolecules* **1988**, *21*, 294.

(2). Mates, T.; Ober, C. K.; Angelopous, B.; Martin, H. Mat. Res. Soc. Symp. Proc. 1990, 123, 167.

(3). Matsuura, T.; Yamada, N.; Nishi, S.; Hasuda, Y. *Macromolecules* **1993**, *26*, 419.

(4). Bidar, H.; Fabre, C.; Geniès, E. M. *New J. Chem.* **1987**, *11*, 721.

(5). Hara, S.; Mitani, M.; Toshima, N. *J. Macromol. Sci.-Chem.* **1990**, *A27*, 1431.

(6). Yamamoto, K.; Nishide, H.; Tsuchida, E. *Polym. Bull.* **1987**, *17*, 163.

(7). Chaturvedi, V.; Tanaka, S.; Kaeriyama, K. *Macromolecules* **1993**, *26*, 2607.

(8). Tour, J. M.; Stephens, E. B. *J. Am. Chem. Soc.* **1991**, *113*, 2309.

(9). Kallitsis, J. K. ; Rehahn, M.; Wegner, G. *Makromol. Chem.* **1992**, *193*, 1021.

(10). Wallow, T. I.; Novak, B. M. *J. Am. Chem. Soc.* **1991**, *113*, 7411.

(11). Yamamoto, T.; Hayashi, Y.; Yamamoyo, Y. *Bull. Chem. Soc. Jpn.* **1978**, *51*, 2091.

(12). Gin, D. L.; Conticello, V. P.; Grubbs, R. H. *J. Am. Chem. Soc.* **1992**, *114*, 3167.

(13). Crivello, J. P.; Lam, J. H. W. *J. Polym. Sci., Polym. Lett. Ed.* **1979**, *17*, 759.

(14). Ito, H.; Ueda, M. *Macromolecules* **1990**, *23*, 2885.

(15). Ito, H.; Ueda, M.; Ito, T. *J. Photopolym. Sci. Technol.* **1990**, *3*, 335.

(16). Kim, H. K. Ober, C. K. *Polym. Bull.* **1992**, *28*, 33.

(17). Zhong, X. F.; Francois, B. Makromol. Chem., Rapid Commum. **1988**, *9*, 411.

(18). Chu, J. Y. C.; Stolka, M. *J. Polym. Sci.: Polym. Chem.* **1975**, *13*, 2867.

(19). Pearson, J. M.; Stolka, M. *Poly(N-vinylcarbazole)*, Gordon and Breach Science Publishers, New York, NY, 1981.

(20). Feiring, A. E.; Auman B. C.; Wonchoba, E. W. *Macromolecules* **1993**, *26*, 2779.

RECEIVED September 13, 1994

Chapter 24

Charge-Carrier Generation and Migration in a Polydiacetylene Compound

Studied by Pulse Radiolysis Time-Resolved Microwave Conductivity

G. P. van der Laan[1], M. P. de Haas[1], J. M. Warman[1], D. M. de Leeuw[2], and J. Tsibouklis[3]

[1]Radiation Chemistry Department, Interfaculty Reactor Institute, Delft University of Technology, Mekelweg 15, 2629 JB Delft, Netherlands
[2]Philips Research Laboratories, P.O. Box 80000, 5600 JA Eindhoven, Netherlands
[3]Interdisciplinary Research Center, University of Durham, South Road, Durham DH1 3LE, United Kingdom

The long-lived transient conductivity change resulting from nanosecond pulsed irradiation of (5,7-dodecadiyn-1,12-diol-bis(n-butoxy-carbonyl-methylurethane), 4BCMU, has been studied using the time-resolved microwave conductivity (TRMC) technique. The polymerization of the 4BCMU upon irradiation results in an increase of the end-of-pulse conductivity with increasing dose up to approximately 10 kGy. The observed radiation-induced conductivity leads to a lower limit of 5×10^{-4} m^2/Vs for the sum of the charge carrier mobilities. At higher accumulated dose the conductivity decreases continuously up to the maximum total dose given of 1 MGy. These results are explained in terms of long chain length polymerization being counteracted at high accumulated dose by radiation-induced damage.

Polydiacetylenes (PDA) constitute a class of unusual polymeric materials, since perfect crystals can be prepared by solid-state topological polymerization (*1,2*). 4BCMU is one of the most widely studied diacetylenes and can be polymerized by UV or γ–irradiation (*3*). The C1 and C4 atoms of the adjacent 4BCMU monomers link up to form fully conjugated polydiacetylene chains resulting in high molecular weight polydiacetylene, P4BCMU, (figure 1) even at low monomer conversion (chain length of ca. 2400) (*4,5*). Hydrogen bonds are present in both the monomer and polymer. In the present work polymerization is induced by 3-MeV electrons from a van de Graaff accelerator.

The backbone structure in the single crystal of the polydiacetylenes provides an ideal conjugated system for charge transport, as is indicated by the rather high charge carrier mobilities determined (*6-13*). Most conductivity studies report values between 5×10^{-4} m^2/Vs and 500×10^{-4} m^2/Vs (*6,8-13*) observed in polydiacetylene PDATS (the polymer ofbis(p-toluene sulphonate) ester of 2,4 hexadiyne-1,6 diol). The large spread

0097–6156/94/0579–0316$08.00/0

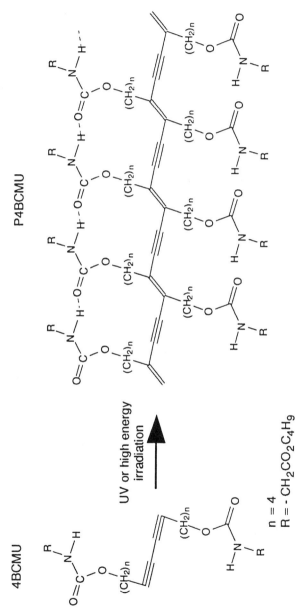

Figure 1. The structure of 4BCMU before and after polymerization.

in the mobility values is due to the as yet unsolved problem of the field dependence of the charge carrier drift velocity. Even a much higher mobility value of 20 m^2/Vs has been reported by Donovan and Wilson (7). This controversial value is based on the conclusion that even at low electric field strength the charge carrier velocity is saturated at the speed of sound. In recent articles of Bässler (9), Fisher et al. (11), and Yang et al. (12) the mobility measurements are reviewed and it is generally concluded that a mobility of no higher than ca. 10^{-2} m^2/Vs is needed to explain the experimental results.

Conductivity measurements in polymeric materials are often complicated by micro-heterogeneities within the sample, chain end effects, field induced space charge problems, saturation of the drift mobility, and the need of good electrode contacts. Recently Blum and Bässler (13) discussed the problems associated with conductivity measurements in crystalline polydiacetylenes caused by charge carriers that are blocked when reaching the chain ends. They concluded that a DC conductivity study was rather a study of chain to chain hopping and suggested that an AC conductivity experiment on a GHz scale was needed to study the charge transport process along the backbone itself.

In this work we present results obtained using such an ultra high frequency AC method, the Time-Resolved Microwave Conductivity (TRMC) technique (14-18). The charge carriers are produced by pulses of 3 MeV electrons and are distributed close to uniformly throughout the sample up to a thickness of several millimeters. Microwaves are then used to monitor the change in conductivity (dielectric loss) of the medium resulting from the formation of mobile charge carriers. The decay kinetics of the radiation-induced conductivity transients are studied on the timescale of nanoseconds to milliseconds.

In an attempt to obtain information on the mobility of the charge carriers along individual polydiacetylene backbones, we have studied partially polymerized 4BCMU. The structure of the polydiacetylene backbone of P4BCMU is similar to that of PDATS. However, the length of the main chain may be an order of magnitude larger (3,19).

It was reported that 100 % crystalline polymers are radiation resistant (20 and references therein), i.e. they undergo very little crosslinking or degradation. Some degradation has been reported by prolonged exposure of P4BCMU to ^{60}Co γ–rays (21) or to intense e-beams of 20 keV electrons (22). We have studied the effect of accumulated irradiation dose on its conductive properties. Charge transport is found to be rather sensitive to radiation-induced defects.

This study indicates that the TRMC method can be a valuable additional tool in the study of solid-state polymerization of diacetylenes. This application of TRMC will be discussed in a future publication in more detail.

Experimental

Materials

Monomer. The 4BCMU monomer was synthesized as follows. 5,7-Dodecadiyn-1,12-diol (2.425 g; 0.0125 mol) was dissolved in 30 ml of dry tetrahydrofuran (THF). Once the thiol had dissolved completely, triethylamine (0.5 ml; dried over KOH) and dibutyltin bis(2-ethylhexanoate) (0.05 g) were added with vigorous stirring followed by the dropwise addition of a solution of butyl isocyanoate (4.325 g; 0.0275 mol) in dry THF (20 ml) over 30 minutes. The temperature of the reaction mixture was not allowed to rise above 25 °C during the addition. The solution was stirred for another 3 hrs and added dropwise to 375 ml of petroleum ether (40-60) °C from which the crude 4BCMU precipitated as a white solid and was isolated by filtration. The product was purified by repeated recrystallizations from THF/petroleum ether (40-60) °C.

Polymerization. All 4BCMU samples were irradiated within a microwave cell under identical irradiation conditions. The microwave cell consisted of a 14 mm length of rectangular (7.1x3.55 mm^2) copper waveguide closed at one end with a metal plate and flanged at the other. For the conversion analysis 170 mg of 4BCMU powder was compressed in a cell, filling it for a length of 7 mm. For the conductivity measurements 35 mg of 4BCMU powder was compressed into a rectangular shaped cavity of 2x6x3 mm^3 (length x width x depth) dimension in a perspex block of 20x7.1x3.5 mm^3. Perspex itself is known to give no TRMC transients upon pulsed irradiation. The filled perspex block was inserted into the microwave cell. The samples were not subjected to any evacuation procedures and were thus "air saturated".

The crystalline monomer was gradually polymerized using trains of 0.5 ns, 4 A pulses of 3 MeV electrons from a Van de Graaff accelerator at a rate of 2 pulses per second, depositing ca. 1 Gy (=1 J/kg) per pulse. Above an accumulated dose of 1 kGy and 10 kGy, 2 ns and 5 ns pulses were used, respectively.

The monomer to polymer conversion upon irradiation was followed by removing a few mg material from the microwave cell after several levels of accumulated dose and determining the maximum of the polymer UV-absorption peak of the irradiated 4BCMU dissolved in chloroform. The extinction coefficient of the polymer absorption in chloroform is 17,500 dm^{3}mol^{-1}cm^{-1} at 468 nm (*23*).

Conductivity measurements. For the conductivity measurements the same sample was used for the whole dose range. At various intervals of total accumulated dose the sample was irradiated with 0.5 or 2 ns pulses of 3-MeV electrons with beam charges of 2 and 8 nC, respectively. A typical 0.5 ns 4 A pulse deposits a dose of 1 Gy, creating a concentration of charge carrier (electron-hole, e$^-$-h$^+$) pairs less than 6x10^{20} m^{-3} (<1 μM). For dose calculations the density of 4BCMU is taken to be 1 kg/dm^3 so that 1Gy equals 1 kJ/m^3. The total beam charge in the pulse was monitored routinely by deflection onto a 50 Ohm target. The energy deposition was close to uniform throughout the sample and equal to 550 J/m^3 per nC beam charge. A more detailed discussion of the dose deposition in the pulse-radiolysis TRMC setup can be found elsewhere (*25*).

The time resolved microwave conductivity, TRMC, technique. The transient change in conductivity on irradiation was monitored by measuring the change in the microwave power reflected by the sample using the time resolved microwave conductivity, TRMC, technique (figure 2). In the present experiment microwaves in the Ka band (26.5-38 GHz) were used. For small changes the relative change in absorbed microwave power ΔP/P is proportional to the change in conductivity of the sample, Δσ. The quantitative relation between the microwave loss and Δσ was determined using computational and data fitting procedures described previously (*14-18*). The radiation-induced conductivity changes measured in the present study are between 10^{-4} and 10^{-3} S/m.

Changes in the output of the microwave detector diode were monitored using either a Tektronix SCD 1000 digital oscilloscope or a tandem combination: a Tektronix 2205

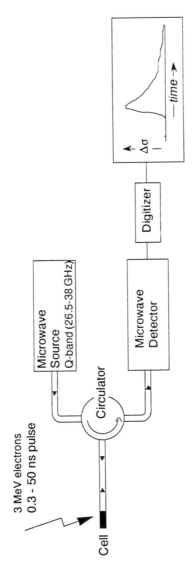

Figure 2. The pulse-radiolysis TRMC technique.

oscilloscope (7A13 plug-in) of which the output is measured on a Sony/Tektronix RTD 710 digitizer. Using the former, the time response is approximately 1 ns. The latter combination has a rise time of 5-10 ns, but is capable of registering the data using a pseudo logarithmic time base. This allowed recording of transient data from 10 ns to 5 milliseconds using a single accelerator pulse. In the subsequent text and Figures the conductivity transients are presented in a dose normalized form, $\Delta\sigma/D$ with D the energy absorbed in J/m^3. The value of $\Delta\sigma/D$ at the end of the pulse is denoted $\Delta\sigma_{eop}/D$. The significance of this parameter is explained in the discussion section.

In order to reduce spurious conductivity signals due to electrons produced in the irradiated air inside the waveguide, the waveguide was flushed with sulfur hexafluoride.

Results

Pulsed irradiation of the 4BCMU at room temperature results in readily measurable microwave conductivity transients. In Figure 3 the conductivity transients are shown for increasing accumulated dose. All transients are much longer lived than the pulse length and have highly disperse decay kinetics. There is no evidence of a short lived conductivity component during the pulse. The transients in Figure 3 show an increase of $\Delta\sigma_{eop}/D$ with increasing the accumulated dose from 0.11 to 0.80 kGy while the decay kinetics of the conductivity are almost unchanged. At higher accumulated dose $\Delta\sigma_{eop}/D$ decreases gradually. At the same time the lifetime of the mobile charge carriers decreases considerably.

In Figure 4 the dependence of $\Delta\sigma_{eop}/D$ on accumulated dose is shown for an accumulated dose increasing from 10 to 10^6 Gy. $\Delta\sigma_{eop}/D$ increases steadily with increasing dose until a maximum is reached at an accumulated dose of ca. 10 kGy. At higher dose it decreases.

In order to relate the conductivity to the amount of polymer present in the sample we determined the monomer conversion at several levels of accumulated dose. In Table I some properties for the 4BCMU irradiated with different levels of accumulated dose are shown. The monomer conversion was similar to that reported for ^{60}Co γ-irradiation in vacuum at the same total dose (23). The values presented for the charge carrier mobility in Table I will be discussed below.

Table I. Properties for 4BCMU irradiated with 3-MeV electrons

Accumulated Dose	$\Delta\sigma_{eop}/D$	monomer conversion P (%)		Charge carrier mobility[a]
(kGy)	(Sm^2/J)	3-MeV electron	^{60}Co γ(23)	$\Sigma\mu_{min}$ $(m^2V^{-1}s^{-1})$
0.11	7×10^{-7}	3.6		5.0×10^{-4}
0.80	18×10^{-7}	14.4		3.1×10^{-4}
10	25×10^{-7}	33	50	1.9×10^{-4}
512	8×10^{-7}	66	66	0.3×10^{-4}

[a] Normalized to the amount of polymer present (see below).

The change in decay kinetics is described in terms of the first half-life of the conductivity transients. In figure 5 the dose dependence of this half-life is shown. Up to an accumulated dose of 1 kGy no effect is observed. At higher doses the lifetime of the mobile charge carriers decreases considerably.

Discussion

In the low accumulated dose range (< 0.1 kGy) the dependence of the conductivity on dose is close to linear (figure 4) and would appear to be governed by the increase in the polymer fraction in the sample. From the radiation-induced conductivity transients in this dose range we can derive a lower limit to the mobility of the mobile charge carriers on a single chain of P4BCMU. At higher accumulated dose the charge transport mechanism is perturbed by radiation-produced defects as is shown by the decrease in both the magnitude and the lifetime of the conductivity transients.

Firstly, we will discuss the conductivity data obtained in the low accumulated dose range and the relation between $\Delta\sigma_{eop}/D$ and the charge carrier mobility. Secondly, we will discuss the high dose range data and its relationship to the radiation-induced damage in P4BCMU. Finally we will comment on the lifetime of the mobile charge carriers.

End-of-pulse conductivity. The radiation-induced conductivity is proportional to the product of the sum of the mobilities of the (positive and negative) charge carriers, $\Sigma\mu$ (m^2/Vs), and the number of mobile electron-hole e$^-$-h$^+$ pairs present per unit volume, N_p (m^{-3}).

$$\Delta\sigma_{eop} = e\ \Sigma\mu\ N_p \tag{1}$$

In equation 1 $\Delta\sigma_{eop}$ is the conductivity at the end of the pulse and e is the elementary charge (1.6x10^{-19} C). With the TRMC technique it is only possible to measure the sum of the mobilities and not the individual contributions. However, using a time of flight method, Fisher (11) has reported that the major charge carrier in PDATS is the electron.

The value of Np in equation 1 is related to the energy deposited per unit volume, D (J/m^3), by

$$N_p = \frac{D}{I_p}\ W\ P \tag{2}$$

where W is the fraction of the initially formed e$^-$-h$^+$ pairs that are still present after the irradiation pulse as mobile charge carriers, P is the fraction of the monomer converted to polymer in the sample and I_p is the energy in Joule required to form one (e$^-$-h$^+$) pair. From equations 1 and 2 $\Sigma\mu$ will be given by

$$\Sigma\mu = \frac{\Delta\sigma_{eop}}{D}\ \frac{I_p}{e\ W\ P} \tag{3}$$

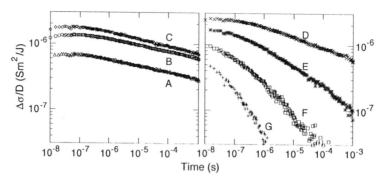

Figure 3. Dose normalized conductivity transients at room temperature for 4BCMU irradiated with 3-MeV electrons up to an accumulated dose in kGy of A: 0.11; B: 0.40; C: 0.80; D: 6.4 ; E: 45 ; F: 224; and G: 512.

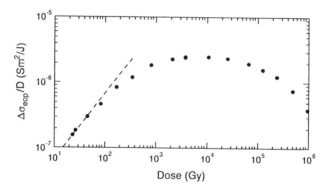

Figure 4. Dose normalized end-of-pulse conductivity for 4BCMU versus accumulated dose. The dashed line corresponds to linear dependence on dose.

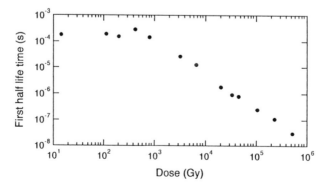

Figure 5. The dependence of the first half-life, $t_{1/2}$, of the microwave conductivity transients on the accumulated dose.

We are only interested in those $e^- \text{-} h^+$ pairs that become localized on the polydiacetylene chain since only these charge carriers will attribute to the observed conductivity. A considerable fraction of the geminate ion pairs may be trapped on the carboxyl or nitrogen groups or geminate recombination may have occured. The nanosecond time resolution of the present detection method is not sufficient to study geminate decay. Therefore we have to make an assumption about the value of W. The situation in which all $e^- \text{-} h^+$ pairs formed survive until the end of the pulse corresponds to W = 1. Using this value an upper limit for Np is obtained, which yields a lower limit to the sum of the mobilities, $\Sigma\mu_{min}$

$$\Sigma\mu_{min} = \frac{\Delta\sigma_{eop}}{D} \frac{I_p}{e\,P} \qquad (4)$$

It is apparent from equation 4 that if I_p is known, $\Sigma\mu_{min}$ can be determined. For organic materials the average energy for pair formation I_p is approximately 4×10^{-18} J (25 eV). Using the experimental value of $\Delta\sigma_{eop}/D$ for 4BCMU equation 4 yields the values of $\Sigma\mu_{min}$ given in Table I. For the lowest accumulated dose of 0.11 kGy the value of $\Sigma\mu_{min}$ is 5×10^{-4} m^2/Vs. This is close to the low field mobility values published by Bässler (9,26), Moses (10), and Fisher (11) for PDATS.

An upper limit for the mobility, $\Sigma\mu_{max}$, can be determined (27) if we use in equation 3 the value for the free ion yield of $W \approx 0.025$, which is a lower limit for organic materials at room temperature (28). This results in $\Sigma\mu_{max} = 2\times10^{-2}$ m^2/Vs, which is still within the range of most measured values with the exception of the estimates of 20 m^2/Vs made by Wilson et al. (7).

Domain boundary effects are avoided in an AC conductivity technique if the extent of the domains, d, fulfills the condition

$$d \gg \mu E/\omega \qquad (5)$$

Where E is the field strength of the applied field and ω is the radian frequency of the probing electric field (ca. 2×10^{11} s^{-1} in the present work). For a mobility of 2×10^{-2} m^2/Vs and a maximum field strength of the microwave field of 10^4 V/m, d must be considerably larger than 1 nm according to equation 5. This is amply fulfilled for P4BCMU for which the chain length is ca. 1000 nm. We conclude that domain boundary problems are absent in the present results. Thus the mobility obtained will be the mobility of charge carriers along the polymer backbone itself.

Dose dependence of $\Delta\sigma_{eop}/D$. The monomer conversion increases with increasing dose over the whole dose range studied. A gradual increase of the conductivity would therefore be expected if the polymer, once formed, is not damaged by irradiation. However figure 4 demonstrates that the conductivity actually decreases significantly at higher levels of accumulated dose. This effect could be caused by either a decrease in mobility or a decrease in the number of charge carriers that become localized on the polymer backbones or both.

It was claimed in previous studies that 100 % crystalline polymers would be radiation resistant (20 and references therein); i.e. they undergo very little crosslinking or degradation. In other studies on P4BCMU (21,22) it was concluded that while the unsaturated backbone remains intact, defects are formed in the side chain regions on exposure to high energy radiation. In a study using high intensity, 20 keV e-beam radiation on thin P4BCMU films (22) changes have been observed in the N-H and C=O bonds of the urethane group. This damage will effect the hydrogen bonding and thus the alignment of the polymer backbone may be perturbed. This kind of radiation-induced effects may reduce the conjugation of the main chain which in turn could reduce the charge carrier mobility. We conclude that in determining the mobility of the charge carriers in P4BCMU one should not use samples that have been irradiated with a dose of more than a few kGy. This is much lower than the dose of 500 kGy normally applied to obtain polymeric material from solid-state polymerization using high energy irradiation.

Some of the chemical species formed in the side chain region upon irradiation may be highly reactive towards the initially formed charge carriers. At high doses the concentration of these trapping sites could be so high that the number of charge carriers that become localized on the polymer backbone decreases, hence leading to a reduction in $\Delta\sigma_{eop}/D$.

Decay kinetics. The conductivity transients observed in P4BCMU are dispersive and very long lived; i.e. with a first half-life of longer than 100 μs at low doses (figure 3). This is much longer than reported from DC conductivity measurements in PDATS (*6-13*), in which the conductivity transients decay within 10-1000 ns. However, in a recent study of Donovan et al. (*29*) also a dispersive, long lifetime has been observed for the light induced charge carriers in PDATS. In this experiment the electric field to collect the light induced free charge carriers was switched on at a time t after the excitation pulse. The charge carriers extracted upon application of the field at time t=100 μs and 10 ms were 60 % and 40 %, respectively, of the amount collected when the field was applied continuously.

The mean square distance, $<x^2>$, that a charge carrier may travel by diffusive motion along a (1-D) chain in a time interval t is given by

$$<x^2> = 2 \mu kT t / e \qquad (6)$$

At room temperature kT equals 0.025 eV. Thus for a mobility of 10^{-3} m^2/Vs it takes 20 ns to reach the end of the 1 μm long chain by diffusion alone. Thus within the halflife of 100 μs the charge carriers will have reached the end of the chain several thousand times. Therefore we conclude that the charge carriers are not preferentially trapped at the chain ends.

To explain the long lifetime in these systems we propose that the mobile charge carriers observed are derived from those e$^-$-h$^+$ pairs for which the electron and hole have escaped very rapid (picosecond time scale) geminate recombination and have become localized on separate conducting main chains (*30,31*) or on scavenging sites such as the urethane or carboxyl groups. While localized on the polymer backbone, the charge carriers are free to undergo rapid, one-dimensional motion along the chain. Charge recombination is, however, retarded by the intervening *n*-butoxy-carbonylmethyl urethane side chains and can only occur via long-distance tunneling of electrons from chain to chain.

Dose dependence of the decay kinetics. In Figure 5 is shown that the half-life of the conductivity transients does not change up to an accumulated dose of ca.1 kGy. At higher accumulated dose the half-life decreases steadily with increasing dose. Some of the chemical species formed in the side chain region upon irradiation may be reactive towards the mobile charge carriers located on the polymer backbones. As shown above the lifetime of the mobile charge carriers is so long that they will travel along the polymer backbone many times before becoming trapped or before recombination occurs with a carrier of opposite charge. Increasing the number of trapping sites by high energy radiation could, therefore, reduce the lifetime, as is observed for a total dose of 1 kGy or more.

We cannot rule out chain scission in the conjugated backbone as a source of the decrease in lifetime. This would contribute to the decrease in the lifetime if the chain breaks were of a free radical character and could act as charge trapping sites.

Conclusions

Using the pulse-radiolysis time-resolved microwave conductivity (PR-TRMC) method to study the low field mobility of the charge carriers in a polydiacetylene chain, a lower limit of 5×10^{-4} m^2/Vs is found for radiation-polymerized 4BCMU at low conversion. This is of the same order of magnitude as earlier published values for thermally polymerized PDATS. An upper limit of the mobility is estimated to be ca. 2×10^{-2} m^2/Vs.

The decrease in the magnitude and decay time of the radiation-induced conductivity transients found for doses above a few kGy is attributed to the radiation damage in the side chain regions, which results in the formation of trapping sites and the possible perturbation of the conjugation of the polymer backbone.

Acknowledgements

The present investigation was supported by the Dutch Ministry of Economic Affairs Innovation-Oriented Research Programme on Polymer Composites and Special polymers (IOP-PCBP). Financial support from the ECC under ESPRIT program 7282 TOPFIT is gratefully acknowledged.

References

(1) Wegner, G. Z. Naturforschung **1969**, *24B*, 824.
(2) Bässler, H. *Adv. in Polym. Sci.* **1984** *63*, 49.
(3) Eckhardt, H.; Prusik, T. Chance, R.R. In *Polydiacetylenes*; Bloor, D.; Chance, R.R., Eds.; Martinus Nijhoff publishers: Dordrecht, The Netherlands, 1985, pp 25-39.
(4) Patel, G.N.; Walsh, E. *J. Polym.Sci.: Polym. Lett. Ed.* **1979**, *17*, 203.
(5) Chance, R.R.; Patel, G.N.; Witt, J.D. *J. Chem. Phys.* **1979**, *71*, 206.
(6) Reimer, B.; Bässler, H. *Chem. Phys. Lett.* **1976**, *43*, 81.
(7) Donovan, K.J.; Wilson, E.G. *J. Phys. C* **1979**, *12*, 4857.
(8) Spannring, W.; Bässler, H. *Chem. Phys. Lett.* **1981**, *84*, 54.
(9) Bässler, H. In *Polydiacetylenes*; Bloor, D.; Chance, R.R., Eds.; Martinus Nijhoff publishers, Dordrecht: The Netherlands, **1985**, pp 135-153.

(10) Moses, D.; Heeger, A.J. *J.Phys.: Condens. Matter* **1989**, *1*, 7395.

(11) Fisher, N.E.; Willock, D.J. *J. Phys.: Condens. Matter* **1992**, *4*, 2517.

(12) Yang, Y.; Lee, J.Y.; Kumar, J.; Jain, A.K.; Tripathy, S.K.; Matsuda, H. ;Okada, S.; Nakanishi, H. *Synth. Met.* **1992**, *49-50*, 439.

(13) Blum, T., Bässler, H. *Chem. Phys.* **1988**, *123*, 431.

(14) Warman, J.M.; de Haas, M.P. In *Pulse Radiolysis of Irradiated Systems*; Tabata, Y., Ed.; CRC: Boca Raton, 1991, pp 101-133.

(15) de Haas, M.P. In *The measurement of electrical conductance in irradiated dielectric liquids with nanosecond time resolution;* PhD Thesis University of Leyden, Delft University Press: Delft, 1977.

(16) Infelta, P.P.; de Haas, M.P.; Warman, J.M. *Radiat. Phys. Chem.* **1977**, *10*, 353.

(17) Warman, J.M. In *The Study of Fast Processes and Transient Species by Electron Pulse Radiolysis*; Baxendale, J.H.; Busi, F., Eds.; Reidel: Dordrecht, 1982, pp 129-161.

(18) Warman, J.M.; de Haas, M.P.; Wentinck, H.M. *Radiat. Phys. Chem.* **1989**, *34*, 581.

(19) Bloor, D. In *Polydiacetylenes*; Bloor, D.; Chance, R.R., Eds.; Martinus Nijhoff publishers: Dordrecht, The Netherlands, 1985, pp1-24.

(20) Patel, G.N.; *Radiat. Phys. Chem.* **1980**, *15*, 637.

(21) Se, K.; Ohnuma, H.; Kotaka, T. *Polymer. J.* **1982**, *14*, 895.

(22) Colton, R.J.; Marrian, C.R.K.; Snow, A.; Dilella, D. *J. Vac. Sci. Technol.* **1987**, *B5*, 1353.

(23) Patel, G.N.; Khanna, Y.P.; Ivory, D.M.; Sowa J.M.; Chance, R.R.; *J.Polym. Sci.: Polym. Phys. Ed.* **1979**, *17*, 899.

(25) Schouten, P.G.; Warman J.M.; de Haas, M.P. *J. Phys. Chem.* **1993**, *97*, 9863.

(26) Blum, T.; Bässler, H. *Phys. Status Solidi (b)* **1989**, *153*, K57.

(27) Schouten, P.G.; Warman, J.M.; de Haas, M.P.; van der Pol, J.F.; Zwikker, J. *J. Am. Chem. Soc.* **1992**, *114*, 9028.

(28) *Handbook of Radiation Chemistry*; Tabata, Y.; Ito, Y.; Tagawa, S., Eds.; CRC Press, Boca Raton, 1991, p 409.

(29) Donovan, K.J.; Elkins, J.W.P.; Wilson, E.G. *Mol. Cryst. Liq. Cryst.* **1993**, *236*, 157.

(30) Warman, J.M.; de Haas, M.P.; van der Pol, J.F.; Drenth, W. *Chem. Phys. Lett.* **1989**, *164*, 581.

(31) Schouten, P.G.; Warman, J.M.; de Haas, M.P.; Fox, M.A.; Pan, H-L. *Nature* **1991**, *353*, 736.

RECEIVED September 13, 1994

Chapter 25

Excitation Dynamics in Conjugated Polymers

H. Bässler[1], E. O. Göbel[2], A. Greiner[1], R. Kersting[3], H. Kurz[3], U. Lemmer[2], R. F. Mahrt[1], and Y. Wada[2]

Fachbereiche[1] Physikalische Chemie und [2]Physik und Zentrum für Materialwissenschaften, Philipps-Universität Marburg, 35032 Marburg, Germany
[3]Institut für Halbleitertechnik II, Rheinisch-Westfälische Technische Hochschule, Aachen, Germany

Time resolved fluorescence spectroscopy has been employed to gain information on the nature of optical excitations in conjugated polymers of the PPV type as well as on their dynamics. Both the bathochromic shift of the fluorescence spectra with increasing delay time and the spectrally dependent fluorescence decay support the notion that the elementary excitations are neutral excitons that undergo an incoherent random walk among subunits of the chain with different excitation energies.

The potential application of conjugated polymers in opto-electronic devices has stimulated intense research concerning the dynamics of elementary excitations in this class of materials. The general notion has been to treat a π - conjugated polymer chain as a onedimensional semiconductor ignoring Coulomb and electron correlation effects yet invoking strong electron phonon interaction (1). A photon is considered to raise an electron from the valence to the conduction band. The charges rapidly self-localise to form polarons (or bipolarons) that can subsequently recombine radiatively or non-radiatively. Recent steady state photoluminescence work as well as theoretical studies (4,5) have provided evidence, though, that the neglect of Coulomb-interaction is unjustified and that the primary excitation is exciton-like as expected for a molecular solid with a low dielectric constant. Notably site-selective fluorescence work (3) on members of the

0097–6156/94/0579–0328$08.00/0
© 1994 American Chemical Society

poly(phenylenevinylene) (PPV) family substantiated the notion that the absorbing elements are ordered subunits of the polymer backbone comprising several monomer units. Because of the spatially random distribution of topological faults that determine the length of these units (the "effective conjugated length") their transition energies are subject to a distribution which is the origin of inhomogeneous band broadening. In this concept the red shift between absorption and emission - previously considered as evidence in favor of polaron formation - is explained in terms of a loss of electronic energy the excitations suffer while migrating incoherently within a distribution of hopping states whose width is much larger than kT. Time resolved studies will be reported that substantiate this conclusion. A preliminary account of part of the experimental facts has already been published (6).

Experimental and Results

The temporal evolution of the photoluminescence spectra of PPV, its phenyl-substituted soluble counterpart (PPPV), as well as PPPV blended with polycarbonate has been studied employing a streak-camera and upconversion techniques allowing a temporal resolution of $\simeq$ 20 ps and $\simeq$ 200 fs, respectively. Two types of experiments were performed. Firstly, the evolution of the luminescence spectra was studied at variable delay time between excitation and spectra recording and, secondly, the temporal decay of the luminescence was followed within different spectral windows encompassing the spectral region of the low energy absorption band considered to be the inhomogeneously broadened $S_1 \leftarrow S_0$ 0-0 transition.

The absorption spectrum of PPV, prepared via the precurser route, shows poorly resolved vibronic structure (Fig. 1). Fluorescence was excited at room temperature by 150 fs pulses of 3.12 eV photons with a spectral width of 15 meV. Its decay was monitored at selected energies within the inhomogeneously broadened $S_1 \leftarrow S_0$ 0-0 transition at a monochromator limited spectral resolution of 20 meV. Fig. 2 shows that (i) except for low detection energies the fluorescence rises instantaneously indicating that vibronic relaxation is completed on a 100 fs scale, (ii) the subsequent decay is nonexponential and strongly dependent on the monitor energy, and (iii) a fast initial rise is observed at low detection energies. The initial

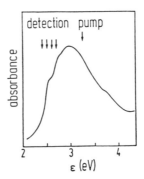

Figure 1. Absorption spectrum of PPV prepared by the precursor method. Arrows indicate the pump energy and the energies at which the photoluminescence decay has been monitored employing upconversion techniques.

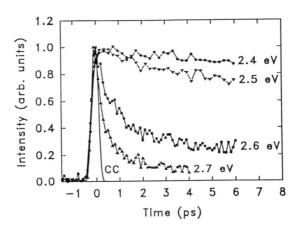

Figure 2. Normalised luminescence decay in PPV at different detection energies within the inhomogeneously broadened $S_1 \leftarrow S_0$ 0-0.

decay process monitored at 2.7 eV yields a minimum decay time of order 250 fs. Performing the same experiment with a 1% blend of PPPV in polycarbonate yields a similar result except that the entire time frame is shifted by a factor of roughly 5. (Fig. 3). The 10K absorption and cw-photoluminescence spectra of PPPV are presented in Fig. 4. Because of the larger inhomogeneous broadening that is due to the steric hindrance imposed by the phenyl substituent the individual vibronic bands are resolved only in emission yet not in absorption. Measuring the temporal evolution of the fluorescence spectra at two different pump energies reveals an interesting localisation phenomenon. Upon exciting above a demarcation energy E_{loc} the emission spectra are inhomogeneously broadened and experience a bathochromic shift with increasing delay time. On the other hand, excitation close to E_{loc} generates spectra that are substantially narrower and time-independent (Fig. 5). This confirms the conclusion drawn from previous site-selective fluorescence studies that E_{loc} is the energy that separate states participating in energy transport from states that do not. The shift of the spectra excited above E_{loc} occurs on a logarithmic time scale (Fig. 6). This spectral behavior translates into a luminescence decay that becomes slower as the monitor energy gets lower, in analogy to the results shown in Fig. 2, and approaches an exponential with a decay time of about 0.5 ns. The above features are independent of the pump intensity, indicating that no molecular phenomena are involved, and vanish if the PPPV in the blend systems is replaced by an oligomer containing three phenylene-moieties.

Discussion

The above results can be rationalised in terms of a concept that treats the conjugated polymers as an array of subunits of statistically varying length (the effective conjugation length L_{eff}) separated by topological faults. These "sites" act like chromophores in an organic glass among which incoherent hopping transport of neutral (excitons) or charged excitations occurs. The statistical variation of L_{eff} translates into a distribution of excitation energies of the sites giving rise to inhomogeneous broadening of both energy and charge transporting levels. The dynamics of excitations in such energetically random systems has been studied both theoretically (8,9) and experimentally (10). The key results are as follows: (i) Upon random initial population of the sites, the excitations relax towards the tail states of

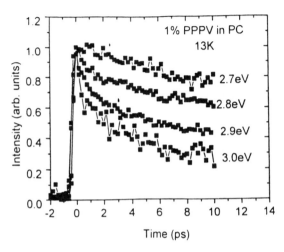

Figure 3. Normalised luminescence decay in 1% PPPV/polycarbonate blend at different detection energies.

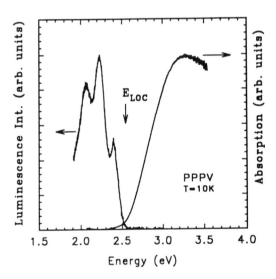

Figure 4. Low temperature (10 K) absorption and cw-luminescence spectra of PPPV. E_{loc} denotes the demarcation energy below which incoherent excitation transport is blocked. Reproduced with permission from reference 7.

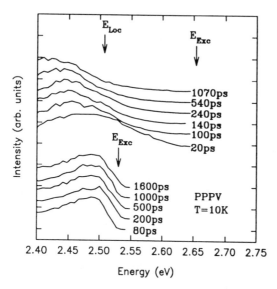

Figure 5. Time resolved $S_1 \to S_0$ 0-0 luminescence spectra of PPPV after excitation with 7 ps pulses of different photon energies E_{exc}. Reproduced with permission from reference 7.

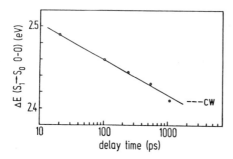

Figure 6. Decay of the maximum of the $S_1 \leftarrow S_0$ 0-0 fluorescence band (upper set of spectra in fig. 5) as a function of delay time.

the distribution of states (DOS) in course of their random walk, the mean energy loss being a logarithmic function of time. Time resolved photoluminescence spectroscopy allows to monitor the temporal relaxation pattern of the ensemble. (ii) If the system contains extra traps outside the distribution of intrinsic localised states capturing of excitations competes with relaxation. The trapping process is non-exponential in time. (iii) At low temperatures characterized by σ /kT $>>$ 1,σ being a measure of the width of the DOS-function, excitations with finite lifetime are subject to localisation because resonant or quasi-resonant emission must occur once the dwell of an excitation exceeds its intrinsic lifetime. Owing to the distribution of states at the tail of the DOS a demarcation energy must therefore exist that separates states that participate in incoherent transport from states that do not (11). Site-selective fluorescence spectroscopy on conjugated polymers provided evidence for this phenomenon (12).

Were the absorption spectrum of PPV and its analoques due to a valence to conduction band transition, no emission that overlaps with the absorption spectrum should be observed because of rapid "intra-band" scattering and accumulation of the carriers at the band edges. (i) The occurrence of such photoluminescence, (ii) its non-exponential temporal decay pattern, (iii) the strong increase of the decay time with decreasing monitor energy (iv) the observation that dilution causes extension of the temporal decay pattern (v), the localisation phenomenon borne out by Fig. 4, and (vi) the asymptotic approach of exponential decay behavior of the photoluminescence upon probing at low emission energies provide compelling evidence that the elementary excitations are neutral exciton like entities that migrate incoherently among subunits of the polymer chains. We therefore conclude that conjugated polymers behave like a random array of oligomers of different length. This concurs with recent theoretical work concerning the nature of excited states in these systems. (4,5)

Acknowledgments

This work was supported by the Deutsche Forschungsgemeinschaft, the Stiftung Volkswagenwerk, the Fond der Chemischen Industrie and the Alfred-Krupp-Stiftung. Discussions with B. Molley and Prof. H. Kauffmann are gratefully acknowledged.

Literature Cited

1 Su, W. P.; Schrieffer, J. R.; Heeger, A. J.
  Phys.Rev.Lett **1979**, *42*, 1698

2 Rauscher, U.; Bässler, H.; Bradley, D.D.C.; Hennecke, M.
  Phys.Rev.B. **1990**, *42*, 9830

3 Heun, S.; Mahrt, R. F.; Greiner, A.; Lemmer, U.; Bässler, H.; Halliday, D.
  A.; Bradley, D.D.C.; Burn, P.L.; Holmes, A.B.
  J.Phys.Condens.Matter **1993**, *5*, 247

4 Mukamel, S.; Wang,H.X. *Phys.Rev.Lett.* **1992**, *69*, 65

5 Abe, S.; Schreiber, M.; Su, W.P.; Yu, J. *Phys.Rev.B.* **1992**, *45*, 9432

6 Kersting, R.; Lemmer, U.; Mahrt, R.F.; Leo, K; Kurz, H.; Bässler, H.; Göbel,
  O.E. *Phys.Rev.Lett.* **1993**, *70*, 9820

7 Lemmer, U.; Mahrt, R.F.; Wada, Y.; Greiner, A.; Bässler, H.; Göbel, O.E.
  Chem.Phys.Lett. **1993**, *209*, 243

8 Movaghar, B.; Grünewald, M.; Ries, B.; Bässler, H.; Würtz, D.
  Phys.Rev.B. **1980**, *42*, 9830

9 Rudenko, A.I.; Bässler, H. *Chem.Phys.Lett.* **1991**, *182*, 581

10 Richert, R.; Bässler, H.; Ries, B.; Movaghar, B.; Würtz, D.
  Phil.Mag.Lett. **1989**, *59*, 95

11 Bässler, H. In *Optical Techniques to Characterize Polymer Systems*
  Bässler, H., Ed.; Elsevier: Amsterdam, 1989, p. 181

12 Elschner, A.; Mahrt, R.F.; Pautmeier, L.; Bässler, H.;
  Stolka, M.; Mc Grane, K. *Chem.Phys.* **1991**, *150*, 81

RECEIVED September 13, 1994

Chapter 26

Application of Polyaniline Films to Radiation Dosimetry

Yuichi Oki, Takenori Suzuki, Taichi Miura, Masaharu Numajiri, and Kenjiro Kondo

Radiation Safety Control Center, National Laboratory for High Energy Physics (KEK), 1-1 Oho, Tsukuba, Ibaraki 305, Japan

Radiation-induced doping of polyaniline was applied to a measurement of the integrated radiation dose. Samples consisting of undoped polyaniline and polyvinylchloride powder were irradiated with ^{60}Co γ-rays. The conductivity increased along with an increase in the dose in the range of 10 to 10^5 Gy. It was shown that this method in the polyaniline system is promising for radiation dosimetry. The relationship between the amount of doped chlorine and the conductivity increase of polyaniline is discussed.

Conducting polymers have been expected to be applied to various electrical devices and sensors. In this work, radiation-induced doping was applied to the measuremet of the integrated radiation dose in a polyaniline system. It has been reported that radiation results in an increase in the electrical conductivity of conducting polymers due to radiation-induced doping in electron irradiation for polyacetylene (1, 2) and polythiophene (3–5) and in neutron irradiation for polyacetylene (6). Yoshino et al. (3, 7) have pointed out that radiation-induced doping can be applied to measurements of radiation in a polythiophene–SF$_6$ system.

In high-energy accelerator facilities, the beam-line components are exposed to high radiation fields. A device for measuring the integrated radiation dose is very important in estimating the radiation damage of these accelerator components. Such devices are required to be highly stable to radiation and to have the capability of measuring a very wide range of doses. Polyaniline is known to exhibit a drastic increase in its conductivity, by about 10 orders of magnitude, by doping it with a protonic acid (8). A polyaniline device using the radiation-induced doping technique is therefore one of the candidates for dosimeters used in such facilities.

0097–6156/94/0579–0336$08.00/0

The prepared polyaniline samples consisted of a strip of an undoped polyaniline and the source material of the dopant, which were sealed in a glass-ampoule *in vacuo*. Halogenated materials were used as the source material, from which gases formed through radiolysis were supplied. The halogen-containing dopant gas reacts with the film on its surface. This radiation-induced doping results in an increase in its surface conductivity.

It is considered that the increase in electrical conductivity can be expressed as a function of the concentration of the dopant absorbed on the surface. The amount of dopant depends on the G-value of the dopant gas liberated from the source material, the weight of the source material and the integrated radiation dose. Therefore, the relationship between these parameters and conductivities must be elucidated in order to apply this method as a radiation dosimeter.

This paper reports on the characteristics of a polyaniline sample consisting of a polyaniline strip and polyvinylchloride (PVC) powder, while focusing on the relationships among the conductivity change, the integrated radiation dose and the content of chlorine introduced in the polyaniline film through radiation-induced doping.

Experimental

Samples.   The polyaniline used in this experiment was chemically synthesized from a solution of aniline dissolved in 1 M perchloric acid. The resulting polyaniline was converted to the emeraldine base by repeated washing with a NaOH solution. The thus-prepared powder was dissolved in *N*-methyl-2-pyrrolidone; the solution was then poured onto a glass plate. The film was prepared by thoroughly evaporating the solvent at 80 °C under a vacuum for about 24 hours.

The polyaniline sample, as a prototype polyaniline dosimeter, consisted of a polyaniline strip (0.5 × 4 cm, *ca.* 40 μm thick) and PVC powder, which were sealed in a Pyrex glass tube (volume: *ca.* 2.5 cm^3) under vacuum or in air at 1 atm. In each glass tube, quartz wool was put between the strip and the powder so as not to contact them directly. The degree of polymerization of the PVC used was 1100, and its mesh size was about 150.

Irradiation.   The samples containing a given amount of PVC were irradiated with γ-rays from ^{60}Co sources at room temperature. Irradiation above 100 Gy was carried out at the ^{60}Co irradiation facilities of Japan Atomic Energy Research Institute.

Measurement of Electrical Conductivity.   The resistance of the polyaniline film was measured with an ultra-high resistance meter (Advantest Co., Model

R8340A) in a pure nitrogen atmoshere at room temperature. The original volume and surface conductivities of the undoped polyaniline film were measured with Au electrodes. After γ-ray irradiation, the glass ampoule of the sample was open, and the resistance of the irradiated strip was measured.

Neutron Activation Analysis of Chlorine.   In order to examine the relationship between the conductivity change and the chlorine content of the strip, the content was determined by a method involving neutron activation analysis. After γ-ray irradiation, a piece of the strip was irradiated in a pneumatic tube of the JRR4 reactor in Japan Atomic Energy Research Institute. The intensity of the 2168-keV γ-ray of ^{38}Cl formed by the (n, γ) reaction of ^{37}Cl was measured with a Ge semiconductor detector system. The chlorine content was determined from its γ-ray intensity, compared with reference samples.

Results and Discussion

Conductivity Change of Irradiated Polyaniline.   The volume conductivity of the undoped polyaniline film used in this experiment was 1×10^{-13} S/cm and the surface conductivity was 1×10^{-13} S. Figure 1 shows the change in the conductivity of the polyaniline film, itself, due to γ-ray irradiation. The abcissa represents the absorbed dose for thick polyaniline, and the ordinate shows the ratio of the conductivity of irradiated and non-irradiated polyaniline.

As Figure 1 shows, the film was found to be very resistant to radiation. Almost no conductivity change was observed up to 3.6 MGy in a film irradiated *in vacuo*. However, the conductivity of the film irradiated in air increased above *ca.* 0.1 M Gy, and reached 10^5 times the original conductivity at 3.6 MGy. This suggests that certain dopant gases are produced from the air, itself, in a high radiation field. Nitric acid is known to be formed in air due to radiation (*9, 10*). Nitric acid and/or NO_x gases are considered to be possible candidates to serve as dopants.

The polyaniline–PVC powder system.   Figure 2 shows the conductivity change of irradiated polyaniline strips for polyaniline samples *in vacuo* with three different contents of PVC powder (10mg, 50mg, and 150mg). The conductivity increases almost linearly on a log–log plot along with an increase in the dose over a wide region of about 10 to 0.1 MGy; the maximun conductivity reached 10^{10} times its original value. This conductivity change is due to only a radiation-induced doping effect, because, as shown above, there was no conductivity change when the polyaniline was irradiated alone *in vacuo*. This linear relationship is one of the important characteristics required for an integrated radiation dosimeter. The conductivity showed a maximum at around 0.1 M Gy, and then decreased at

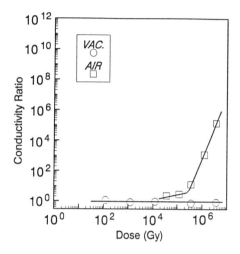

Fig. 1 Conductivity change of polyaniline film due to γ-ray irradiation

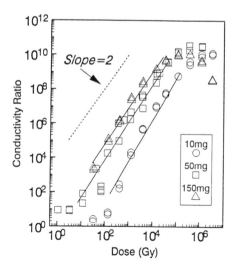

Fig. 2 Relationship between the dose and conductivity change of polyaniline film in the polyaniline–PVC system

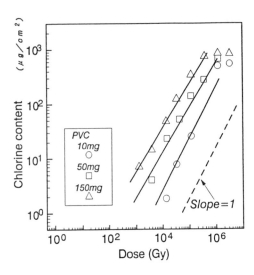

Fig. 3 Chlorine content on the surface of polyaniline film

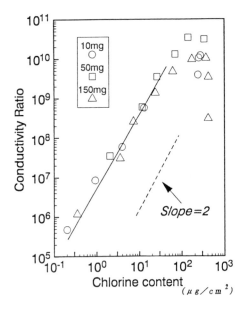

Fig. 4 Relationship between the chlorine content of polyaniline film and its conductivity change

higher doses. The slopes of the fitting lines in the figure were about two, showing that the conductivity is proportional to the square of the dose.

On the other hand, the conductivity in polyaniline samples sealed with 50 mg of PVC powder in air increased by only a factor of 2 at 10^4 Gy. This result is considered to be due to a suppression of the diffusion of the heavy dopant gas by air.

The cross section of the doped polyaniline films was observed with a microscope. The micrographs indicated that the surface of the film was uniformly doped with dopants.

The gas produced from the γ-ray irradiated PVC powder was analyzed with a mass spectrometer. The dopant gas in this system was confirmed to be HCl, because only H_2 and HCl were detected in the mass spectra. Figure 3 shows the relationship between the content of chlorine in the irradiated polyaniline and the dose. The chlorine originally contained in undoped polyaniline was also determined to be *ca.* 10 μg per 1 cm^2 of the film based on a neutron activation analysis. The ordinate of Figure 3 represents the content in $\mu g/cm^2$, which was obtained by subtracting the original content from the content in the doped polyaniline. The content and dose exhibited a linear relationship with a slope of unity. This result indicates that the chlorine content is directly proportional to the dose. It was therefore found that the radiation-induced doping proceeded quantitatively along with an increase in the dose.

From Figures 2–3, the relationship between the chlorine content and the conductivity can be derived. There is a clear linear relationship with a slope equal to two, as shown in Figure 4. This exhibits that the conductivity is proportional to the square of the chlorine content. Figures 2–4 show that the decrease in the conductivity at a higher dose region than 0.1 MGy is not caused by a decrease in the chlorine content due to radiation effects. This decrease in the conductivity may be attributed to a saturation of the dopants on the surface and/or radiation effects on a surface which is highly doped with chlorine, in addition to a decrease in the production rate of the dopant gas due to the decomposition of PVC at a high-dose region.

The present results show that the polyaniline–PVC powder system is promising for wide-range integrated radiation dosimetry. A further study concerning the physico-chemical properties of the polyaniline film has been extensively carried out. The details will be reported elsewhere.

Acknowledgments

The authors wish to express their appreciation to Dr. Toshihiro Ohnishi and Mr. Shuji Doi of Tsukuba Research Laboratory, Sumitomo Chemical Co. for supplying the polyaniline powder. They are also grateful to the members of the

Inter-University Laboratory for the Common Use of Japan Atomic Energy Research Institute Facilities for their assistance in γ-irradiation.

Literature Cited

1 Yoshino, K.; Hayashi, S.; Ishii, G.; Inuishi, Y. *Jpn. J. Appl. Phys.* **1983**, *22*, L376–L378.
2 Hayashi, S.; Ishii, G.; Yoshino, K.; Okube, J.; Moriya, T. *Jpn. J. Appl. Phys.* **1984**, *23*, 1488–1491.
3 Yoshino, K.; Hayashi, S.; Kohno, Y.; Kaneto, K.; Okube, J.; Moriya, T. *Jpn. J. Appl. Phys.* **1984**, *23*, L198–L200.
4 Hayashi, S.; Takeda, S.; Kaneto, K.; Yoshino, K.; Matsuyama, T. *Jpn. J. Appl. Phys.* **1986**, *25*, 1529–1532.
5 Yoshino, K.; Hayashi, S.; Kaneto, K.; Okube, J.; Moriya, T.; Matsuyama, T.; Yamaoka, H. *Mol. Cryst. Liq. Cryst.* **1985**, *121*, 255–258.
6 Butler, M. A.; Ginley, D. S.; Bryson, J. W. *Appl. Phys. Lett.* **1986**, *48*, 1297–1299.
7 Yoshino, K.; Hayashi, S.; Ishii, G.; Kohno, Y.; Kanato, K.; Okube, J.; Moriya, T. *Kobunshi Ronbunshu* **1984**, *41*, 177–182.
8 Chiang, J. C.; MacDiarmid, A. G. *Synth. Metals* **1986**, *13*, 193–205.
9 Less, L. N.; Swallow, A. J. *Nucleonics* **1964**, *22(9)*, 58–61.
10 Tokunaga, O.; Nishimura, K.; Suzuki, N.; Washino, M. *Radiat. Phys. Chem.* **1978**, *11*, 117–122.

RECEIVED September 13, 1994

Chapter 27

Present and Future of Fullerenes
C$_{60}$ and Tubules

Katsumi Tanigaki

NEC Corporation, 34 Miyukigaoka, Tsukuba 305, Japan

A general concept of new types of carbon, C$_{60}$ and tubules, is presented with reference to the relevant literature. Conductivity and superconductivity found for alkali C$_{60}$ fullerides are described and discussed on the basis of their crystal structures. The potential utility of the inner space of both C$_{60}$ and tubules is also addressed from the nano material science point of view.

New types of carbon, both spherical-shaped C$_{60}$ (Fig.1c) [1,2] and rod-like tubule (Fig.1d) [3,4], have recently been found. Like conventional graphite and diamond (Fig.1a,b), these carbon materials are making very strong impacts on material science. C$_{60}$ solids in combination with other elements exhibit very high conductivity [5] and superconductivity [6] with quite high transition temperature (Tc), the latter of which is surpassed only by that of copper oxides [7-11]. Unique magnetic properties are also reported for the charge transfer complex salts of C$_{60}$ and organic donors [12]. For the tubules, theoretical calculations predicted electronic properties varying from metal to semiconductor, which can be regulated by the diameter and the helical arrangement of hexagons in the tubule layer [13-15]. It can be expected that these unique properties will offer the possibility to make advanced electronic devices in the future. In this paper are described general concepts and ideas which can be made for these new types of carbon, with reference to the relevant literature. Electronic properties of fullerenes are presented focusing on the crystal structure and superconductivity of alkali C$_{60}$ fullerides. The potential utility of the inner spaces inside the hollow of fullerenes is also addressed in terms of nano science of future electronics.

Structure of C$_{60}$ and tubules

C$_{60}$ is a cluster-type molecule consisting of sixty carbon atoms and has a spherical shape with I$_h$ symmetry and radius of about 0.8 nm as shown in Figs.2 (graphics) and 3 (image by electron microscopy). In terms of geometry this is categorized as a closed polyhedron consisting of twenty hexagons and twelve pentagons. We can construct various closed polyhedrons according to Euler's theorem by changing the number of hexagons, keeping the number of pentagons to

0097–6156/94/0579–0343$08.00/0

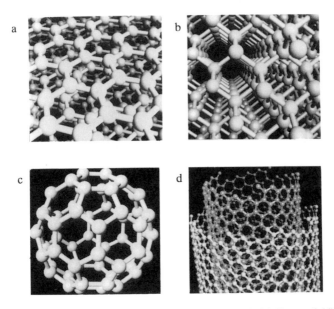

Fig.1 Four types of carbon; (a) graphite, (b) diamond, (c) C$_{60}$ and (d) tubule.

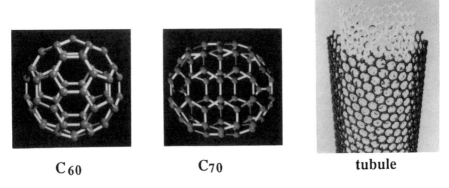

C$_{60}$ C$_{70}$ tubule

Fig.2 Structure of spherical-shaped fullerides (C$_{60}$ and C$_{70}$) and tubule.

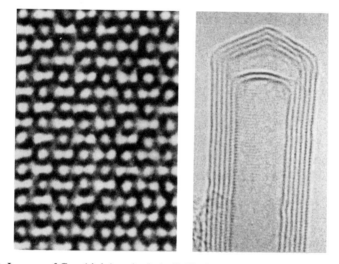

Fig.3 Images of C$_{60}$ (right) and tubule (left) observed by transmission electron microscope.

twelve . As a result, in principle we can expect quite a large number of C_n analogues to C_{60}. Actually the existence of such C_n compounds has been experimentally confirmed (see Fig.2, here C_{70} is displayed as the second majority C_n product). However, the magic number of n is known to be required. For instance, so far only C_{70}, C_{76}, C_{78}, C_{82}, C_{84} --- have been detected and isolated [16], but the C_{62}-C_{68} clusters are missing in experiments and no evidence has been reported for supporting that C_{80} can be formed. The reason for these magic numbers should be related to the formation mechanism and still remains as one of very interesting research items in these new materials.

Two final directions of the C_n cluster growth can be imagined when the number of n is increased for the C_n materials described above. One is the Cn clusters with round shapes, and the other is the rod-shaped clusters. Although the final destination of the formation of C_n materials with increasing n in a gas phase, where C_{60} is usually formed, is not fully understood yet, rod-shaped C_n materials are found in the deposits in the vicinity of the cathode carbon rod. These tubules have a hexagon network in the side surface (Fig.2) and their edges are generally closed by including pentagons as shown in Fig.3. In this sense these rod-shaped tubules also belong to the C_{60} family. The two new types of carbon, spherical shaped molecules like C_{60} and tubules, having hexagon and pentagon networks are called fullerenes after the name of Buckminster Fuller who designed the world's largest geodestic dome [17].

Electronic states and properties of C_{60} and tubules

In order to understand the electronic properties of C_{60} and tubules, one must have the conceptual stand-point that C_{60} is a molecular-type cluster and tubule is a crystal-type cluster. In the case of C_{60}, crystals are made from the C_{60} units and they show various interesting solid properties. On the other hand, one tubule itself is a crystal having a regulated structure and shows unique electronic properties as discussed later.

The energy levels of C_{60} [18,19] are shown in Fig.4a. As shown in this figure, the HOMO level of the C_{60} molecule has five degenerate levels with h_u symmetry and the LUMO level with t_{1u} symmetry is triply degenerate. Since the five HOMO levels are completely occupied by ten electrons, C_{60} has a closed-shell electronic structure [18]. The orbitals forming these levels are p-type and the electrons delocalize over the molecule. Therefore, in principle the properties of the C_{60} molecule are determined by the π-electron characteristics.

Because of the sphericity of C_{60}, a crystal with closed packing structure of either hexagonal closed packing (h.c.p.) or face centered cubic (f.c.c.) can be expected for C_{60} solids. In these two choices pristine C_{60} is shown to be f.c.c. [20] as is the solid crystal structure of other closed-shell elements such as Ar, Kr and Xe. The electronic states of the f.c.c. C_{60} solid are also shown in Fig.3a. In solids h_u HOMO and t_{1u} LUMO levels form band structures. The higher edge of the valence band consists of the h_u-derived levels and the lower edge of the conduction bands is made of the t_{1u}-derived levels. The band gap of the C_{60} solid is about 1.8 eV and is categorized as a typical semiconductor. Reflecting the fact that C_{60} is a closed shell molecule with a relatively large gap between HOMO and LUMO, the C_{60} solid is reported to be a van der Waals type crystal. However, it should be noted that the covalent character of C_{60} solids is relatively stronger than that of the conventional van der Waals organic crystals.

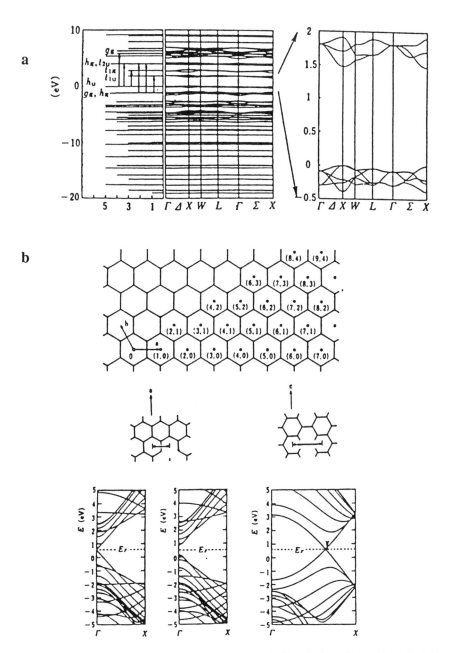

Fig.4 Band structures of (a) C60 solid and (b) tubules. Reproduced with permission from references 13 and 19. Copyright 1991 and 1992 American Physical Society.

In contrast to C_{60}, tubules themselves are crystals and band calculations have been performed by several researchers [13-15]. As shown in Fig.4b, single shell tubules show a variety of electronic properties depending on their diameter as well as their helical structure, which is determined by how the graphitic sheet rolls up to make tubules. In general the electronic properties of tubules can be related to their structures described by the annotation $A(n_1,n_2)$, where $A(n_1,n_2)$ means the tubule made by rolling up a graphitic sheet so that the origin $A(0,0)$ can be overlapped on the $A(n_1,n_2)$ point, as follows: (1) $n_1-2n_2=0$: metal (2) $n_1-2n_2=3m$ (m=1,2,...): narrow gap semiconductor (3) the other cases: wide gap semiconductor. As the diameter of tubules becomes larger and approaches to infinity their electronic properties will be the same as that of graphite having a semimetallic property.

The electronic properties expected between C_{60} and tubules are greatly different. As for C_{60} the high molecular symmetry of C_{60} is responsible for the electronic properties. For tubules, although carrier injection might be effective, the precise control of structural parameters such as diameter and helix is much more important for using these as nano-electronic materials. Additionally, high conductivity as well as superconductivity with high transition temperatures are observed for f.c.c. C60 solids doped with other elements. These properties will be discussed in more detail in the paragraphs that follow.

Conductivity and superconductivity in C_{60} fullerides

As mentioned earlier, the C_{60} solid is semiconducting. There are two methods to be considered for carrier injection as shown in Fig.5. One is the replacement of some of C_{60} by electron-rich or electron-poor molecules. This is the same technology as that used in silicon. For example, B and P doping replaces some of Si thereby generating P-type and N-type semiconductors, respectively. As a candidate molecule for replacing C_{60}, BC_{59} and NC_{59} could be considered for hole and electron injection and endohedral materials such as $La^{3+}C_{82}^{3-}$ [21-23] could also be used. However, there are no such experimental reports at the present stage. The other promising method is intercalation like that used for graphite. Many different elements can be intercalated into the spaces of graphite layers. For example, one of the graphite intercalations KC_8 is known to be superconducting. In contrast to graphite having two-dimensional character, C_{60} solids have three dimensional character. Instead of the interlayer spacings of graphite, the interstitial site spacings can be used for C_{60}. Such examples are actually reported in combination with alkali metals and alkali-earth metals.

The most attractive electronic properties of C60 have so far been found for the intercalations with alkali metals. In Fig.6 the variation in conductivity is shown as a function of the doping time of K. The picture (right) in Fig.6 is one of a typical apparatus used for measuring the conductivity of doped C_{60}. The conductivity increases first with the concentration of K and then decreases. The maximum conductivity is observed at x=3 for K_xC_{60} and the temperature dependence of conductivity shows that K_3C_{60} solid is metallic. This can be interpreted if we consider the band structure of C_{60} solids. Three electrons can be injected into the t_{1u}-derived conduction band for this structure and accordingly the band is half filled, rendering this composition to be metallic.

The carrier injection into C_{60} was described ealier in terms of the intercalation of graphite. In another word, the f.c.c. K_3C_{60} can also be considered as an ionic salt. At present A^+ and C_{60}^{n-} are known to make various ionic solids, where A denotes the

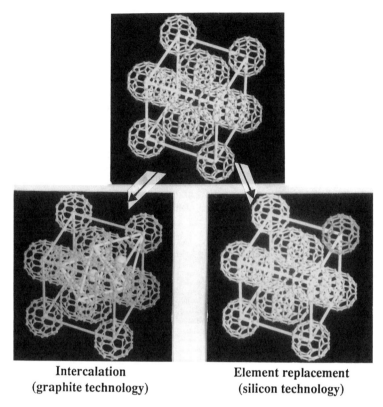

Intercalation
(graphite technology)

Element replacement
(silicon technology)

Fig.5 Two types of method for carrier injection. Intercalation is successfully used for graphite and element replacement is used for silicon.

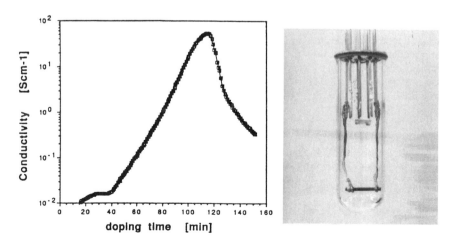

Fig.6 Conductivity of the C_{60} thin film doped with K as a function of the doping time. The picture (right) is a set-up for experiments.

alkaline metal. This is partially because C_{60} can be easily reduced into several stages of anions due to the degeneracy of the t_{1u} level and has multi valences from -1 to -6 [24]. So far f.c.c. A_1C_{60} [25], f.c.c. A_2C_{60} [26], f.c.c. A_3C_{60} [27], body centered cubic (b.c.t.) A_4C_{60} [28] and body centered cubic (b.c.c.) A_6C_{60} [29] have been reported, depending on the number of A with some exceptions in the case of Na doping. In the case of K the stable phases at room temperature are K_3C_{60}, K_4C_{60} and K_6C_{60}, and only K_3C_{60} is metallic among these. Therefore, considering these existing crystal phases and their electronic properties, the observed conductivity change can be reasonably understood.

High T_c superconductivity is also reported for the ionic solids having the stoichiometric composition of A_3C_{60} with f.c.c. lattice. In this structure both crystallographic interstitial sites, one larger octahedral (O-) and two smaller tetrahedral (T-) sites per C_{60}, are occupied by A elements, and the alkali-metals are completely ionized to be $A^+{}_3C_{60}{}^{3-}$. Since the f.c.c. A_3C_{60} crystals differ from the other existing crystal phases from the viewpoint of electronic properties, the detailed studies of the crystal structure of this phase are crucial. In the following paragraph the lattice parameters of A_3C_{60} are simply discussed in connection with the observed superconductivity.

Lattice parameters in f.c.c. A_3C_{60}

The lattice parameters (a_0) of these A_3C_{60} compounds are shown in Fig.7 as a function of the total volume of A^+ cations in the unit cell [30]. In general a_0 would have a good correlation with the ionic radii of alkali-metals. Actually at first glance the lattice parameters (a_0) seem to vary according to the size of alkali-metal cations in a simple manner as shown in this figure. Looking a bit closer, however, we can see a slight difference in lattice contraction in the small lattice parameter region. For instance, Na_3C_{60} and Na_2KC_{60} do not follow the line of the lattice parameters. The other thing to be pointed out is that KCs_2C_{60} is not plotted in this figure, since the phase stability of this composition is much less than that of the other A_3C_{60} fullerides [31,32]. The discrepancies in lattice parameter observed here are considered to be encountered in the following two cases. The former discrepancy is the case that the O-site is occupied by alkali-metals with small ionic radius. Owing to the mismatching between the cation size and the O-site spacing, the C_{60}^{3-}-A^+ coulombic interactions would become less relative to the central field repulsive interactions and the lattice cannot contract to be closely packed. The latter phase instability is the case that the T-site is forced to expand by the occupation of alkali-metals with large ionic radius. In this case the stress in the lattice will cause the f.c.c. structure to another stable crystal structure having energy minimum.

In order to discuss the lattice parameters in detail, a_0 for $A(T)_2Cs(O)C_{60}$ and $A(T)_2Rb(O)C_{60}$ (where A=Li, K, Rb and Cs) are plotted as a function of the alkali-metal ionic radius (r_A^+) in the T-site in Fig.8. In this figure we can see that a_0 varies reasonably with the $A(T)$ ionic radius. It is clear that the lattice parameters are controlled by the r_A^+ in T-site and the alkali-metal in O-site has only a small influence, although other factors have to be taken into account in order to explain further details [30].

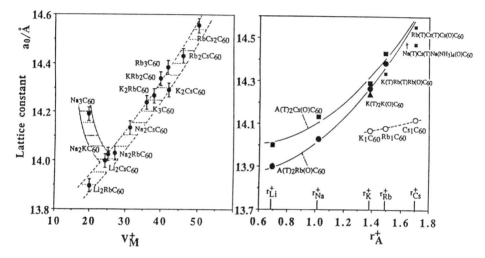

Fig.7 The relationship between a_0 and the total volume of the intercalated alkali cation. The f.c.c. cell is taken as an equivalent b.c.t. cell with $a_{b.c.t.} = a_{f.c.c.}/\sqrt{2}$. P shows the turning point previously supposed. Reproduced with permission from reference 30 with copyright 1993 Elsevier Science Publishers.

Fig.8 The lattice parameters of a_0 as a function of the alkali-metal ionic radius in T-site. The dotted line is the lattice parameters of rock salt type f.c.c. A_1C_{60} for comparison. Reproduced with permission from reference 30 with copyright 1993 Elsevier Science Publishers.

Superconductivity

As explained earlier, A_xC_{60} fullerides have a variety of crystal phases depending on the number of alkali-metals. Among these crystal phases, it has been clarified that only f.c.c. A_3C_{60} is metallic and superconducts under certain temperatures, and that the other b.c.t. A_4C_{60} and b.c.c. A_6C_{60} phases are nonsuperconducting [28]. The superconductivity can be checked by either transport or diamagnetic susceptibility measurements. However, so far, the preparation of high quality single crystal or thin films of f.c.c. A_3C_{60} containing binary alkali-metals is still not well established. Therefore, magnetic diamagnetic shielding measurements are generally employed for studying T_c's using polycrystalline powder samples. The Tc of the prepared samples can be evaluated from the on-set temperatures of the shielding curves as shown in Fig.9.

The T_c values of these compounds are plotted as a function of lattice parameter in Fig.10 [32,33]. In this figure the data of high pressure experiments and the expectation from the simple BCS theory are also shown. There are two important aspects that should be noted. One is that the actual Tc decreases very far from both the pressure data and the simple BCS expectation in small lattice parameters [33-35]. The other is that the T_c increase seems to saturate in the large lattice parameter region [32]. These are very important and should be understood in order to explain what controls Tc in this new superconductor family. The highest Tc so far among this superconductor family is 33 K for $RbCs_2C_{60}$.

Importance of inner space

Another aspect which should be considered for these new carbon materials is the inner space of the cages. C_{60} has an open spherical space of about 0.8 nm inside the hollow. Tubules also have a cylindrical shaped space inside the hollow. These nano-cavities will offer many profound thoughts of nano science to us.

If other electron-donating elements can be introduced into the hollow of C_{60} as shown in Fig.11, molecules with a negatively charged exterior and a positively charged interior would be obtained. This could look like atomic elements whose size is extremely large. New crystals might be achieved from these quasi-atomic elements [36]. On the other hand, the tubules have a quite narrow and long space inside the tubule that grows in the center of multi-layered tubules as also shown in Fig.11. This inner space could accommodate a variety of materials from inorganic to organic compounds. Nowadays quantum confinements are of large interest. However, experimental examinations related to these phenomena are generally difficult and such studies are limited mainly to theoretical considerations. The encapsulating tubules might open a way to materials science studying the influence of such confinements on electrical and optical properties of materials.

A lot of experimental data show that some elements such as Sc, Y, and La can actually be introduced into the hollow of the C_{82} molecule [21-23], but so far no success has been reported for C_{60}. Depending on the size of the elements relative to the inner-space of C_{82}, $Sc_n@C_{82}$ (n=1-3), $Y_n@C_{82}$ (n=1,2) and $La@C_{82}$ are spectroscopically found [21-23], where @ denotes the endohedral state. These are now being isolated with the final structural confirmation. Similar endohedral modifications on the tubules have recently also been reported [37]. As can be seen in Fig.12, lead oxides can successfully be introduced into the inmost tubule after breaking up the edge of tubules. Furthermore a method of opening the edge of nano tubules using oxidation was found [38], and now we have possibilities for confining a variety of materials in the nano cavity world. Another item to be addressed with

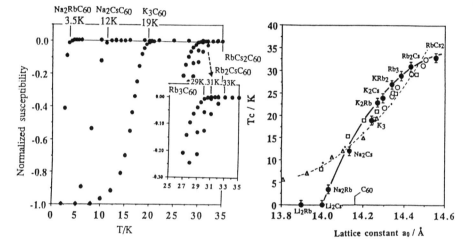

Fig.9 Magnetic shielding of typical f.c.c. A_3C_{60} compounds. The observed Tc varies from 33 K for $RbCs_2C_{60}$ to 3.5 K for Na_2RbC_{60}. Li_2CsC_{60} and Li_2RbC_{60} are not superconducting down to 2 K.

Fig.10 The relationship between Tc's and a_0's for M_3C_{60} (M=Li, Na, K, Rb, Cs and their binary alloys) superconductors in a wide range of a_0. Data indicated by open and closed circles are experimental data. The solid line is a fitting curve of both experimental data. The open triangles and squares are the relationships for K_3C_{60} and Rb_3C_{60} obtained from the pressure experiments and the dotted line is a Tc-a_0 relationship expected from simple BCS theory.

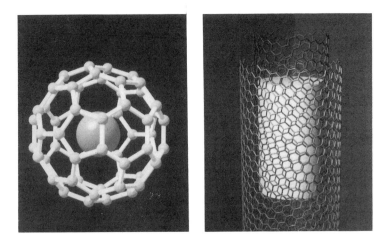

Fig.11 Graphics of encapsulating C$_{60}$ (left) and tubule (right).

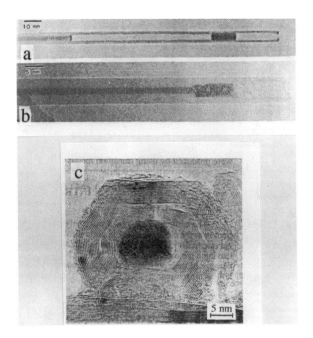

Fig.12 Electron microscope images of metal encapsulating tubules (upper a and b) and a large fullerene containing metal (lower c). Pictures (a and b) are reproduced with permission from reference 37. Copyright 1993 Nature. Picture (c) is from courtesy of Yahachi Saito.

regard to the inner space of fullerenes is the encapsulation of inorganic crystals in the large size C_n fullerenes usually with n=1000. Using the encapsulating large C_n fullerenes the air-sensitive substances can be protected. Such an example is shown in Fig.12, where LaC_2 is encapsulated in the large fullerenes without degradation [39,40]. A lot of examples are also reported to date regarding such encapsulations [21-23, 37-40]. These might open the way to medical applications. Substances that are sensitive to oxygen or other outer environments can be protected by encapsulation and be released in a special position of the body.

Concluding remarks

At present two new types of carbon are added to the conventional carbon materials of graphite and diamond. As described the new forms of carbon show very interesting electrical properties. Also the inner space inside the hollow is of great importance in nano science. A new stage of material science including applications will be advanced using fullerenes as an exotic material.

Acknowledgements

The author is grateful to T. W. Ebbesen, J. Mizuki, J.-S. Tsai, I. Hirosawa, S. Kuroshima, M. Kosaka, T. Manako and Y. Kubo for collaborations in this work.

References

[1] Kroto H. W., Heath J. R., O'Brien C., Curl R. F. and Smalley R. E., Nature, 318, 162 (1985).
[2] Krätschmer W., Lamb L. D., Fostiropoulos K and Huffman D. R., Nature 347, 354 (1990).
[3] Iijima S., Nature354, 56 (1991).
[4] Ebbesen, T. W. and Ajyan P. M., Nature, 358, 220 (1992).
[5] Haddon R. C., Hebard A. F., Rosseinsky M. J., Murphy D. W., Duclos S. J., Lyons K. B., Miller B., Rosamilia J. M., Fleming R. M., Kortan A. R., Glarum S. H., Makhija A. V., Muller A. J., Elick R. H., Zahurak S. M., Tycko R., Dabbagh G. and Thiel F. A., Nature, 350, 320 (1991),
[6] Hebard A. F., Rosseinsky M. J., Haddon R. C., Murphy D. W., Glarum S. H., Palstra T. T. M., Ramirez A. P. and Kortan A. R., Nature, 350, 600 (1991)
[7] Holczer K., Klein O., Huang S.-M., Kaner R. B., Fu K.-J., Whetten R. L. and Diederich F., Science, 252,1154 (1991)
[8] Rosseinsky M. J., Ramirez A. P., Glarum S. H., Murphy D. W., Haddon R. C., Hebard A. F., Plastra T. T. M., Kortan A. R., Zahurak S. M. and Makhija A. V., Phys. Rev. Letters, 66,2830(1991)
[9] Tanigaki K., Ebbsen T. W., Saito S., Mizuki J., Tsai J. S., Kubo Y. and Kuroshima S., Nature, 352,222(1991)
[10] Murphy D. W., Rosseinsky M. J., Fleming R. M., Tycko R., Ramirez A. P., Haddon R.C., Siegrist T., Dabbagh G., Tully J. C. and Walstedt R. E., J. Phys. Chem., Solids, 53, 1321 (1992)
[11] Kortan A. R., Kophylov N., Glarum S., Gyorgy E. M., Ramirez A. P., Fleming R. M., Thiel F. A. and Haddon R. C., Nature, 355, 529 (1992)
[12] Allemand P.-M., Chemani K. C., Koch A., Wudl F., Holczer K, Donovan S., Grüner G and Thompson J. D, Science253, 301 (1991).

[13] Hamada N., Sawada S. and Oshiyama A., Phys. Rev. Lett., 68, 1579 (1992).
[14] Mintmire J. W., Dunlop B. I. and White C. T. Phys. Rev. Lett. 68, 631 (1992).
[15] Tanaka K., Okawara K., Okada M and Yamabe T., Chem. Phys. Lett., 191, 469 (1992).
[16] Wakabayashi T. and Achiba Y., Chem. Phys. Lett. 190, 465 (1992).
[17] "Fullerenes--Synthesis, Properties and Chemistry of Large Carbon Clusters", ACS Symposium Series 481 (American Chemical Society, Washington D. C.), Hammond G. S. and Kuck V. J. eds. , 1992.
[18] Haddon R. C., Acc. Chem. Res. 25, 127 (1992).
[19] Saito S. and Oshiyama A. Phys. Rev. Lett. 66, 2637 (1991).
[20] Fleming R. M., Siegrist T., Marsh P. M., Hessen B., Kortan A. R., Murphy D. W., Haddon R. C., Tycko R., Dabbagh G., Mujsce A. M., Kaplan M. L. and Zahurak S. M., Mater. Res. Soc. Symp. Proc. 206, 691 (1991).
[21] Chai Y., Guo T., Jin C., Haufler R. E., Chibante L. P. F., Fure J., Wang L., Alford J. M. and Smalley R. E., J. Phys. Chem. 95, 7564 (1991).
[22] Shinohara H. Sato H., Ohkohchi M., Ando Y., Kodama T., Shida T., Kato T and Saito Y., Nature 357, 52 (1992).
[23] Jhonson R. D., Varies M. S. Salem J., Bethune D. S.and Yannoni C. S., Nature 355, 239 (1992).
[24] Xie Q., J. Am. Chem. Soc., 114, 3978 (1992).
[25] Zhu Q., Zhou O., Bykovetz N., Fischer J. E., McGhie A. R., Ramanow W., Lin C. L., Strongin R. M., Cichy M. A. and Smith III A. B., Phys. Rev. B., in press.
[26] Rosseinsky M. J., Murphy D. W., Fleming R. M., Tycko R., Ramirez A. P., Siegrist T., Dabbagh G. and Barrett S. E., Nature, 356 416 (1992)
[27] Stephens P. W., Mihaly L., Lee P. L., Whetten R. L., Huaang S.-M., Kaner R. and Deideerich F., Natture, 351, 632 (1991)
[28] Fleming R. M., Rosseinsky M. J., Ramirez A.P., Murphy D. W., Tully J.C., Haddon R.C., Siegrist T., Tycko R., Glarum S.H., Marsh P., Dabbagh G., Zahurak S.M., Makhija A. V. and Hampton C., Nature, 352, 701 (1991)
[29] Zhu Q., Cox D. E., Fischer J. E., Kniaz K., McGhie A. R. and Zhou O., Nature 355, 712 (1992).
[30] Tanigaki K., Hirosawa I., Mizuki J. and Ebbesen T. W., Chem. Phys. Lett. 213, 395 (1993).
[31] Hirosawa I, Tanigaki K., Ebbesen T. W., Shimakawa Y., Kubo Y. and Kuroshima S., Solid State Commun. 82, 979 (1992).
[32] Tanigaki K, Mater. Sci. Eng., B19, 135 (1993).
[33] Tanigaki K. Hirosawa I., Ebbesen T. W. and Mizuki J., J. Phys. Chem. Solids, 54, 1645 (1993).
[34] Tanigaki K., Hirosawa I., Ebbesen T. W., Mizuki J., Shimakawa Y., Kubo Y., Tsai J. S. and Kuroshima S., Nature, 356,419 (1992)
[35] Tanigaki K, Ebbesen T. W., Tsai J.-S., Hirosawa I., and Mizuki J., Euro. Phys. Lett., 23, 57 (1993).
[36] Oshiyama A., Saito S., Hamada N. and Miyamoto Y, J. phys. Chem. Solids, 53, 1457 (1992),
[37] Ajayan P. M. and Iijima S., 361, 333 (1993).
[38] Ajayan P. M., Ebbesen T. W. Ichihashi T., Iijima S., Tanigaki K. and Hiura H., Nature 362, 522 (1993).
[39] Ruoff R. S. Lorents D. C., Chan B, Malhotra R. and Subramoney S., Science, 259, 346 (1993).
[40] Tomita M., Saito Y. and Hayashi T., Jpn. J. Appl. Phys. 32, L280 (1993).

RECEIVED October 21, 1994

Chapter 28

Photoresponsive Liquid-Crystalline Polymers

Holographic Storage, Digital Storage, and Holographic Optical Components

Klaus Anderle and Joachim H. Wendorff

Fachbereich Physikalische Chemie und Wissenschaftliches Zentrum für Materialwissenschaft, Philipps-Universität Marburg, D–35032 Marburg, Germany

A storage process is described which is based on light-induced trans-cis-trans-isomerization reactions of azobenzene chromophores attached to a polymer backbone as side groups in liquid crystalline polymers. The chromophores are able to rotate in the glassy state if subjected to linearly polarized light: the azobenzene units approch a saturation orientation which is perpendular to the polarization direction of the light. The contribution discusses the molecular mechanism of this process as well as possible applications.

In the past decade optical recording and the development of suitable recording media have become a subject of extensive scientific and industrial interest *(1)*. High optical sensitivity, an enhanced storage density and short switching as well as access times are as important as, in certain applications, reversibility and high signal to noise ratio after many write-erase cycles.

The contribution describes the principles of a storage process based on the liquid crystalline state of organic polymeric materials *(2-6)*. Such materials are capable of forming anisotropic glasses, which can be obtained as thin films. By suitable means one is able to align the optical axis of the uniaxial

0097–6156/94/0579–0357$08.00/0
© 1994 American Chemical Society

system within the film along a given direction or parallel to the film normal. A storage process is made possible by incorporating into the polymer a suitable dye - such as azobenzene - characterized by the fact that it is able to undergo a light-induced isomerization process   even in the solid glassy state. This in turn leads to a reorientation of the optical axis within the film and thus to strong modifications of the optical properties. The information written-in in this way can be erased either by heating to temperatures above the glass transition temperature or by light. The paper describes the physical processes involved in the storage process and the capability of such materials to store holograms. Of particular interest is that the novel type of storage material is able to store information not only on amplitude and phase (scalar properties) of the light but also on the state of polarization. It may thus be used for polarization holography.

Materials Used and Preparation of the Liquid Crystalline Polymer (LCP)-based Recording Films

The compounds used are predominantly   nematic side chain polymers with acrylate or polyester-type chain backbone (Figure 1). The polymers contain azobenzene-units in different concentrations. This allows to tailor the absorption and optical and dielectric properties more or less at will, depending on the choice of the copolymer. The synthesis has been the topic of many previous publications (2-6). To prepare the films used for storage the liquid crystalline polymers were routinely annealed above their glass transition temperature in a vacuum oven to remove residual solvent. Prefabricated liquid crystalline display (LCD)-cells of various sizes and thicknesses were filled at temperatures well above the clearing temperature by capillary action. The cells consist usually of two quartz substrates which were covered by spin coating with a thin polyamide layer. This layer was subsequently structured by a buffing technique. Defect free and large size (as large as 25 cm^2) homogeneous monodomain films of high optical quality can be achieved by annealing the cells at temperatures close to but still below the clearing temperature. Figure 2 displays the birefringence of such a sample as a function of the temperature within the nematic and isostropic phase.

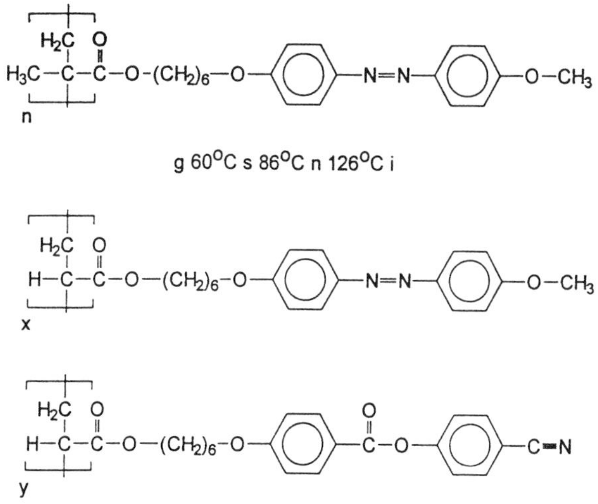

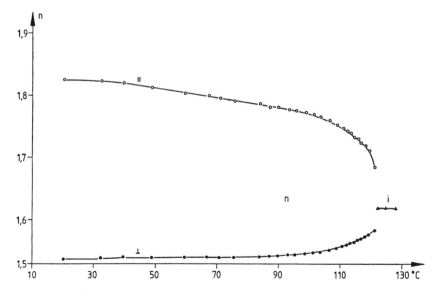

g 60°C s 86°C n 126°C i

g 28°C n 117°C i

Figure 1
Chemical structure of some of the polymers used for optical storage. The transition temperatures (in °C) are also given in the usual notation with the symbols: g: glass transition; s :smectic transition; n nematic transition; i isotropic melt

Figure 2
Variation of the ordinary and extraordinary index of refraction with temperature for a side chain liquid crystalline polymer in the nematic and isotropic state

Basic Mechanism Controlling the Storage Process: Reorientation by Photoselection *(4)*

The modulation of the optical properties of monodomains of liquid crystalline polymers is achieved by the trans-cis isomerization which is caused by the interaction of light with the covalently attached azobenzene groups of the polymer. This type of reaction is of particular interest in a liquid crystalline state: the cis isomer is bent and nonmesogenic and hence cannot form a liquid crystalline phase.

An effective storage process results if linearly polarized light is used and if the illumination is done in such a way that the liquid crystalline state is not destabilized *(2-6)*. Typically intensities of the order of mW/cm^2 are used. In particular the illumination can be performed within the liquid crystalline glassy state i.e. using solid films. The experimental finding is that by using polarized light one is able to reorient the director significantly. This is apparent from the polar diagram of extinction, as displayed in Figure 3. The saturation value of the reorientational angle amounts to 90 deg with respect to the polarization direction of the light. This result is independent of the relative orientation of the polarization direction with respect to the director. This modulation of the direction of the nematic director is directly reflected in the variation of the birefringence: the birefringence is reduced and the sign of the birefringence is reversed due to the induced director reorientation.

Grating Experiments *(3)*

The simplest hologram one can think of is obtained by superimposing two planar waves. This leads to a characteristic periodic intensity modulation in the plane of the recording film and thus to an optical grating in the material. Its characteristic features which can be read out by a planar wave (periodicity of the grating and the amplitude of modulation of the optical properties) allow to obtain valuable information of the storage capability of the recording material (diffraction efficiency, resolution, etc, see below). A second type of a simple hologram is obtained by the superposition of a planar and a spherical wave which gives rise to a Fresnel type interference pattern. The hologram stored in a recording film can be considered as an optical element manufactured by holography (HOE: holographic

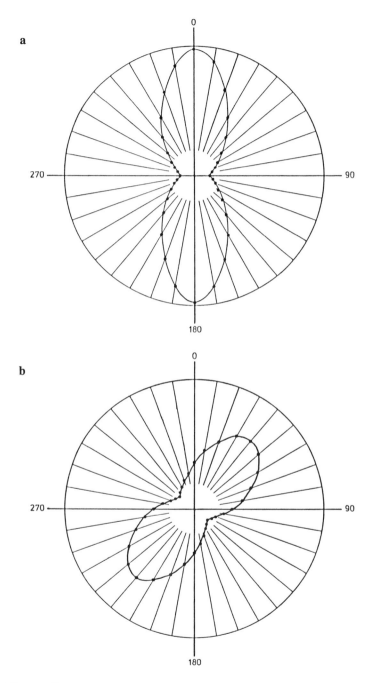

Figure 3
Polar diagram of the extinction as obtained for a nematic
monodomain for a side chain liquid crystalline polymer
a) before and b) after irradiation with polarized light

optical component). It can be used as a Fresnel lens to enlarge the size of an object or to decrease ist size *(5)*.

In order to test the capability of the liquid crystalline polymers with respect to the storage of holograms we performed grating experiments. The result is that a very high diffraction efficiency which can be achieved combined with an excellent resolving power which makes this type of liquid crystalline polymer an attractive novel recording medium (Figure 4). Actually we were able to store real holograms reversibly in such materials and we were also able to prepare holographic optical components, such as a Fresnel lens, using azobenzene containing liquid crystalline side chain polymers.

Polarization   Holography   *(6)*

The characteristic feature of holographic recording is the transformation of phase information into intensity using the interference of the beam carrying the information (object beam) with a reference beam. The reason is that optical storage media are not able to record phase information directly. Optical recording media have still another disadvantage. The recorded pattern does not contain information on the state of polarization of the incident light. So if the surface of the object to be recorded has the property of changing the polarization of the reflected light one will not be able to store this information in the hologram. Sometimes this kind of information may be crucial. Polarization holography is the answer.

It has become apparent that the material described here is to a certain extent able to store information on the polarization of the light: it tends to change the direction of its main optical axis which approachs a saturation value of 90⁰ relative to the polarization direction (Figure 5). So in principle this material may be capable of storing polarization holograms. To test this capability we performed grating experiments (6). The result is that the material is able to store to a certain extent information related to polarization and that one is able to retrieve such information using an unpolarized beam. The pattern obtained are, however, quite complex. Figure 6 gives an example for the angle of maximum diffraction efficiency as a function of the irradiation time for a nonparallel polarization of the intersecting beams.   We are currently performing experiments to try to develop an understanding of the basic

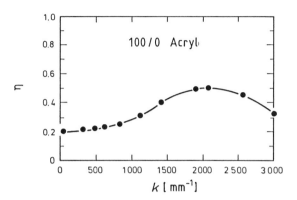

Figure 4
First order diffraction efficiency as a function of the spatial frequency

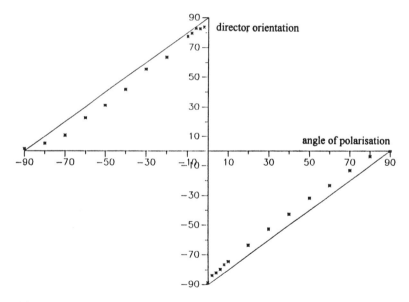

Figure 5
Dependence of the orientation of the director after irradiation with polarized light on the original relative orientation of the polarization of the light and the nematic director

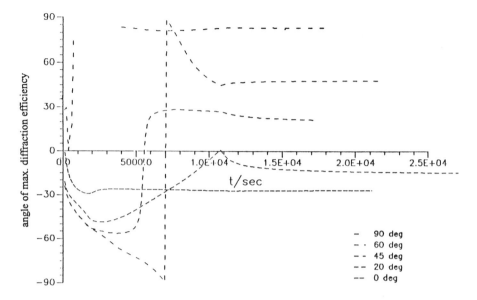

Figure 6
Angle of maximum diffraction efficiency as a function of the irradiation time, nonparallel polarization of intersecting beams

features in relation to the processes happening during storage in the liquid crystalline polymers.

Literature   Cited

1.) Hartmann, M; Jakobs, B.J.A.; .Braat, J.J.M. Philips Techn. Rev. **(1985)**,42, 37; Gravesteijn,D.J.; van der Poel,C.J.; Scholte, L.O.; van Uitjen, C.M.J. Philips. Techn. Rev. **(1987)** 44, 250

2) Eich,M; J.H.Wendorff,J.H. Makromol. Chem. Rapid Commun. **(1986)** 8,467

3.) Eich. M; Wendorff,J.H. J.Opt. Soc. Am. B, 7, 1428 (199o)

4.) Birenheide, R.; Wendorff J.H. Proc.SPIE-Int. Soc. Opt. Eng. **(1990)**, 1213

5.) Anderle, K.; Birenheide,R.; Wendorff,J.H. Liq. Cryst. **(1991)** 5,691

6.) Anderle,K; Wendorff,J.H. Mol. Cryst. Liq. Cryst, in press; Birenheide, R.PhD-Thesis, Darmstadt **(1991)**

RECEIVED September 13, 1994

Chapter 29

Azimuthal Alignment Photocontrol of Liquid Crystals

Kunihiro Ichimura

Research Laboratory of Resources Utilization, Tokyo Institute
of Technology, 4259 Nagatsuta, Midori-ku, Yokohama 227, Japan

Novel ways to achieve in-plane alignment photocontrol of
nematic liquid crystals are presented. A surface of a silica
plate was treated with an azobenzene derivative having a
triethoxysilyl residue at the o-position. Exposure of a liquid
crystal cell fabricated with the modified plate to linearly
polarized blue light resulted in a homogeneous alignment.
The alignment direction was perpendicular to the electric
vector of the actinic light and was regulated at will by
changing the polarization plane. Modification of a thin film
of poly(vinyl alcohol) to introduce azobenzene units
selectively on the surface was an alternative way to perform
the in-plane alignment regulation. The efficiency of the in-
plane alignment photocontrol was influenced prominently by
the nature and position of substituents on the azobenzene
moiety.

The alignment of liquid crystals (LCs) is governed by the nature of the
substrates. When substrate surfaces are covered with monolayered
photochromic molecules like azobenzenes to prepare command surfaces,
a reversible alignment change of nematic LCs between homeotropic and
planar modes is caused upon alternate photoirradiation with UV and
visible light for the surface photochromic reactions (*1*) (Figure 1). This
type of surface-assisted alignment control has been extended to the
photoregulation of the in-plane LC orientation with the use of linearly
polarized UV light through the n-π^* electronic transition of azobenzenes
to give a cis-rich photostationary state (*2*) and for a photochromic

0097–6156/94/0579–0365$08.00/0

reaction of a spiropyran (*3*) on silica surfaces, resulting in the formation of homogeneous alignment. A closely related alignment photoregulation was observed in other systems (*4,5*) where thin films of polymers doped with dye molecules are used for photoactive layers. These photoresponsive LC cells driven by the action of polarized light are of potential value for erasable optical memories as well as display devices without any electric field application.

An alternative photoresponsive system coupled with linearly polarized light involves the molecular reorientation of azo-chromophores which are attached to backbones of liquid crystalline (*6-8*) as well as amorphous polymers (*9*). In these cases, the azobenzenes are excited by polarized visible light through the $n-\pi^*$ electronic transition to yield a trans-rich photostationary state which leads to the molecular reorientation of the chromophores in polymeric matrices. Our recent paper has described that a persistent homogeneous alignment of nematic LCs is obtained by the action of a command surface covered with a p-cyanoazobenzene upon exposure to linearly polarized visible light (*10*).

This paper deals with modifications of silica plates and poly(vinyl alcohol) thin films with azobenzenes which are attached onto the surfaces at their o-position to realized a in-plane reorientation of nematic LCs by irradiation of the substrates with linearly polarized visible light.

Photoactive Silica Surfaces

A series of our works on the command surfaces suggest that a specific interaction between command molecules linked upon substrate surfaces and liquid crystalline molecules at a boundary region plays an essential role with respect to inducing an alignment alteration between homeotropic and planar modes. The working principle of the alignment regulation has been interpreted as follows. The common feature of command molecules exhibiting geometrical photoisomerizations involves a rod-like molecular shape in the trans-isomers which are subjected to a drastic transformation into the corresponding cis-isomers exhibiting bent structures. The rod-like trans forms are so similar to LC molecules that they are assimilated into the LC mesophase to result in a homeotropic alignment. The molecular shape conversion into the V-shaped cis-forms induces distortion of the molecular arrangement of the nearest neighbor LC molecules and leads to the alignment alteration into the planar mode (Figure 1).

These considerations had led us to an idea that an azimuthal alignment control of LC molecules is induced by a molecular reorientation of azo-chromophores on a substrate plate surface when the plate is exposed to linearly polarized visible light which results in trans-rich isomers during the photoirradiation. Thus, the molecular design of the azobenzene moiety for modification of silica surfaces has been undertaken to take account of the following conditions; the chromophore is substituted with alkyl residues at the p-positions to mimic the size and shape of nematic LC molecules, and the chromophore is linked covalently and laterally on the substrate surface so that the long axis of the photoisomerizable units is oriented preferentially parallel with the surface.

Eventually, we have designed an azobenzene (1) derivative which bears hexyl and hexyloxy residues at the opposite p-positions and a carboxylic acid group at a lateral position of the chromophore through decamethylene spacers (*11*). A triethoxysilyl residue is introduced at the o-position of the molecules through an amide bond in order to attach the command molecules laterally onto a substrate quartz plate according to the conventional way. It was confirmed that irradiation of the azo-modified plate with visible light of wavelength longer than ca. 430 nm brought about a trans-rich photostationary state.

A hybrid cell was assembled by sandwiching a nematic LC, DON-103 (Rodic) of $T_{NI} = 74°C$, between the azo-modified plate and a plate which was treated with lecithin to give homeotropic alignment. Linearly polarized light was incident perpendicularly to the cell surface. The evaluation of the photoinduced homogeneous alignment was made by passing a polarized He-Ne laser beam as a probe light through the cell and a crossed polarizer to monitor the transmitted light intensity as a function of the rotational angle (ϕ) of the cell around the optical axis (Figure 2). If a birefringent homogeneous alignment is induced, the transmitted light intensity is related with $\sin^2 2\phi$.

Photobirefringence was observed after prolonged polarized irradiation at room temperature. It was found that the temperature plays an important role in the homogeneous alignment photoregulation, just as in the case of the cell commanded by a p-cyanoazobenzene (*10*). The photoresponse was markedly enhanced at temperatures higher than T_{NI}. Figure 3 shows the birefringence patterns of the cell after irradiation with linearly polarized visible light of various polarization angles (θ) at 100°C. Here θ is defined as an angle contained by the polarization plane and a cell axis which is tentatively defined as the direction parallel to that of the rectangular cell. The results indicate clearly that the direction of homogeneous alignment is freely controlled by changing θ.

One of the characteristic features of this cell is the anomalous stability of the photoinduced alignment toward heat and unpolarized light. No essential modification of the alignment was detected even after heating the cell at 100°C for 1hr. Furthermore, a simultaneous irradiation with visible light at 100°C for 1hr had no effect on the alignment unless the light was polarized. Such a behavior is rather surprising since a thermal randomization of the orientated surface azobenzene may readily take place by heating above T_{NI} as observed for

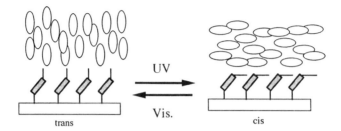

Figure 1   Illustrative representation of surface-assisted alignment photocontrol of liquid crystals.

DON-103 (n,m= 3,2–4,5–5,1)

1

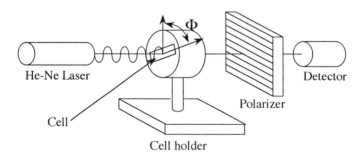

Figure 2   Experimental setup for measuring photobirefringence of an LC cell.

the cases of p-cyanoazobenzene and a spiropyran. This stability of the present azobenzene implies that the two-dimensional free volume at the interface is so small for the azobenzene having two long substituents to reduce the thermal randomization remarkably.

The exposure energies required for the in-plane reorientation were estimated according to the following procedure. The cell was again irradiated with polarized light at $\theta = 0°$ to induce a homogeneous alignment and subsequently exposed to the light at various temperatures after rotating the polarization plane angle $\theta = 45°$. Figure 4 shows rotational angle of the direction of homogeneous alignment as a function of exposure energies. The rotational angles reached a saturation value of about 35° at about 200 mJ/cm^2 when the cell was heated above $T_{NI} = $ 74°C during the irradiation. Lower cell temperature resulted in slower photoresponses and smaller rotational angles. It is noteworthy here that the rotational angle after the photoinduced reorientation of LC is smaller than 45°. This phenomenon can be interpreted as follows. The azobenzene chromophores on the surface may rotate 90° toward the polarization plane after photoirradiation because the polarized light is absorbed favorably by the chromophores with the transition moment parallel to the polarization plane. Although LC molecules align side by side to the azobenzene, the direction of the LC molecules is not always precisely parallel to that of the surface azobenzene units. A gap between both directions may arise from the existence of the long alkyl substituent at the p-positions; the direction of the alkyl chains is tilted from the azobenzene molecular axis.

The photoinduced homogeneous alignment is illustrated in Figure 5. Polarized light results in a molecular reorientation of azobenzene units on the surface which trigger the reorientation of nearest neighbor LC molecules to modify the alignment of bulk mesophase. This was supported by the fact that a photodichroism of the surface azobenzene emerged in the polarized electronic absorption spectra although a dichroic ratio was relatively small ($A_\perp$ / $A_{//}$ = 1.18).

Photoactive PVA Films Surfaces

Alignment photocontrol of LCs with photochromic units tethered to polymeric surfaces is of significance from practical as well as fundamental viewpoints. It follows that poly(vinyl alcohol) (PVA) thin films were employed because of their available surface-selective modification through esterification with azobenzene acid chlorides (*12*).

The esterification of PVA films with chlorides of carboxylic acids (**2**, **3** and **4**) was carried out in hexane at 25°C in the presence of pyridine. Figure 6 shows the average density of the azobenzene derived from **2** on PVA films as a function of reaction time. Although the amount of the

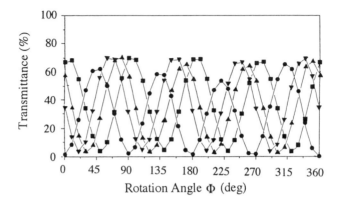

Figure 3   Angular (ϕ) dependence of transmittance of probing He-Ne laser beam through a cell upon exposure at 100°C to linearly polarized visible light with polarization plane of a)$\theta = 0°$ (-●-), b) - 30° (-▼-), c) - 45° (-▲-) and d) - 60° (-■-), respectively.

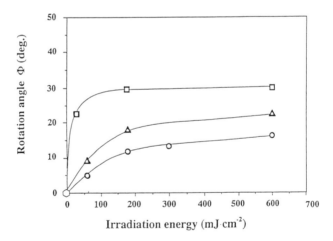

Figure 4 Rotation angles of a homogeneous alignment direction of the cell as a function of exposure energies of polarized visible light at 20°C (-○-), 50°C (-△-) and 100°C (-□-), respectively, after changing the polarization plane angle 45°.

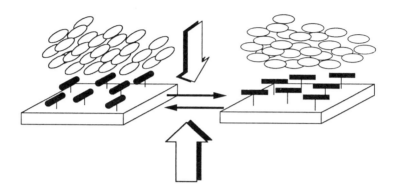

Figure 5 Illustrative representation of azimuthal photocontrol of LC alignment.

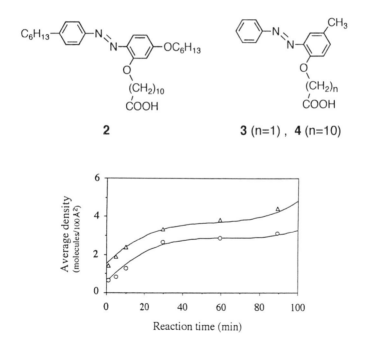

Figure 6 Surface-selective esterification of films of fully saponified PVA (-○-) and 80 % saponified PVA (-△-) with an acid chloride of **2**.

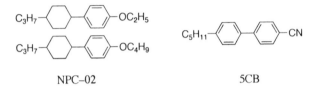

NPC–02   5CB

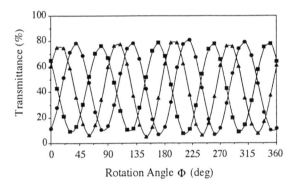

Figure 7   Angular (ϕ) dependence of transmittance of probing He-Ne laser beam through a cell upon exposure to linearly polarized visible light with polarization plane of a)$\theta = 0°$ (-●-), b) - 30° (-▲-) and c) - 60° (-■-), respectively.

azobenzene attached on a film surface of partially saponified PVA (saponification degree = 80%) increased markedly after prolonged reaction, there appeared a plateau at an average density of slightly larger than 2 units of **2** per 100Å^2. This value is approximately equal to that reported as determined by means of carbamoylation to estimate the number of hydroxyl groups which are localized on an uppermost polymer surface layer (*13*). This implies that the azo-chromophores are introduced selectively onto the polymeric surface through esterification of hydroxyl groups which are exposed to the organic solvent.

Nematic LCs were sandwiched between the plate and a plate modified with lecithin. The extent of homogeneous alignment induced by irradiation with linearly polarized light was influenced by the nature of the LCs. Among the LCs tested, a mixture of NPC-02 and 5CB gave the most distinct result for a reversible photocontrol of the azimuthal orientation (Figure 7). On the contrary, PVA surfaces which were modified with an azobenzene having no long alkyl groups (**3** and **4**) showed no alignment regulation even upon prolonged irradiation with polarized light.

Conclusion

The side-on type attachment of azobenzene units on surfaces of a silica plate and of a PVA films is an effective way to regulate the direction of a homogeneous alignment of nematic LCs by using linearly polarized visible light.

Literature Cited

1) Ichimura, K. In *Photochemical Processes in Organized Molecular Systems*; Honda, K. Ed.; Elsevier: Amsterdam, the Netherland, 1991, pp 343-357
2) Kawanishi, Y.; Tamaki, T.; Seki, T.; Sakuragi, M.; Ichimura, K. *Mol. Cryst. Liq. Cryst.* **1992**, *218*, 153
3) K. Ichimura, Y. Hayashi and N. Ishizuki, *Chem. Lett.* **1992**, 1063
4) Gibbons, W. M.; Shannon, P. J.; Sun, S.-T.; Swetlin, B. J. *Nature* **1991**, *351*, 49
5) Iimura, Y.; Kusano, J.; Kobayashi, S.; Aoyagi, T.; Sugano, T. *Jpn. J. Appl. Phys. Part 2* **1993**, *32*, **L93**
6) Anderle, K.; Birenheide, R.; Werner, M. J. A.; Wendorff, J. H. *Liq. Cryst.* **1991**, *9*, 691
7) Ivanov, S.; Yokovlev, I.; Kostromin, S.; Shibaev, V.; Laesker, L.; Stumpe, J.; Leysig, D. *Makromol. Chem., Rapid Commun.* **1991**, *12*, 709

8) Eich, M.; Wendorff, J. H.; Reck, B.; Ringsforf, H. *Makromol. Chem., Rapid Commun.* **1987,** *8,* 59

9) Rochon, P.; Gosselin, J.; Natansohn, A.; Xie, S. *Appl. Phys. Lett.* **1992,** *60,* 4

10) Ichimura, K.; Hayashi, Y.; Kawanishi, Y.; Seki, T; Tamaki, T.; Ishizuki, N. *Langmuir* **1993,** *9,* 857

11) Ichimura, K.; Hayashi, Y.; Akiyama, H.; Ikeda, T.; Ishizuki, N. *Appl. Phys. Lett.* **1993,** *63,* 449

12) Ichimura, K.; Suzuki, Y.; Seki, T.; Kawanishi, Y.; Aoki, K. *Makromol. Chem., Rapid Commun.* **1989,** *10,* 5

13) Matsunaga, T.; Ikada, Y. *ACS. Symp. Ser.* **1980,** *1212,* 391

RECEIVED September 13, 1994

SILICON-CONTAINING POLYMERS

Chapter 30

Electronic Structure and UV Absorption Spectra of Permethylated Silicon Chains

Harald S. Plitt, John W. Downing, Hiroyuki Teramae[1], Mary Katherine Raymond, and Josef Michl[2]

Department of Chemistry and Biochemistry, University of Colorado, Boulder, CO 80309-0215

We report the present state of our experimental and computational investigations of the conformational dependence of the electronic structure and absorption spectra of permethylated linear oligosilanes. So far, we have unraveled the individual spectra of oligosilane conformers with four and five silicon atoms in the chain. *Ab initio* calculations suggest dihedral angles close to 90 ° in the gauche links, and close to 165 ° in the anti links. The conformational properties are important for detailed understanding of the nature of excitation localization in polysilane high polymers. We concentrate on: (i) the remarkable differences between the properties of molecules differing merely by a single bond rotation, reproduced by a very simple "ladder C" model at the Hückel level of theory, but not by the usual even simpler Sandorfy C model, and (ii) ominous indications from *ab initio* calculations and initial results for the six-silicon chain that even the ladder C model is still too simple, and that an explicit consideration of electrons located in bonds to the chain substituents will be unavoidable to describe correctly the gauche conformers ("ladder H" model).

Perhaps the most striking property of polysilanes $(RR'Si)_n$, polymers with a long backbone of silicon atoms, is their intense near-UV absorption band. As discussed at length in a series of reviews (1),(2),(3),(4),(5),(6), there is hardly any doubt nowadays that this is due to excitation of a molecular type within the silicon backbone, described quite adequately as a transition between σ_{SiSi} and σ_{SiSi}^* orbitals localized in a chain segment extending over a few or as many as 20 - 30 SiSi bonds, and differing by one in the number of nodes. While the absorption properties extrapolate smoothly to those of hexamethyldisilane as the chain becomes shorter, it has been noted recently that the emission properties of the permethylated chain change abruptly when the chain length reaches six or seven silicon atoms, depending

---

[1] Permanent address: NTT Basic Research Laboratories, Morinosato, Atsugi, Kanagawa 243-01, Japan
[2] Corresponding author

0097-6156/94/0579-0376$08.00/0

on temperature (7),(8),(9),(10). The emission becomes weak, strongly red-shifted and Franck-Condon forbidden, and the radiative lifetime increases anomalously. These results agree with the proposal (11) that the short segments in a peralkylated polysilane have an anomalously small fluorescence rate constant, and the report (12) that isolated nine-silicon chain segments carrying a more complex substituent pattern and incorporated into a long polymer exhibit a broadened fluorescence with an anomalously long lifetime, attributed to a self-trapped exciton. This assignment has since been confirmed by the near chain-length independence of the energy of the anomalous emission, and the nature of the distortion responsible for the trapping has been proposed to be the stretching of one of the SiSi bonds (7). Since the present review concentrates on the absorption properties of silicon chains, the reader interested in additional detail on emission spectra is referred elsewhere (13).

Permethyloligosilane Conformers

The first direct investigation of a conformational equilibrium in an oligosilane was performed on decamethyltetrasilane (14). The authors investigated the temperature dependence of the solution IR and Raman spectra and concluded that two conformers are present. The slightly (about 0.5 kcal/mol) more stable one was assigned the anti structure, the other the gauche structure. The assignment was based on the presumed presence of a center of inversion symmetry in the former and the IR-Raman mutual exclusion rule. The SiSiSiSi dihedral angles were not determined, and UV spectra were not measured.

Several empirical and semiempirical methods have been used to calculate the equilibrium geometries of individual oligosilane conformers, and particularly the dihedral angles (3). We have now performed completely optimized RHF 3-21 G* *ab initio* calculations of the equilibrium geometries of the permethylated chains up to Si_5Me_{12} and find the dihedral angles to be about 165 ° in the anti and about 90 ° in the gauche arrangement. These values are nearly the same in the conformers of Si_4Me_{10} (15) and Si_5Me_{12}. The anti arrangement is computed to be about 0.6 - 0.8 kcal/mol more stable than gauche, and the conformational energies are roughly additive, except for the g^+g^- conformer of Si_5Me_{12}, which is hindered and less stable, and for which we have not found a minimum. The computed dihedral angles are in fairly good agreement with those of prior MM (16) and MNDO/12 (17) calculations. Since the computed energy difference between the planar zig-zag anti geometry (dihedral angle 180 °) and the helical anti geometry (dihedral angle 165 °) is minute, it is difficult to state with certainty that the all-anti chains of permethylated oligosilanes are helical at their equilibrium geometry in solution, but it appears likely. The closest experimental evidence available originates in the single-crystal X-ray structure of the macrocyclic analogue, $Si_{16}Me_{32}$, which is dominated by angles close to 92 ° and 165 ° (18). However, the dihedral angles in the high molecular weight polymer, $(SiMe_2)_n$, are 180 ° (19). An experimental confirmation of the anticipated helicity in solution would be highly desirable.

For want of better labels, we shall refer to the bond conformations in an oligosilane chain as anti and gauche, and shall assume in the following that the dihedral angles are about 165 ° and about 90 °, respectively. The exact values of the dihedral angles will have little importance for our arguments.

Absorption Spectra

The IR and UV absorption spectra of the oligosilanes with more than three silicon atoms show multiple peaks attributable, at least in part, to the presence of conformational isomers. This has been demonstrated by recording the spectra under conditions of matrix isolation, where high viscosity prevents conformer interconversion, and destroying individual conformers selectively with monochromatic radiation. Computer subtraction then yields the spectra of the individual conformers. Results for the anti and gauche conformers of permethyltetrasilane have already been reported (20) and results for the anti-anti, anti-gauche, and gauche-gauche conformers of permethylpentasilane have now been obtained as well. If the anti dihedral angle deviates from 180 °, this pentasilane will have a larger number of conformers defined by the sign of the deviation, but we have not been able to find evidence for them in our spectra so far.

While the mid-IR spectra of all the conformers are quite similar, the positions of their first intense peaks in the UV spectra are vastly different. We show these in Figure 1A not only for the chains with up to five silicon atoms, for which we have a detailed analysis, but also for the longer chains, where we only have UV spectra in organic glasses so far. By extrapolation from the shorter chains, we attribute the lowest-energy peaks of these longer chains to their all-anti conformers, and assume that the small peaks at higher energies correspond to conformers with some gauche links, whose structure we cannot assign at present.

If we assign the intense peak of each conformer to its first σ-σ^* transition, Figure 1A suggests that σ conjugation in the all-anti chain is strong, as the transition energy drops rapidly with the increasing length of the chain. In contrast, in the all-gauche series the transition energy does not drop at all from Si_3Me_8 to Si_5Me_{12}, and we are forced to conclude that in this series there is no σ conjugation. The properties

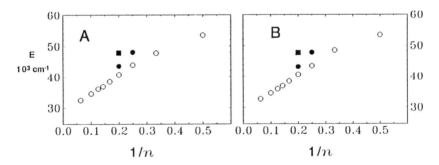

Figure 1: $\sigma\sigma^*$ Excitation energies of permethylated oligosilanes Si_nMe_{2n+2}. A: Observed in argon matrix. B: Computed from the ladder C model. Open circles: all-anti conformers. Dark circles: conformers with one gauche link. Dark square: conformer with two gauche links.

of the Si_5Me_{12} conformer with one anti and one gauche bond are intermediate, again as if the anti link conjugated and the gauche link did not.

The Ladder Model

The simple Sandorfy C model (21), which has seen much use in the analysis of the photophysics of polysilanes and oligosilanes, assumes that only the two valence orbitals on each silicon atom that are used for the formation of SiSi bonds need to be considered (the backbone orbitals), and that they interact only in the nearest-neighbor fashion, yielding a topologically strictly linear chain of interacting σ orbitals in conjugation, analogous to the linear chain of carbon $2p$ orbitals in a polyene. The two parameters of the model are the vicinal (β_{vic}) and the geminal (β_{gem}) resonance integral (Figure 2), and the solution is usually found at the Hückel level, although the Pariser-Parr-Pople approximation has also been used (22).

The simple Sandorfy C model is clearly incapable of reproducing any conformational differences in excitation energies, since neither of its parameters is a function of the dihedral angle. If we wish to develop an ability to model the absorption properties of the various conformers of long polysilane chains in a simple way, we need to adopt a model that is more realistic.

In the simplest such model, we keep the restriction to the backbone orbitals (Figure 2), but introduce all resonance integrals between either of the two valence orbitals on a silicon atom and either of the two valence orbitals on the one or two neighboring silicon atoms (20). This adds two new resonance integrals, β_{13} and $\beta_{14}(\omega)$, to the β_{vic} and β_{gem} resonance integrals already present in the Sandorfy C model. The β_{13} resonance integral is still independent of the SiSiSiSi dihedral angle ω, but $\beta_{14}(\omega)$ clearly is not, and corresponds to the syn-periplanar (0 ° dihedral angle) and anti-periplanar (180 ° dihedral angle) perivalent interactions of organic chemistry, and to the full range of values in between. The $\beta_{14}(\omega)$ integral was identified as responsible for the higher excitation energy of the gauche compared to anti links in polysilanes a long time ago (23). Reduced conjugation in the gauche conformer of $(SiH_2)_n$ was also deduced from energy band calculations (24).

In effect, we have thus introduced three new adjustable parameters, β_{13}, $\beta_{14}(anti)$, and $\beta_{14}(gauche)$. We have estimated physically realistic values of the new parameters in two ways. First, we have represented the valence orbitals by hybrids constructed from Slater orbitals, and assumed the resonance integrals to be proportional to overlaps. Second, we have calculated the Fock matrix elements between natural hybrid orbitals (25) obtained from an RHF 3-21 G^* wavefunction of parent Si_4H_{10}. Both methods yielded the same qualitative results: β_{vic} and β_{gem} should both be negative, with the latter about 6 - 7 times larger, and β_{13} should be quite close to zero. None of these should depend perceptibly on the dihedral angle ω. In contrast, the sign of $\beta_{14}(\omega)$ should change as the dihedral angle is varied from 0 ° (syn planar geometry, $\beta_{14}(syn)$ negative) to 90 ° (gauche geometry, $\beta_{14}(gauche)$ approximately zero) and 180 ° (anti planar geometry, $\beta_{14}(anti)$ positive and nearly twice larger than $\beta_{14}(syn)$). Since attempts to fit the values of the parameters by non-linear least-squares fitting all yielded very small values for β_{13}, we have decided to reduce the number of adjustable paramaters by setting $\beta_{13} = 0$ exactly. This produces a Hückel model with two parameters that are independent of the dihedral angle (β_{vic} and β_{gem}), and one that depends on the dihedral angle ($\beta_{14}(\omega)$). However, since we only have data for two values of the dihedral angle at present, we need only the values β_{14}(gauche) and β_{14}(anti) and not the whole function, $\beta_{14}(\omega)$.

The topology of the valence orbital interactions corresponds to a ladder: each of the two side-pieces consist of linearly conjugated chain of orbitals interacting through alternating β_{vic} and $\beta_{14}(\omega)$ resonance integrals, and the rungs correspond to β_{gem} interactions between orbitals of the two side-pieces. For this reason, we have proposed to call this model the ladder model of σ conjugation (20). Since it now appears that it will be neccessary to go beyond this model in the future, as discussed below, we propose to call this specific version of the ladder model of σ conjugation, in which only the SiSi backbone orbitals are included, the ladder C model, in analogy to the Sandorfy C model.

Non-linear least squares fitting of the four parameters of the ladder C model to the excitation energies of Figure 1A at the Hückel level produced the values β_{vic} = -3.32 eV, β_{gem} = -0.62 eV, β_{14}(gauche) = -0.40 eV, and β_{14}(anti) = +0.78 eV. These values produce the predicted energies shown in Figure 1B, in essentially perfect agreement with the experimental values of Figure 1A. Although an improvement relative to the simple Sandorfy model clearly must be expected in view of the increased number of adjustable parameters, it was not obvious a priori that such perfect agreement for all the observed conformers could be achieved.

In the ladder model, multiple interaction paths exist between the basis set orbitals, and provide opportunities for destructive or constructive interference. Closer inspection of the properties of the HOMO and the LUMO shows that the values of the resonance integrals are such that in an all-anti chain, the paths involving only side-piece conjugation and those also involving rung conjugation interfere constructively. This leads to strong increase of the excitation energy upon chain shortening, and thus strong σ conjugation. In contrast, in an all-gauche chain, they interfere destructively, leading to no increase of the excitation energy upon chain shortening, and therefore to no σ conjugation (20). Thus, it is possible to rationalize the striking differences between the conformers at an intuitive level.

The parameterized ladder C model performs far better than standard semiempirical methods (INDO/S, MNDO, and AM1) that were only parameterized against the properties of other compounds. Although these reproduce some of the qualitative trends among conformers, they do so rather poorly, and the computed excitation energies are far off (13).

At this point, it thus appears that our goal has been met, and that we now have a simple parameterized ladder C model based on backbone orbitals only and suitable for application even to very long permethylated silicon chains, and thus to the interpretation of the spectra of polysilane high polymers, discussion of excitation localization on conformationally defined chain segments, etc.

However, all is not well: First, the optimized value of β_{14}(gauche), -0.40 eV, is much more negative than our simple a priori estimates suggested (close to zero), and corresponds to expectations for β_{14}(syn). This appears unphysical, and is quite disturbing. Second, a preliminary analysis of the results for Si_6Me_{14} does not fit very well the expectations based on the ladder model. Until the analysis of the experimental results for the six-silicon chain is complete, this is perhaps not too disturbing. Finally, our *ab initio* calculations suggest that the situation in short oligosilanes is far more complicated, and that the introduction of additional orbitals into the simple model is probably inevitable. These would be the valence orbitals used by silicon to hold the methyl substituents, and their addition corresponds to going in the direction of Sandorfy model H (21). We now suspect that a ladder model H will be the simplest reliable description of σ conjugation in a saturated chain if chains of all lengths are to be described simultaneously.

Beyond the Backbone Orbitals

As discussed in more detail elsewhere (3), the two silicon valence orbitals on each silicon atom that are used to hold the methyl substituents can be combined pairwise into a sum (SiC$^+$) and a difference (SiC$^-$). They interact strongly with their respective counterparts on the methyl carbon atoms and yield bonding and antibonding combinations.

In a planar arrangement of the silicon atoms, either the anti or the syn geometry, SiC$^+$ is of σ symmetry and SiC$^-$ of π symmetry. The former will mix with the backbone orbitals formed from the SiSi valence orbitals, all of which are of σ symmetry as well, and shall modify them and their energies somewhat. Such mixing is far stronger between the LUMO of the backbone and the antibonding combinations of the SiC$^+$ and carbon orbitals than between the HOMO of the backbone and their bonding combinations, for reasons that are well understood (3). The effects of the mixing are presumably absorbed in the empirically fitted parameters of simple models of the Sandorfy C and ladder C type, in which only the backbone orbitals are considered explicitly.

At the planar backbone geometries, both the bonding and the antibonding combinations of the SiC$^-$ orbitals and the methyl carbon orbitals combine into molecular orbitals of π symmetry. Because of the much higher electronegativity of carbon, the Si-C bonding combinations are quite low in energy and do not affect the spectroscopic properties of interest here. However, the energies of the unoccupied Si-C antibonding π^* orbitals are comparable to those of the backbone σ^* orbitals, and one might expect σ-π^* transitions to occur at energies comparable to those of σ-σ^* transitions. Indeed, *ab initio* calculations at various levels of approximation (26) suggest that at planar backbone geometries, the shortest Si_nH_{2n+2} chains, up to perhaps a trisilane or a tetrasilane, have a σ-π^* state below the lowest σ-σ^* state. Since the former decreases in energy much more slowly than the latter as the chain length is extended, in the longer chains there is no contest, and the σ-σ^* state clearly

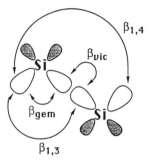

Figure 2:   Backbone valence orbitals in an oligosilane chain and definition
             of the resonance integrals of the ladder C model.

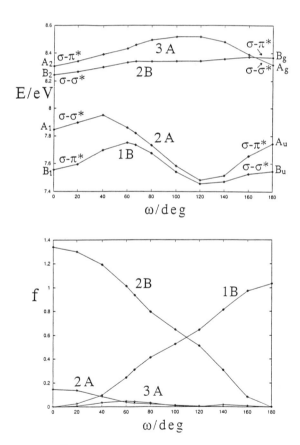

Figure 3:   6-31G** CIS (13×26) excitation energies (A) and intensities (B)
             of the low-energy transitions of Si_4H_{10} as a function of the
             dihedral angle ω.

is the lowest. Similar calculations showed that the lowest Rydberg states lie below the valence states discussed here in the shortest chains, but not in the longer ones, since their energy drops only as the ionization potential as the chain gets longer, while the σ-σ* excitation energy drops faster.

No reliable calculations have been performed for the permethylated chains, but it is probable that their behavior is similar to that of the parent chains, and that the shortest ones have a weak and unobserved σ-π* state (and Rydberg states) below the lowest σ-σ* state. There is no direct experimental evidence for low-lying σ-π* states, but this is perhaps not surprising, given their small expected transition moments. This is the justification for ignoring the σ-π* states in the spectral analyses performed so far, and for the use of C type models for the interpretation of the strong bands in the optical spectra. It is however possible that the failure of Si_2Me_6 and Si_3Me_8 to fluoresce detectably (7), unlike Si_4Me_{10} and the longer chains, is related to a change in the nature of the lowest excited state.

We have now performed many additional *ab initio* calculations on Si_4H_{10} as a function of the dihedral angle ω, all of which yielded qualitatively similar information. A typical result is shown in Figure 3. First of all, we note that in addition to the long-axis polarized transitions into states of B symmetry, the only ones dealt with so far, the calculations predict the presence of much weaker short-axis polarized transitions into states of A symmetry at comparable energies. These involve excitations between orbitals whose node number differs by two, and it is qualitatively obvious why they should be weak. At the anti geometry, the lowest σ-σ* state of A symmetry lies quite far above the lowest σ-σ* state of B symmetry, but it is greatly stabilized as the dihedral angle is reduced, and for the syn geometry the A state lies below the B state. The A states have not been observed, presumably because of their low intensity, and we shall leave them out of the following discussion.

At the syn (0 °) and anti (180 °) planar geometries, the first long-axis polarized σ-σ* transition into a B state is the only one of any significant calculated intensity, and the other states should be difficult to observe. It would appear quite reasonable to expect the ladder C model to work for this σ-σ* transition, and thus for the interpretation of the observed spectrum.

Note, however, that the large increase in the excitation energy of the intense σ-σ* transition upon going from the anti to the syn geometry is not accompanied by a similar increase in the energy of the lowest σ-π* transition. Quite to the contrary, the σ-π* state drops in energy and becomes the lowest excited singlet state of the syn form. Thus, in a simple correlation diagram, we would expect the σ-σ* and the σ-π* states to cross. However, at geometries intermediate between anti and syn, the σ versus π symmetry distinction disappears. The crossing between the intensely allowed σ-σ* state and the extremely weakly allowed σ-π* state is therefore avoided. The calculated energy of the resulting lowest two states of B symmetry hardly changes at all as the dihedral angle is varied, while the calculated intensity is transferred smoothly from the lowest long-axis polarized transition to the second lowest one (Figure 3). This spells disaster for attempts to use C type models with backbone orbitals only, since they do not contain σ-π* states and are therefore inherently incapable of describing the avoided crossing. Also the standard semiempirical models such as MNDO, AM1, and INDO/S, which can describe the σ-π^* states but place them at incorrect energies, fail to reproduce the observations.

At the gauche geometry, the calculation for Si_4H_{10} suggests that the transition to the lower state resulting from the avoided crossing should have lower intensity, and that to the higher state should have higher intensity. Assuming that the situation is qualitatively similar in Si_4Me_{10}, the observed transition is presumably the one to the higher state. It is clear that the use of its energy, which is strongly modified by the avoided crossing, for the parameterization of the ladder C model is incorrect. In a model that only contains backbone orbitals, the proper energy to use would be that of the diabatic σ-σ^* state before the crossing is avoided. One would then have to accept a disagreement between the observed and the calculated excitation energy of the gauche form of Si_4H_{10} or Si_4Me_{10} and attribute it to interaction with a state of σ-π^* type, absent in the model, but at least the magnitude of the derived resonance integrals would be correct. One might then hope that in longer chains, in which the σ-σ^* state will presumably be below the σ-π^* state at all geometries, the avoided crossing will not occur, and that for these longer and more important chains the agreement between the ladder C model and experiment would be good. It is possible that certain conformations, such as all-gauche, will have a σ-π^* state below a σ-σ^* state even for the longest chains, but these are unlikely to be of much importance in the modelling of high polymers. Thus, the conclusion from the *ab initio* calculations is that it is not correct to use the excitation energies of the gauche conformers of the shortest chains to parameterize the ladder C model. This is most unfortunate, since these are just the chains for which the spectral separation of all the conformers is feasible.

Our use of the energy of the intense transition of gauche Si_4Me_{10} in deriving the resonance integrals that were employed for the calculations in Figure 1B in effect used an excitation energy appropriate for the syn conformation (Figure 2), where the diabatic and adiabatic σ-σ^* energies agree. This accounts for the fact that the strange value of β_{14}(gauche) that resulted was approximately equal to that expected for β_{14}(syn). Also, if the σ-π^* states of Si_6Me_{14} already are sufficiently high in energy and do not perturb the energies of the σ-σ^* states much, it would be understandable that the predictions of the ladder C model parameterized on the shorter chains would be incorrect.

We do not yet know whether the *ab initio* results are correct, but since they are quite invariant with the level of approximation used (basis set, size of CI), we suspect that they are. It would be highly desirable to measure the spectral properties of a peralkylated tetrasilane constrained to a planar geometry, and to verify some of the *ab initio* results directly. If these are confirmed, we can either try to parameterize the ladder C model on longer chains, or give it up, and attempt to parameterize the ladder H model. Because of the large number of new parameters, this would best be avoided, but it is no longer likely that it can be.

Conclusions

We have measured separately the IR and UV spectra of the individual conformers of Si_4Me_{10} and Si_5Me_{12}, and calculated their optimized geometries at a fairly reliable level of *ab initio* theory, which also gives a good account of the IR spectra. The dihedral angles are close to 90 ° in the gauche links, and close to 165 ° in the anti links.

The interpretation of the UV spectra is not straightforward. The standard Sandorfy C model is hopelessly inadequate. The optimally parameterized ladder C model reproduces very well the excitation energies of the first intense σ-σ^* transitions of all the all-trans conformers available, up to $Si_{16}Me_{34}$. The parameterization derived from the spectral properties of the various gauche conformers of Si_4Me_{10} and Si_5Me_{12} reproduces their own spectra very well. However, all *ab initio* calculations we have performed suggest that the spectra are most likely distorted by σ - π interactions, and that the agreement obtained with a method that only considers the σ electrons of the backbone is fortuitous. The β_{14}(gauche) parameter value derived from the spectra apparently actually corresponds to the syn rather than the gauche geometry, and is not expected to give good results for the various gauche conformers of longer chains, in which the σ - π interactions would be weaker or absent.

According to the *ab initio* results, the degree of σ conjugation in the oligosilane chain is a very sensitive function of the chain conformation, and is the strongest for the all-anti geometry and the weakest for the (hypothetical) all-syn geometry. It is still too early to tell just which conformational feature separates the individual chromophoric segments of the polysilane chain.

Acknowledgement

This work was supported by National Science Foundation (CHE 9020896). HSP is grateful to the Deutsche Forschungsgemeinschaft for a scholarship, and HT to the Nippon Telephone and Telegraph Co. for financial support for a stay at the University of Colorado.

References

1.   West, R. J. *Organomet. Chem.* **1986**, *300*, 327.

2.   Zeigler, J.M. *Synth. Met.* **1988**, *28*, C581.

3.   Miller, R.D.; Michl, J. *Chem. Rev.* **1989**, *89*, 1359.

4.   Michl, J. *Synth. Met.* **1992**, *50*, 367.

5.   Michl, J; Sun, Y.-P. ACS Symposium Series 527, Reichmanis, E., O'Donnell, J. H., and Frank, C.W., Eds.; American Chemical Society: Washington, DC, **1993**;Chapter 10, p. 131.

6.   Kepler, R.G.; Soos, Z.G., in "Relaxation in Polymers", Kobayashi, T., Ed.; World Scientific, Singapore, 1994.

7.   Plitt, H. S.; Michl, J. *Chem Phys. Lett.*, **1993**, *213*, 158.

8.   Sun, Y.-P.; Michl, J., *J. Am. Chem. Soc.* **1992**, *114*, 8186.

9.   Sun, Y.-P.; Veas, C.; Raymond, M.K.; West, R.; Michl, J. (to be published).

10.   Sun, Y.-P.; Hamada, Y.; Huang, L. M.; Maxka, J.; Hsaio, J.-S.; West, R.; Michl, J. *J. Am. Chem. Soc.* **1992**, *114*, 6301.

11.   Sun, Y.-P.; Miller, R.D.; Sooriyakumaran, R.; Michl, J. *J. Inorg. Organomet. Polym.* **1991**, *1*,3.

12.   Thorne, J. R. G.; Williams, S. A.; Hochstrasser, R. M.; Fagan, P. J. *Chem. Phys.*, **1991**, *157*, 401.

13.   Plitt, H. S.; Downing, J. W.; Raymond, M. K.; Balajki, V.; Michl, J. *J. Chem. Soc. Faraday Trans.*, **1994**, *90*, 1653.

14.   Ernst, C.A.; Allred, A. L.; Ratner, M. A. *J. Oranomet. Chem.*, **1979**, *178*, 119.

15.   The value of 180° previously reported[7] for the anti conformer of Si_4Me_{10} on the basis of an extensive but not ocmplete geometry optimization is now known to be wrong.

16.   Welsh, W. J.; Debolt, L.; Mark, J. E. *Macromolecules*, **1986**, *19*, 2987.

17.   Welsh, W. J.; Johnson, W. D. *Macromolecules*, **1986**, *23*, 1881.

18.   Shafiee, F.; Haller, K. J. *J. Am. Chem. Soc.*, **1986**, *108*, 5476.

19.   Lovinger, A. J.; Davis, D. D.; Schilling, F. C.; Padden, Jr., F. J.; Bovey, F. A.; Ziegler, J.M., *Macromolecules*, **1991**, *24*, 132.

20.   Plitt, H. S.; Balaji, V.; Michl, J., *Chem. Phys. Lett.*, **1993**, *213*, 158.

21.   Sandorfy, C., *Can. J. Chem.*, **1955**, *33*, 1337.

22.   Herman, A., *Chem. Phys.*, **1988**, *122*, 53; Kepler, R. G.; Soos, Z. G., *Phys. Rev. B*, **1991**, *43*, 12530.

23.   Klingensmith, K. A.; Downing, J. W.; Miller, R. D.; Michl, *J. Am. Chem. Soc.*, **1986**, *108*, 7438.

24.   Teramae, H.; Takeda, K., *J. Am. Chem. Soc.*, **1989**, *111*, 1281.

25.   Reed, A. E.; Curtiss; L.A. ; Weinhold, F., *Chem. Rev.* **1988**, *88*, 899.

26.   Balaji, V.; Michl, J.; *Polyhedron*, **1991**, *10*, 1265.

RECEIVED September 13, 1994

Chapter 31

Electronic Properties of Polysilanes

R. G. Kepler[1] and Z. G. Soos[2]

[1]Sandia National Laboratories, Department 1704, M/S 0338, P.O. Box 5800, Albuquerque, NM 87187–0338
[2]Department of Chemistry, Princeton University, Washington Road, Princeton, NJ 08544

The results of recent studies of the electronic properties of polysilanes are reviewed. The electronic states can be described by the Hückel model if coulomb interactions are included using the Pariser-Parr-Pople approximation. The long polymer chains appear to be divided into random length, short, ordered segments by conformational defects, with the energy of the excited states depending on the length of the segments. In isolated polymer chains energy is transferred from high-energy, short segments to longer, lower energy segments but the distance and time during which transfers take place is very limited. In solid films the excitons become highly mobile and remain mobile throughout their lifetime, even at low temperatures. Holes are quite mobile in solid films and the characteristics of transport are the same as those of charge carrier transport in molecularly doped polymer films.

Soluble polysilanes are a relatively new class of polymers that have recently come under intense scrutiny for a variety of possible applications. They are excellent photoconductors (1), their third order nonlinear optical coefficients are quite large for materials which are transparent in the visible (2), potentially useful structures have been produced photochemically in thin films (3), and they may prove to be useful in photoresist applications (4), particularly those which tend to be environmentally benign and do not require high resolution.

Since polysilanes are σ-bonded polymers which contain no π-bonds, it initially appears that their electronic states would be quite different from those in the more familiar π-conjugated polymers. It has now been shown, however that the σ-electrons are delocalized and play the same role as the π-electrons. The theory that can be used to describe the electronic states, as a first approximation, is the Hückel theory modified by the inclusion of coulomb interactions using the Pariser-Parr-Pople approximation (5-7). In both types of material the transfer integrals t alternate, t+δ and t-δ, but they alternate for

0097–6156/94/0579–0387$08.00/0

different reasons. In the σ-conjugated polysilanes one integral is between σ-orbitals on neighboring silicon atoms and the other between σ-orbitals on the same silicon atom. In π-conjugated polymers the alternation is typically caused by different bond lengths. The magnitude of δ is generally larger in the polysilanes also.

This theory and the experimental evidence that supports it (8) points up a strong similarity between the states of solid polysilane films and those of molecular crystals. In Figure 1 we indicate the energy, relative to the ground state, of the various excited states of poly(di-n-hexylsilane), PDHS, films and of the same states in anthracene single crystals.

Solution Studies

The absorption and fluorescence spectra of poly(methylphenylsilane), PMPS, in a dilute cyclohexane solution are shown in Figure 2. Two noteworthy aspects of these spectra will be discussed. The absorption spectrum is broad and featureless and the fluorescence spectrum is narrower and on the low energy edge of the absorption spectrum.

The broad featureless absorption spectrum is believed to result from inhomogeneous broadening. The long polysilane molecule is thought to be divided into random length, short, ordered segments on the order of 20 silicon atoms long. The defects which divide the molecule have been left undefined but are presumed to be conformational defects. The energy of the excited state of these segments depends on the length, being higher for shorter lengths. Experimental evidence for this hypothesis is provided by studies of the temperature dependence of the absorption spectrum in dilute solutions. The energy at the peak of the absorption spectrum decreases as the temperature is lowered. The temperature dependence can be modeled quite well by assuming that it takes a specific amount of energy to create the conformational defects which divide the segments, leading to more defects and shorter segments at higher temperatures (9). Further evidence is provided by the fact that in dilute solutions of PDHS an order-disorder transition is observed below room temperature. Schweizer (9) has shown that the solvent dependence of the transition temperature can be explained by assuming that the transition results from a competition between entropy favoring shorter segments created by more defects and energetic stabilization provided by van der Waals interaction between the segments, the polarizability of which is greater for longer segments, and the polarizable solvent molecules. The transition temperature is observed to increase with the polarizability of the solvent molecules, as predicted by this model, and does not depend on the solvent dielectric constant.

The fact that the fluorescence spectrum is narrower than the absorption spectrum and is on its low energy edge is generally interpreted as evidence that the excited state moves from the shorter, higher energy segments to longer, lower energy segments before a fluorescence photon is emitted. Time-dependent fluorescence depolarization experiments (10) show that the amount of depolarization of the fluorescence spectrum depends on the energy of the exciting light, being greater for higher energies, and occurs only for a short time following excitation. These results have led to the conclusion that the

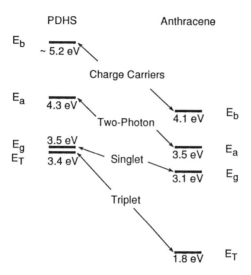

Figure 1. Threshold energies, in eV, for one-photon (E_g), two-photon (E_a), triplet (E_T) and charge carrier (E_b) excitations of PDHS films and anthracene crystals. (Reproduced with permission from reference 8)

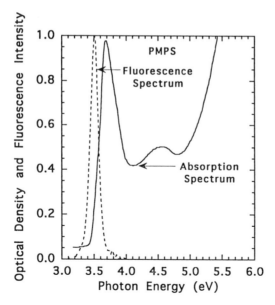

Figure 2. The absorption and fluorescence spectra of PMPS in dilute cyclohexane solution.

excited states quickly become trapped in the longer segments. Measurements of the absorption coefficient and the fluorescence lifetime have led to the conclusion that the fluorescing segments are 30-40 silicon atoms long.

Studies of the fluorescence spectra excited by light in the red edge of the absorption spectrum (11) provide additional evidence for the segment and exciton trapping model, and also show that the Stokes shift of the excited state is very small. If a nearly monochromatic beam of light with energy in the low-energy edge of the absorption band is used to excite PDHS molecules in 3-methylpentane at 1.4 K, the emission parallels the energy of the exciting light, as shown in Figure 3, indicating a Stokes shift of at most 10 cm^{-1}. If the energy of the exciting photons is higher, so that shorter segments are excited, the broad fluorescence spectrum is observed. These results also show that in isolated strands, the excitation energy is trapped in the longer, low-energy segments.

Two-photon absorption spectra can be observed in PDHS solutions by observing the fluorescence induced by two-photon transitions as a function of the energy of the exciting light (12). Such experiments locate the 2^1A_g state and the results are in agreement with the predictions of the theoretical model outlined above (13).

Solid State Studies

The absorption, fluorescence and two-photon absorption spectra of solid films of polysilanes are essentially the same as those in solution except for small solid-state energy shifts. Therefore, it is generally believed that in the solid state the absorption spectrum is still inhomogeneously broadened as it is in solution but it has been shown that excitons are no longer trapped in the longer segments, even at low temperature. In addition, photoconductivity studies have shown that positive charge carriers, holes, are quite mobile in a wide variety of polysilanes and the temperature and electric field dependence of the mobility is the same as the mobility of holes in molecularly doped polymers, suggesting that the holes are hopping among the segments discussed above.

Photoconductivity. Early time-of-flight studies of photoconductivity in polysilanes showed that at room temperature hole transport is nondispersive, i. e. the hole mobility is well defined, with very little trapping and that the mobility is strongly temperature dependent (1). Studies of the temperature and electric field dependence of the mobility (14) have shown that hole mobility in polysilanes can be described by the empirical relation

$$\mu = \mu_0 \exp\left[-\frac{\varepsilon_0 - \beta E^{\frac{1}{2}}}{kT_{eff}}\right] \tag{1}$$

where ε_0 is the activation energy for hopping, μ_0 is the limiting value at high temperatures, $1/T_{eff} = 1/T + 1/T_0$ with T_0 being defined empirically and β gives the field dependence. Such behavior was first noted by Gill (*15*) for hole transport in polyvinylcarbazole and has subsequently been observed in many molecularly doped polymers, where transport involves hopping among the dopant molecules. A theoretical description of this behavior still does not exist. However it is clear that it is characteristic of transport by hopping among molecular states and clearly suggests that hole transport in polysilanes is dominated by hopping among the ordered segments of the polysilane chains.

Carrier generation experiments in PMPS have shown that charge carriers are generated when photons with energies in the lowest energy, strong absorption band are absorbed (*1*). This observation was initially surprising since it is known that absorption in this band creates excitons, but it was shown that the carriers are generated by excitons diffusing to the film surface and injecting carriers from the electrode (*1*). It was concluded that the exciton diffusion length was 50 nm, in agreement with later results obtained in studies of exciton-exciton annihilation (*16*), to be discussed below. Measurements of carrier generation with much higher energy photons show a large increase in the quantum efficiency for carrier generation beginning at 4.6 eV which is interpreted as evidence for band-to-band transitions and a free carrier band gap of 4.6 eV (*17*). Similar measurements in PDHS suggest that the free carrier band gap in that material is 5.2 eV (*18*).

Electroabsorption experiments on PMPS films have supported these assignments (*17*). Some of the results are shown in Figure 4 where the change in transmission ΔT, induced by the application of an electric field, divided by the transmission T, is plotted versus photon energy. The strong signal, roughly proportional to the first derivative of the absorption band at 3.7 eV is indicative of a Stark shifted exciton band. The magnitude of the signal indicates a polarizability change of 1.8×10^{-22} cm^3, in reasonable agreement with the PPP model (*17*). The weaker signal at 4.65 eV is clearly not proportional to the first derivative of the absorption spectrum and is presumably associated with the free carrier band gap. The shape of the feature suggests redistribution of oscillator strength away from the center which could result from electron transfer between segments (*17*). Tachibana *et al.* (*19*) have measured the electroabsorption of PDHS and observed a very large feature beginning near 5.2 eV which may be associated with the band gap and is, therefore, also in agreement with the photoconductivity results.

Exciton Transport. An early potential application of polysilanes was as a self-developing resist, one in which material was removed under deep UV radiation (*4*). Studies of the quantum efficiency for material removal by light indicated that it decreased at high light intensities. This observation led to the hypothesis that exciton-exciton annihilation was responsible for the decrease and to measurements of the exciton-exciton annihilation rate constant.

The rate equation for the concentration of excitons, *n*, with exciton-exciton annihilation is

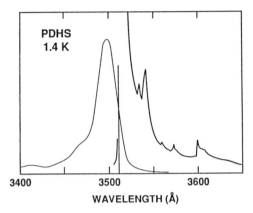

Figure 3   Absorption (thin line) and emission (thick line) spectra of PDHS in 3-methylpentane at 1.4 K. The vertical line at 351.1 nm indicates the wavelength of the exciting light.   (Reproduced with permission from reference 11)

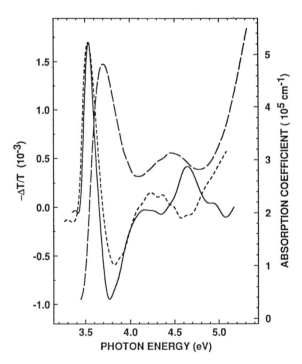

Figure 4. The electroabsorption spectrum at room temperature of PMPS (solid line). The absorption coefficient and its derivative are shown as the long- and short-dashed lines, respectively. The derivative of the absorption spectrum is plotted relative to the left ordinate and normalized to the magnitude of the electroabsorption spectrum at the maximum of the major peak. (Reproduced with permission from reference 17)

$$\frac{dn}{dt} = G(x,t) - \beta n - \gamma n^2 \qquad (2)$$

where t is time, x the distance into the sample, G is the rate of generation of excitons, β the reciprocal of the exciton lifetime τ, and γ is the exciton-exciton annihilation rate constant. It is envisioned that two excitons interact to create one highly excited state which very rapidly radiationlessly decays to a single exciton, thus annihilating one of the two excitons. It has been shown in a variety of experiments that Equation 2 can be used to describe the intensity of fluorescence observed as function of the exciting light intensity using γ as an adjustable parameter and to thus determine γ (16). It has been found that γ is about 1 x 10^{-7} cm^3 s^{-1} in a variety of polysilanes and that it is independent of molecular weight and relatively independent of temperature. This result shows that excitons are extremely mobile in polysilanes. In single crystals of anthracene γ is only about 1 x 10^{-8} cm^3 s^{-1}. If we assume that the excitons are particles, diffusing with a diffusion coefficient D, which annihilate one another at the distance R, the rate constant for annihilation is given (20) by $\gamma = 8\pi DR$. If we assume R to be 0.5 nm, D is found to be 0.08 cm^2 s^{-1}.

In the course of some photoconductivity measurements designed to determine the free carrier band gap in PDHS, we recently found (18) that 0.13% of the annihilation events create a free hole at 2 x 10^5 V cm^{-1}. This observation has provided a new tool for the study of exciton kinetics.

Solution of Equation 2 for a 5 ns laser pulse, a pulse long enough to reach equilibrium conditions since the exciton lifetime in PDHS is 600 ps, shows that the number of carriers generated per absorbed photon by a pulse of light should be, within a factor of about two, a function only of α x I, where α is the absorption coefficient and I is the intensity of the incident laser light pulse. In Figure 5 the number of carriers generated per absorbed photon at many wavelengths and intensities is plotted as a function of α x I. The asymptote at very high exciton concentrations gives the quantum efficiency for carrier generation at an electric field of 2 x 10^5 V cm^{-1}. The dashed lines are the theoretical predictions for $\alpha d = 1$, where d is the sample thickness, and $\gamma = 2$ x 10^{-7} cm^3 s^{-1} for the center, short-dashes line, $\gamma = 1$ x 10^{-7} cm^3 s^{-1} for the lower long-dashes line and $\gamma = 5$ x 10^{-7} cm^3 s^{-1} for the upper long-dashes line. The two solid lines were calculated using $\gamma = 2$ x 10^{-7} cm^3 s^{-1} and $\alpha d \rightarrow 0$ (upper line) and $\alpha d \rightarrow \infty$ (lower line). It is clear that the experimental data are quite consistent with the predictions for $\gamma = 2$ x 10^{-7} cm^3 s^{-1}.

The fact that carriers can be generated by exciton-exciton annihilation provides an opportunity to determine if excitons remain mobile throughout their 600 ps lifetime in PDHS. Since the number of carriers generated depends on the exciton concentration squared, the number of carriers generated by two, equal intensity, 30 ps laser light pulses incident on a sample depends on the time delay between the arrival of the two pulses. If the time delay is long compared to the lifetime of the excitons, the number of carriers generated will be just twice the number of carriers generated by a single pulse. If the two pulses are coincident the number of carriers generated is four times the number of carriers generated by a single pulse. For intermediate delay times, the

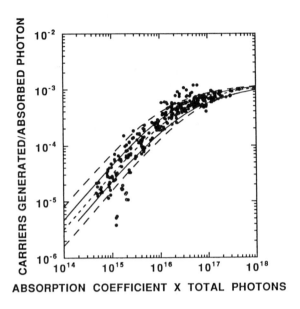

Figure 5. The number of carriers generated per absorbed photon versus the absorption coefficient times the light intensity. See the text for an explanation of the theoretical lines. (Reproduced with permission from reference 18)

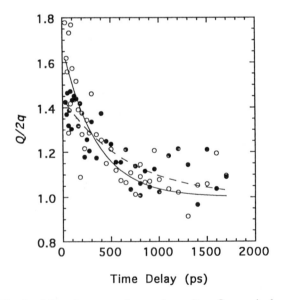

Figure 6. The double pulse experimental results. Open circles for parallel polarization and solid circles for perpendicular polarization. (Reproduced with permission from reference 18)

number of carriers decreases from four times to two times exponentially with the time constant being the exciton lifetime, provided the excitons remain active for carrier generation throughout their life.

Experimental results obtained using polarized light pulses are shown in Figure 6. The total number of carriers generated Q divided by twice the number of carriers generated by a single pulse q is plotted versus the time delay between the arrival times of the two pulses. The open circles are for data taken with the polarization of the two pulses parallel and the solid circles for data taken with the polarization of the two pulses at right angles to one another. The experimental data differ primarily at quite short times, < 300 ps, and the differences are quite reproducible (18). The solid and dashed lines are least square fits of the equation

$$\frac{Q}{2q} = 1 + ae^{-\beta\Delta t} \tag{3}$$

with a and β as adjustable parameters and Δt the time delay. The lifetime obtained from the parallel polarization results, the solid line, was 350 ps and the lifetime obtained from the perpendicular polarizations results, the dashed line, was 650 ps.

We believe that we qualitatively understand these results. PDHS is a crystalline polymer and thus, when the polarization of the two pulses are parallel those crystalline regions in which the molecular transition moment is parallel to the direction of polarization are preferentially excited. Therefore, the exciton concentration can decrease with time both by excitons decaying, either radiatively or nonradiatively, and by excitons diffusing out of the high concentration regions into regions of lower exciton concentration, leading to an apparent shorter exciton lifetime. When the polarizations of the two pulses are perpendicular, the initial exciton distribution is more homogeneous and this effect is minimized.

The double pulse experiments therefore show not only that at room temperature the excitons remain mobile throughout their lifetime and are not trapped in long, low energy segments as they are in isolated chains in dilute solutions, but also that they readily diffuse over distances larger than the crystallite size. The crystallites tend to be lamella on the order of 100 nm thick (21).

The double pulse experiments were performed at room temperature, so the question of whether or not the excitons are trapped in long, low-energy segments at low temperature in PDHS still has not been answered. In an attempt to address that question we have recently measured the change in γ between room temperature and 12 K by measuring the fluorescence intensity versus exciting light intensity. The experimental results are shown in Figure 7. The wavelength of the exciting light, a 5 ns pulse from a dye laser, was 297 nm. In order to calculate the change in γ between the two temperatures we had to measure the change in exciton lifetime and in the absorption coefficient at 297 nm. Within the accuracy of our lifetime measurements, made with a boxcar integrator and a fast photodiode, there was no change in lifetime and the absorption coefficient at 297 nm was found to decrease from 2×10^4 cm^{-1} at

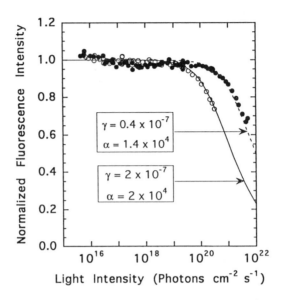

Figure 7. The ratio of the fluorescence intensity to the exciting light intensity, normalized to one at low intensity, versus the exciting light intensity. Open circles are data taken at 295 K and solid circles at 12 K. The solid and dashed lines are theoretical fits.

room temperature to 1.4×10^4 cm^{-1} at 12 K. The solid line in Figure 7 was calculated from Equation 2 using $\alpha = 2 \times 10^4$ cm^{-1}, $\beta = 1.67 \times 10^9$ s^{-1} and $\gamma = 2 \times 10^{-7}$ cm^3 s^{-1}. The dashed line, which fits the 12 K data, was calculated using the same β, changing α to 1.4×10^4 cm^{-1} and using γ as an adjustable parameter. The best fit, shown, was for $\gamma = 0.4 \times 10^{-7}$ cm^3 s^{-1}. Therefore these results show that γ decreases by only a factor of five between room temperature and 12 K, and that excitons are not trapped at low temperature.

Summary

It has now been shown, using a variety of experimental techniques, that the electronic states of polysilanes can be described, to a first approximation, by the Hückel model if coulomb interactions are included using the Pariser-Parr-Pople approximation. The long polymer chains appear to be divided into many random length, ordered segments which are separated by undefined conformational defects. In dilute solutions or glasses, excited states created on shorter, higher energy segments move quickly to longer, lower energy segments and become trapped. Studies in glasses at low temperature show that the Stokes shift in the excited state is less than 10 cm^{-1}. In solid films, when energy transfer to neighboring parallel segments becomes possible, the excited states, excitons, become highly mobile and remain mobile throughout their life, even at low temperatures. Holes are quite mobile in solid films and the mechanism of hole transport has been shown to be the same as that operative in molecularly doped polymers in which the charge carriers hop among the dopants. Presumably the holes hop among the ordered segments of the polymer chain.

Acknowledgments

The technical assistance of P. M. Beeson is gratefully acknowledged. The work at Sandia National Laboratories was supported by the U. S. Department of Energy under Contract No. DE-AC04-76DP00789 and the work at Princeton University was partially supported by NSF Grant No. NSF-DMR-8921072.

Literature Cited

(1)   Kepler, R. G.; Zeigler, J. M.; Harrah, L. A.; Kurtz, S. R. *Phys. Rev. B* **1987**, *35*, 2818-22.
(2)   Kajzar, F.; Messier, J.; Rosilio, C. *J. Appl. Phys.* **1986**, *60*, 3040-4.
(3)   Schellenberg, F. M.; Byer, R. L.; Miller, R. D. *Opt. Lett* **1990**, *15*, 242-244.
(4)   Zeigler, J. M.; Harrah, L. A.; Johnson, A. W. *SPIE Adv. Resist Tech. and Process. II* **1985**, *537*, 166.
(5)   Soos, Z. G.; Hayden, G. W. *Chem. Phys.* **1990**, *143*, 199-207.
(6)   Soos, Z. G.; Ramasesha, S. *Phys. Rev. B* **1984**, *29*, 5410.
(7)   Soos, Z. G.; Ramasesha, S.; Galvao, D. S. *Phys. Rev Lett.* **1993**, *71*, 1609-1612.
(8)   For a more complete review see Kepler, R. G.; Soos, Z. G. In *Relaxation in Polymers*; T. Kobayashi, Ed.; World Scientific: Singapore, 1993; pp 100-133.
(9)   Schweizer, K. S. In *Silicon-Based Polymer Science: A Comprehensive Resource*; J. M. Zeigler and F. W. G. Fearon, Ed.; American Chemical Society: Washington DC, 1990; pp 379-395.
(10)   Kim, Y. R.; Lee, M.; Thorne, J. R. G.; Hochstrasser, R. M.; Zeigler, J. M. *Chem. Phys. Lett.* **1988**, *145*, 75-80.
(11)   Tilgner, A.; Trommsdorff, H. P.; Zeigler, J. M.; Hochstrasser, R. M. *J. Chem. Phys.* **1992**, *96*, 781-96.
(12)   Thorne, J. R. G.; Ohsako, Y.; Zeigler, J. M.; Hochstrasser, R. M. *Chem. Phys. Lett.* **1989**, *162*, 455-60.
(13)   Soos, Z. G.; Kepler, R. G. *Phys. Rev. B* **1991**, *43*, 11908-11912.
(14)   Abkowitz, M. A.; Rice, M. J.; Stolka, M. *Philos. Mag. B* **1990**, *61*, 25-57.
(15)   Gill, W. D. *J. Appl. Phys.* **1972**, *43*, 5033.
(16)   Kepler, R. G.; Zeigler, J. M. *Mol. Cryst. Liq. Cryst.* **1989**, *175*, 85-91.
(17)   Kepler, R. G.; Soos, Z. G. *Phys. Rev. B* **1991**, *43*, 12530.
(18)   Kepler, R. G.; Soos, Z. G. *Phys. Rev. B* **1993**, *47*, 9253-9262.
(19)   Tachibana, H.; Kawabata, Y.; Koshihara, S.; Tokura, Y. *Synth. Met.* **1991**, *41*, 1385-8.
(20)   Chandrasekhar, S. *Rev. Mod. Phys.* **1943**, *15*, 1-89.
(21)   Schilling, F. C.; Bovey, F. A.; Lovinger, A. J.; Zeigler, J. M. In *Silicon Based Polymer Science*; J. M. Zeigler and F. W. G. Fearon, Ed.; American Chemical Society: Washington, DC, 1990; pp 341-378.

RECEIVED October 21, 1994

Chapter 32

UV Photoelectron Spectroscopy of Polysilanes and Polygermanes

Kazuhiko Seki[1], Akira Yuyama[1], Satoru Narioka[1], Hisao Ishii[1], Shinji Hasegawa[2], Hiroaki Isaka[3], Masaie Fujino[3], Michiya Fujiki[3], and Nobuo Matsumoto[3]

[1]Department of Chemistry, Faculty of Science, Nagoya University, Furocho, Chikusa-ku, Nagoya 464–01, Japan
[2]Institute for Molecular Science, Myodaiji, Okazaki 444, Japan
[3]NTT Basic Research Laboratories, Wakamiya, Morinosato, Atsugi 243–01, Japan

UV photoelectron spectra were measured for ten polymers with main chains composed of Si and Ge atoms. The spectra of poly(dialkylsilane)s were described well as the overlap of the spectra of constituent parts, while polysilanes with aryl side groups showed the effect of σ-π mixing. The spectra of polygermanes and Si-Ge block copolymers are fairly similar to those of corresponding polysilanes, while Si-Ge random copolymers show indications of the effect of the presence of many Si-Ge bonds.

Recently polysilanes and polygermanes have attracted attention due to their unique properties. These are mainly due to σ-conjugation along the main chain, with possible applications to photoresists, conducting polymers, and initiators of polymerization, etc. (1,2). Knowledge of their electronic structures forms the basis of understanding these interesting properties. So far there have been reports on the valence electronic structures of several polysilanes studied by UV photoelectron spectroscopy (UPS) (3-5), but they suffered either from low spectral quality or from limited coverage of the valence band. In this work we report the whole valence electronic structures of a series of polysilanes, polygermanes, and Si-Ge copolymers with UPS using synchrotron radiation, and analyze the results by comparison with the spectra of constituent parts.

Experimental

All the sample materials were synthesized at NTT. The samples for UPS measurements were prepared by spin-coating of 0.4 weight % toluene solution of each compound under N_2 atmosphere. The samples

prepared under these conditions   showed no sign of charging during
the   UPS measurement.   The spectra were measured under ultrahigh
vacuum on the order of 10^{-8} Pa by using an   angle-resolving photoelectron spectrometer at the beamline 8B2 of the UVSOR facility at
Institute for Molecular Science.   Monochromatized light from a plane
grating monochromator (5) impinged on the sample, and the energy
distribution of photoelectrons was analyzed with a hemispherical
electrostatic energy analyzer. The overall resolution was about 0.2
eV as deduced from the measurement of the Fermi edge of gold.   No
detectable damage by photon irradiation was observed.   All the measurements were carried out at room temperature.

Polysilanes

Polydialkylsilanes.   In Figure 1(a) the spectrum of poly(methylpropylsilane) $(SiMePr)_n$ at $h\nu = 40$ eV is shown.   We can clearly see
several fine structure peaks   in the upper valence region, in contrast to previously reported spectra (3), which suffered from severe
sample charging and do not show fine structure in this region.   For
comparison, reported UPS and XPS spectra   of methane (7,8), propane
(8,9), and permethyltetrasilane (10) are shown in Figure 1(b)-(d).
They show good correspondence with the spectrum of $(SiMePr)_n$.   From
such a comparison and literature assignments (4,5,10), we can ascribe
the features in 5-6 eV region to Si3p-derived σ-delocalized
states along the chain, those at 9-10 eV   to pseudo-π
states along the Si chain and Si-C bonds, those at 10-15 eV to levels
due to   C2p orbitals of methyl and propyl groups (with weak Si3s-
derived states underneath), and those at 16-25 eV to C2s-derived
states.
    We note that the features associated with   the Si3p levels of
the principal Si chain are in a much lower binding energy region than
the corresponding C2p-derived levels of propane.   Thus the uppermost
part of the valence states is solely derived from the Si main chain
with little mixing with the electronic states of the alkyl side
group.   The alkyl substitution only raises the uppermost valence
states by an inductive effect.
    Still the small electronegativity of silicon results in a small
ionization energy of 5.6_5 eV, comparable to those of π-conjugated
polymers.   This situation is demonstrated in Figure 2 by   comparison
with typical polymers, *i.e.*, polyethylene $(CH_2)_n$ (11), polyacetylene
$(CH)_n$ (12), polytetrafluoroethylene $(CF_2)_n$ (13), poly(dimethylsiloxane) (14), and polythiophene $(C_4H_4S)_n$ (15).   The uppermost valence
state of $(SiMePr)_n$ is at almost the same energy as those of polyacetylene and polythiophene.   Also note that this situation is completely
changed in poly(dimethylsiloxane), where the insertion of oxygen
breaks the σ conjugation, leading to a large ionization threshold of
8.3 eV (15).   This difference between polysilane and polysiloxane can
be regarded as a one-dimensional version of corresponding difference
between Si and SiO_2.
    Unfortunately, the ionization cross-section of Si3p and 3s
orbitals are smaller than that of the C2p orbital, and in the binding energy region above about 9 eV the spectrum in Figure 1 in   is
dominated by the contribution from the carbon-containing   side
groups.
    The spectra of other polydialkylsilanes, namely   poly(di-*n*-

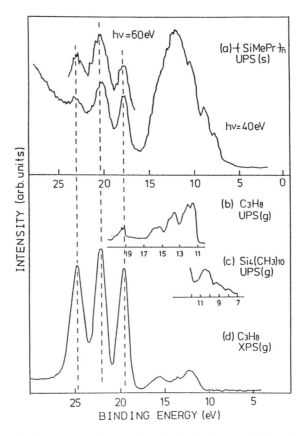

Figure 1.   (a) UPS spectrum of poly(methylpropylsilane) (SiMePr)$_n$ at hν = 40 eV, (b) gas-phase UPS spectrum of propane (9), (c) gas-phase UPS spectrum of permethylbutane (10), and (d) gas-phase XPS spectrum of propane (8). The gas-phase spectra are aligned to (a) at the C2s peaks as shown by the broken line.

butylsilane) $(SiBu_2)_n$ and poly(di-n-hexylsilane) $(SiHex_2)_n$ in Figure 3 can also be described as the overlapping spectra of the constituent parts. We note that the intensity of the uppermost Si3p-derived valence states relative to that of the alkyl-derived deeper levels becomes weaker with increasing side chain length, simply due to the change in the Si/C ratio. At the same time, the right-hand edge of the intense feature due to the alkyl part shifts towards a lower binding energy, because of the σ delocalization in the alkyl chain.

Polysilanes with Aromatic Side Groups. In Figure 4(a) we show the spectrum of poly(methyphenylsilane) $(SiMePh)_n$ and the spectra of constituent parts (Figure 4(b)-(e)) (7,8,10,16). Again the spectral quality is much better than previously reported (3). The spectral features of $(SiMePh)_n$ in the higher binding energy region is generally fairly well described as the overlap of the constituent parts, as in $(SiMePr)_n$, except for the following.

Unlike polyalkylsilanes, in the uppermost part of the valence states of $(SiMePh)_n$ the π-levels in the phenyl group overlap with the delocalized Si-derived states, resulting in σ-π mixing. This is illustrated in Fig. 5 (4) by comparison with phenylpentamethylsilane (17), which can be regarded as the repeating unit of $(SiMePh)_n$. One of the doubly degenerate π highest occupied molecular orbitals (HOMOs) of benzene (left-hand side in Fig. 5), with a node at the Si-C bond, will not be affected by the phenyl substituent. The peak B in the spectrum of $(SiMePh)_n$ at 7.7 eV should come from this unaffected HOMO of benzene. The other HOMO of benzene (right-hand side of Fig. 5), with a large amplitude at the Si-C bond, is mixed with the σ levels delocalized over the main chain to a degree that depends on the angle between the phenyl group and the main chain.

This mixing results in both a bonding and an antibonding combination of these levels. The weak feature C at about 8.3 eV (arrowed in Fig. 6) seems to originate in the bonding combination, with the main contribution due to the HOMO of benzene. The highest valence state feature in the spectrum PMPS, labeled A in Fig. 4, seems to correspond to the antibonding combination of the σ delocalized orbital and the HOMO of benzene, with the main contribution from the former.

Such a situation is confirmed by the comparison with the density of states calculated for $(SiMePh)_n$ shown in Fig. 6, which was obtained by a LCAO band calculation with Slater and Koster type formulation (4). The energy axis is contracted by a factor of 1.2 and shifted to align with the feature B. The correspondence between theory and experiment is fairly good. The levels derived from Si-C bonds, which in permethyltetrasilane appear around the energy of feature C, is calculated to lie in a higher binding energy region.

The mixing is expected to raise the ionization threshold energy of PMPS compared to those of alkylsubstituted polysilanes. Indeed, the observed value of 5.3 eV of $(SiMePh)_n$ is smaller than those of $(SiMePr)_n$ (5.6_5 eV) (K. Seki, T. Takeda, M. Fujino, and N. Matsumoto, unpublished.) and polydimethylsilane $(SiMe_2)_n$ (5.9 eV) (5).

A similar discussion applies to the spectrum of poly(di(4-ethylphenyl)silane) $(Si(Et-Ph)_2)_n$ in Fig. 7. The peak at 6.8 eV is assigned to the nonbonding HOMO of benzene, and the features at 8.0 and ca.5.3 eV can be ascribed to the bonding and antibonding combinations of the other HOMO of benzene and the main chain.

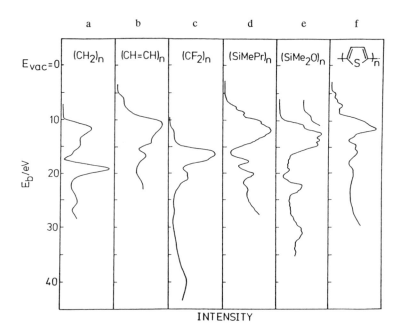

Figure 2. UPS spectra of various polymers. (a) polyethylene (*11*),
(b) polyacetylene (*12*), (c) poly(tetrafluoroethylene) (*13*), (d)
poly(methylphenylsilane) (this work), (e) poly(dimethylsiloxane)
(*14*), and (f) polythiophene (*15*).

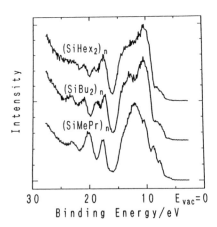

Figure 3.   UPS spectra of poly(dibutylsilane) and poly(dihexylsi-
lane) at hν = 40 eV compared with that of poly(methyl-propylsi-
lane).

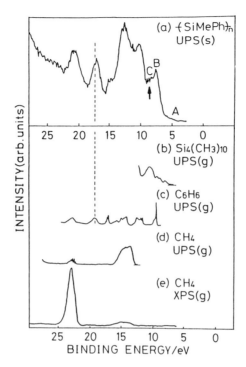

Figure 4. (a) UPS spectrum of poly(methylphenylsilane) (SiMePh)$_n$ at hν = 40 eV, (b) gas-phase UPS spectrum of permethylbutane (*10*), (c) gas-phase UPS spectrum of benzene (*16*), (d) gas phase UPS spectrum of methane (*7*), and (e) gas-phase XPS spectrum of methane (*8*). The gas phase-spectra are aligned to (a) at the C2s peaks as shown by the broken line.

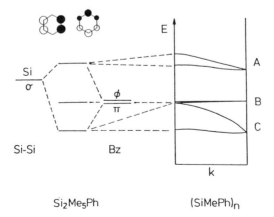

Figure 5. σ-π Mixing in phenylpentamethyldisilane (left) and poly(methylphenylsilane) (right). The two degenerate π HOMOs of benzene are also shown.

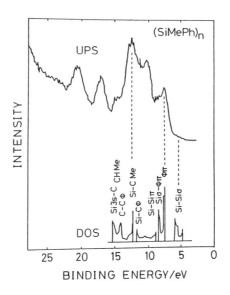

Figure 6. Comparison of the observed UPS spectrum of $(SiMePh)_n$ with the calculated density of states (*4*).

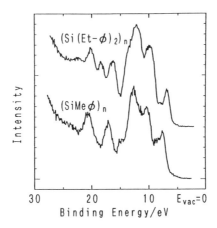

Figure 7. UPS spectra of poly(di(4-ethylphenyl)silane) and poly(methylphenylsilane).

These results show that the σ-π mixing between the main chain and the aromatic substituent will be effective in designing the electronic structure of polysilanes. With larger aromatics, the ionization energy may be further reduced by strong σ-π mixing.

Polygermanes

The spectra of poly(dibutylgermane) $(GeBu_2)_n$ and poly(dihexylgermane) $(GeHex_2)_n$ are shown in Fig. 8. They are similar to the corresponding polysilanes in Fig. 2 except for a small shift towards a lower binding energy. This similarity corresponds well with that observed in the optical spectra (*18*). The reported gas-phase UPS spectra of oligogermanes (*19*) also show fairly good correspondence to those of the silicon analogues.

Silane-Germane Copolymers

In Fig. 9 we show the spectrum of (a) block copolymer poly(dibutyl-germane-co-dibutylsilane) $(GeBu_2)_m(SiBu_2)_n$, (b) random copolymer poly(dibutylgermane-co-tetramethyldisilane) $(GeBu_2)_m(Si_2Me_4)_n$, and (c) random copolymer poly(dibutylgemane-co-dimethylsilane) $(GeBu_2)_m$-$(SiMe_2)_n$. The spectrum is fairly well simulated as the overlap of the constituent parts. In the random copolymers, however, the upper-most spectral feature corresponding to the Si-Ge chain becomes less sharp than other polymers, possibly corresponding to the formation of Si-Ge bonds.

Conclusion

UV photoelectron spectra of ten polymers with Si- and Ge-derived main chains have been measured. They are of much better quality than the previously reported ones, and cover the whole valence state region. They are rather well described as the overlapping contributions from the main chain and the side groups. For compounds with aromatic substituents, an effect of σ-π mixing was observed.

An interesting point in the UPS study of polysilanes, which we could not pursue in this work, is the delicate change of electronic structure related to thermochromism (*19*). This requires accurate determination of the ionization threshold, which is better accomplished with low energy photons. Such a study will also shed more light onto the delicate change of electronic structures between polysilanes, polygermanes, and Si-Ge copolymers. This study is now underway in our group.

Acknowledgments

This work was supported in part by the Grants-in-Aid for Scientific Research (Nos. 05NP0301 and 04403001) from the Ministry of Education, Science and Culture of Japan. This work was carried out as a part of Joint Studies Program of the UVSOR facility of Institute for Molecular Science (No. 5-B209).

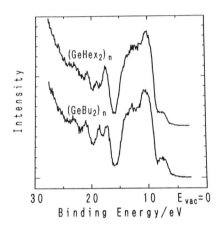

Figure 8.   UPS spectra of poly(dibutylgermane) and poly(dihexyl-germane).

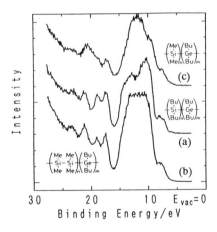

Figure 9. UPS spectra of Si-Ge copolymers. (a) block poly(dibutylgermane-co-dibutylsilane), (b) random poly(dibutylgermane-co-tetramethyldisilane), and (c) random poly(dibutylgermane-co-dimethylsilane).

Literature Cited

1. *Silicon-Based Polymer Science: A Comprehensive Resource*, Zeigler, J.M. and Fearon, F.W.G. Eds. ACS, Washington, 1990.
2. Miller, R.D. and Michl, J. *Chem. Rev.*, 1989, *89*, 1359.
3. Loubriel, G and Zeigler J, *Phys. Rev.*, 1986, *B33*, 4203.
4. Takeda, K., Fujino, M., Seki, K., and Inokuchi, H. *Phys. Rev.*, 1987, *B36*, 8129.
5. Seki, K., Mori, T., Inokuchi, H., and Murano, K. *Bull. Chem. Soc. Jpn.*, 1988, *61*, 351.
6. Seki, K., Nakagawa, H., Fukui, K., Ishiguro, E., Kato, R., Mori, T., Sakai, K., and Watanabe, M. *Nucl. Instrum. Methods*, 1986, *A246*, 264.
7. Potts, A.W, Williams, T.A., and Price, W.C. *Faraday Disc. Chem. Soc.*, 1972, *54*, 104.
8. Pireaux, J.J., Svenson, S., Basilier, E., Malmqvist, P.A, Gelius, U., Caudano, R., and Siegbahn, K. *Phys. Rev.*, 1976, *A14*, 2133.
9. Kimura, K., Katsumata, S., Achiba, Y., Yamazaki, T., and Iwata, S. *Handbook of HeI Photoelectron Spectra of Fundamental Organic Molecules*, (Japan Scientific Societies Press, Tokyo, 1988).
10. Bock, H. and Ensslin, W. *Angew. Chem. Internat. Ed.*, 1971, *10*, 404.
11. Fujimoto, H., Mori, T., Inokuchi, H., Ueno, N., Sugita, K., and Seki, K. *Chem. Phys. Lett.*, 1987, *141*, 485.
12. Tanaka, J., Tanaka, C., Miyamae, T., Kamiya, K., Shimizu, M., Oku, M., Seki, K., Tsukamoto, J., Hasegawa, S., and Inokuchi, H. *Syn. Metals*, 1993, *55-57*, 121.
13. Seki, K., Tanaka, H., Ohta, T., Aoki, Y., Imamura, A., Fujimoto, H., Yamamoto, H., and Inokuchi, H. *Phys. Scripta*, 1990, *41*, 167.
14. Sugano, T., Seki, K., Ohta, T., Fujimoto, H., and Inokuchi, H. *J. Chem. Soc. Jpn.*, 1990, 594 (in Japanese).
15. Fujimoto, H., Nagashima, U., Inokuchi, H., Seki, K., Cao, Y., Nakahara, H., Nakayama, J., Hoshino, M., and Fukuda, K. *J. Chem. Phys.*, 1990, *92*, 4077.
16. Asbrink, L., Edqvist, O., Lindholm, E., and Selin, L.E. *Chem. Phys. Lett.*, 1970, *5*, 192.
17. Pitt, C.G., and Bock, H. *J. Chem. Soc. Chem. Commun.*, 1972, *28*.
18. Tachibana, H., Kawabata, Y., Yamaguchi, A., Moritomo, Y., Koshihara, S., and Tokura, Y. *Phys. Rev.*, 1992, *B45*, 8752.
19. Tokura, T., Mochida, K., Masuda, S., and Harada, Y. *Chem. Lett.*, 1992, 2281.

RECEIVED September 13, 1994

Chapter 33

Radical Ions of Polysilynes

Akira Watanabe[1], Minoru Matsuda[1], Yoichi Yoshida[2], and
Seiichi Tagawa[2]

[1]Institute for Chemical Reaction Science, Tohoku University, Katahira,
Aoba-ku, Sendai 980, Japan
[2]Institute of Scientific and Industrial Research, Osaka University, 8–1
Mihogaoka, Ibaraki, Osaka 567, Japan

The radical ions of polysilynes which have silicon network structure
were studied by the pulse radiolysis. The polysilynes showed
broad absorption and emission spectra due to a quasi-two dimensional
Si skeleton. The Si skeleton of polysilynes was investigated by the
far-IR spectra. The far-IR spectra were compared with the calculated
vibrational bands of cyclic silicon model compounds. As the Si
skeleton of the polysilynes, the localized silicon rings connected by
linear silicon chains were presumed. The transient absorption
spectra of radical ions of poly(n-propylsilyne) and poly(n-hexylsilyne)
showed two characteristic absorption bands in the visible and the IR
region. A time-dependent spectral change of the radical ions was
explained by the formation of the charge resonance (CR) state
where the radical ion site is stabilized after the geometric change of
the σ-conjugated polysilane chain.

Polysilane is an organometallic polymer that has a Si-Si main chain and organic side
chains. The discoveries of soluble polysilanes with high molecular weight have
caused considerable attention to the properties of the σ-conjugated polymers (1-3).
It is a new type of conjugated system for π−conjugation. For π−conjugated
polymers, characteristic absorption bands related to the conduction state were observed.
For example, a soliton model, a polaron model, and a bipolaron model were developed
for polyacetylene, polyaniline, polypyrrole, polythiophene, and so on (4-10). Figure
1 shows energy-band diagrams of polaron and bipolaron. When conjugated polymer
is doped with an electron acceptor, new energy levels are formed between the valence
band and the conduction band. Doping with an electron donor also forms new
energy levels. In a chemical sense, these states can be interpreted as radical cation
and radical anion, respectively (7-10). The bipolaron can be interpreted as dication.
The formation of the new energy levels causes low energy transition as shown in
Figure 1. These low energy transitions which appear in the IR region are closely
related to the conduction state of the conjugated polymer. With an increase in the
conductivity, the transition energy of the absorption band in the IR region decreases
and the spectral shape changes from sharp one to broad one (10). This type of
relationship between the IR band and the conductivity is shown in many π−conjugated
polymers.

No characteristic absorption bands related to the conduction state have been observed for doped polysilane.   One of the reasons is the instability of the Si-Si σ-bond of polysilane to doping with a cationic or anionic dopant.   In such a case, transient techniques are advantageous to observe unstable species.   Pulse radiolysis provides the possibility to observe unstable ionic states of polysilanes (*11-16*).   In the electron beam-induced reaction in solutions, most of the electron beams are absorbed by the solvent and the ionic species of the solvent are produced.   In the next step, the electron transfer between the ionic solvent molecules and the solute takes place (*17*).   When the solvent is tetrahyrofuran (THF), solvated electrons are produced and some of them are transferred to polysilane, and the radical anion of polysilane is produced.   On the other hand, cationic species are produced by the electron beam irradiation to methylene chloride, and electron abstraction from polysilane by the cationic species of the solvent forms a radical cation of polysilane.

Tagawa et al. have reported on the ionic state of polysilanes which have a linear Si-Si chain (*12,13*). The radical ions of polysilanes show two characteristic absorption bands in the UV region and the IR region.   There are some problems concerning the polysilane radical ion.   The first problem is the radical ion site where charge is mainly located.   Is it the side chain type or the main chain type?   The second problem is the assignment of the IR band.   The third problem is the relaxation phenomena of the IR band.   In this paper we report on the radical ion state of polysilynes.   Polysilynes have been considered to be silicon network polymers which have a quasi-two-dimensional Si skeleton (*18-22*). The increase in σ-conjugation along the Si skeleton with an increase in Si-dimensionality is expected, and the absorption band of the radical ion must be influenced by the change of the σ-conjugation state.   In the following, we call polysilanes which have a linear Si-Si chain polysilylene and call polysilanes which have a silicon network structure polysilyne on the basis of the repeating unit.

Experimental

Polysilynes were synthesized by the Kipping reaction of the trichloro-organosilanes using Na in toluene at 110°C under N_2 atmosphere. The molecular weight of polysilynes

R = phenyl, *n*-propyl, *n*-hexyl

was determined by GPC using a monodispersed polystyrenes as standards.   The values of the molecular weight are summarized in Table I.   Fractional reprecipitation

Table I.   The Molecular Weight and the Molecular Weight Distribution of Polysilynes

Polymer	Mw	Mw/Mn
Poly(phenylsilyne)	4,000	2.04
Poly(*n*-propylsilyne)	10,900	4.91
Poly(*n*-hexylsilyne)	12,000	1.23
Poly(*n*-hexylsilyne)	119,000	2.66

from THF solution using methanol as a precipitant was carried out to obtain a high and a low molecular weight poly(n-hexylsilyne).

Absorption and emission spectra were measured by a Hitachi U-3500 and a Shimadzu RF-502A, respectively.   The IR and far-infrared absorption spectra were measured using a JEOL100 FT-IR spectrophotometer.   The Far-IR spectra in the region from 700 to 250 cm^{-1} were measured using a KBr disk.   The Far-IR spectra in the region   from 570 to 50 cm^{-1} were measured using a sample prepared by dispersing a polysilyne into a paraffin matrix and pasting it on a polyethylene film (thickness 0.02 mm).   The pulse radiolysis system has been described in previous papers (23,24).   The electron pulse was with 2 ns duration from a 35 MV electron linear accelerator.   The dose per pulse is evaluated from the absorbance of the hydrated electron at 500 nm in neat deaerated water and determined as 4 - 5 krad per 2 ns pulse.   Solvents, THF and CH$_2$Cl$_2$, were dehydrated and degassed up to 10^{-5} Torr, and transferred to a quartz cell (1 x 1 cm^2 square and 2 cm long) on a vacuum line.   The calculations of vibrational bands were done by MOPAC with the MNDO approximation (25).

Results

Absorption and Emission Spectra of Polysilynes.   Figure 2 shows the absorption and emission spectra of poly(phenylsilyne) and poly(n-propylsilyne).   Poly(phenyl-silyne) shows an intense absorption below 300 nm, which is attributable to the π–π* absorption of the phenyl group.   The absorption above 300 nm can be assigned to the σ-σ* absorption of Si skeleton.   The emission of poly(phenylsilyne) and poly(n-propylsilyne)   can be also assigned to the emission from the Si skeleton. Polysilynes show broad absorption and emission spectra in a lower energy region compared to polysilylene.   This suggests an increase in σ-conjugation due to the silicon network structure.   The effect of the molecular weight on the absorption and emission spectra was investigated for   poly(n-hexylsilyne).   Figure 3 shows the absorption spectra and emission spectra of   the poly(n-hexylsilyne)s with the low molecular weight (Mw = 12,000, Mw/Mn = 1.23) and the high molecular weight (Mw = 119,000, Mw/Mn = 2.66).   The absorption shoulder at 300 nm increases slightly with an increase in the molecular weight.   In the emission spectra, the maximum wavelength shifted from 450 to 466 nm with an increase in the molecular weight.   The emission spectra show a significant effect of the molecular weight compared to the absorption spectra because the emission occurs from the trap site following the energy migration along the Si-Si chain.

Si skeleton of Polysilynes Studied by Far-IR Spectra.   Figure 4 shows the far-IR spectra of polysilanes by KBr disk method.   Poly(methylphenylsilylene) which has a linear Si-Si chain shows two sharp bands assigned to asymmetric (461 cm^{-1}) and symmetric vibrations (382 cm^{-1}) of the silicon-silicon bond.   Polysilynes show broad absorption bands compared to poly(methylphenylsilylene). Poly(phenylsilyne) shows a broad absorption band at 495 cm^{-1} with a peak at 461 cm^{-1}.   The 461 cm^{-1} absorption can be assigned to the asymmetric vibration of a linear Si-Si chain. Poly(n-propylsilyne) shows a broad absorption band at 453 cm^{-1}. Poly(n-hexylsilyne) also shows a broad band at 393 cm^{-1} with shoulders at 496 and 463 cm^{-1}.   The far-IR spectra of polysilanes in the lower energy region from 570 to 50 cm^{-1} were observed using a paraffin matrix (Figure 5).   The polysilanes showed no absorption band below 200 cm^{-1}.   The far-IR spectra are quite different from amorphous silicon (26-28).

Figure 6 shows the calculated vibrational spectra for silicon model compounds. The structure of the model compounds was optimized by MNDO calculation.   The asymmetric (498 cm^{-1}) and symmetric (395 cm^{-1}) vibrations of the silicon-silicon bond were obtained using a linear hexamer model.   The calculated spectrum well

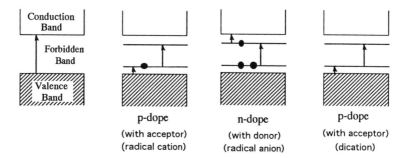

Figure 1.   Energy-band diagram of polaron and bipolaron.

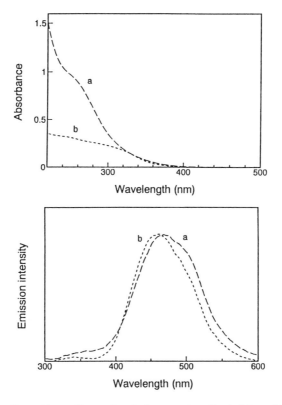

Figure 2.   Absorption and emission spectra of poly(phenylsilyne) and poly(n-propylsilyne) in THF. (a) poly(phenylsilyne); 0.54 mM and (b) poly(n-propylsilyne); 0.56 mM, Optical length; 0.2 cm.

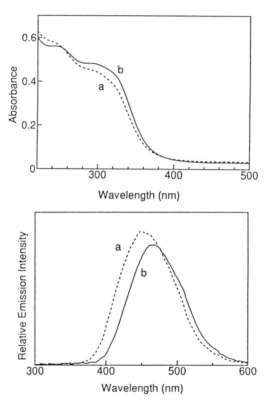

Fugure 3. Absorption and emission spectra of poly(*n*-hexylsilyne) in THF. (a) low molecular weight poly(*n*-hexylsilyne) (Mw = 12,000); 0.50 mM, (b) high molecular weight poly(*n*-hexylsilyne) (Mw = 119,000); 0.50 mM. Optical length; 0.2 cm.

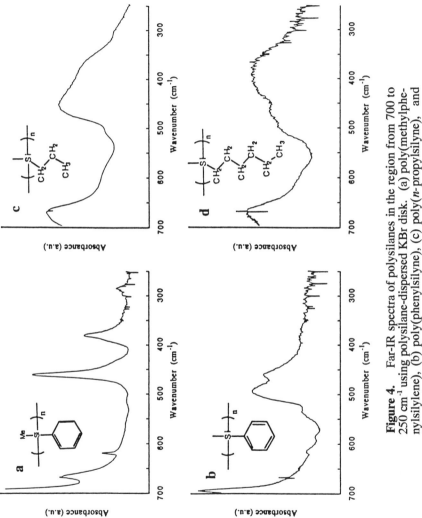

Figure 4. Far-IR spectra of polysilanes in the region from 700 to 250 cm^{-1} using polysilane-dispersed KBr disk. (a) poly(methylphenylsilylene), (b) poly(phenylsilyne), (c) poly(n-propylsilyne), and (d) poly(n-hexylsilyne).

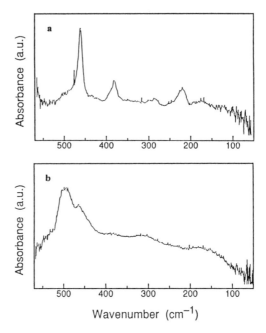

Figure 5.   Far-IR spectra of polysilanes in the region from 570 to 50 cm^{-1} using polysilane-dispersed paraffin matrix. (a) poly(methylphenylsilylene) and (b) poly(phenylsilyne).

corresponds to the observed spectrum of poly(methylphenylsilylene). In Figure 6, the calculated spectra of various cyclic silicon model compounds are displayed. The calculated spectrum of hexacyclosilane shows absorption peaks at 512, 398, 334, and 210 cm^{-1}. The spectral shape is similar to the observed absorption band of poly(phenylsilyne) as shown in Figure 5b. As a main Si skeleton of poly(phenyl-silyne), a silicon 6-membered ring can be expected. The similar far-IR spectrum was reported for perphenylhexasilane (29). In addition, a sharp peak overlapping with the broad band at 461 nm was observed. This band is assigned to a linear Si-Si structure.

Transient Absorption Spectra of Polysilyne Radical Ions. In the previous paper, we reported the transient absorption spectra of the poly(phenylsilyne) radical ions obtained by pulse radiolysis (30). The radical anion of poly(phenylsilyne) shows two characteristic absorption bands at 400 and 900 nm. Similar broad absorption bands were observed for the radical cation of poly(phenylsilyne) in CH$_2$Cl$_2$. Figure 7 shows the transient absorption spectra of the poly(n-propylsilyne) radical anion and radical cation. There are two characteristic absorption bands in the visible and the IR region. Similar spectra were observed for the radical ions of poly(n-hexylsilyne) as shown in Figure 8. In Figures 7 and 8, a noteworthy feature is seen. A significant time-dependent spectral change between delay times of 10 and 100 ns is shown. Figure 9 shows the absorption-time profiles of poly(n-hexylsilyne) radical anion at 460 and 1500 nm. The IR band at 1500 nm increases whereas the visible band at 400 nm is almost constant. Such behavior was more remarkable in the case of the radical cation as shown in Figure 10. The absorption-time profile of the visible band shows significant differences from the IR band. The growth of the absorption band at 1500 nm was observed, which corresponds to the decay of the absorption band at 460 nm.

The radical anion of the high molecular weight poly(n-hexylsilyne) shows a broad spectra similar to the lower molecular weight poly(n-hexylsilyne) as shown in Figure 11. However, the absorption-time profiles are quite different from the low molecular weight poly(n-hexylsilyne) as shown in Figure 12. The absorption-time profile at 1500 nm shows the fast decay immediately after the pulse radiolysis and the following slow growth. The former can be attributed to the decay of the THF-solvated electron. The high molecular weight poly(n-hexylsilyne) must be surrounded tightly by the n-hexyl group. The electron transfer of the THF-solvated electron to the Si-Si chain of the high molecular weight poly(n-hexylsilyne) is reduced by the organic side chain. The radical cation of the high molecular weight poly(n-hexylsilyne) could not be observed due to the insolubility in CH$_2$Cl$_2$.

Discussion

The Si skeleton of polysilynes. A polysilyne has been considered to be a quasi-two-dimensional silicon network polymer. However, there is little data on the silicon skeleton of polysilyne. Far-IR absorption spectroscopy is an effective technique to investigate the Si skeleton (31). For amorphous silicon, the standard TA-LA-LO-TO absorption bands have been reported on the lattice vibration in the range of 550 to 50 cm^{-1}, where TA is the transverse acoustical mode, LA the longitudinal acoustical mode, LO the longitudinal optical mode, and TO the transverse optical mode of silicon lattice vibration, respectively (26-28). The acoustical mode below 200 cm^{-1} is the characteristics of the silicon lattice vibration. As shown in Figure 5, polysilanes show no absorption band below 200 cm^{-1}. This suggests that the silicon network structure of polysilyne is not grown like amorphous silicon. In addition, the band of a linear Si-Si chain was observed for all spectra of polysilynes. The far-IR spectra well correspond to the calculated spectra using the linear and cyclic silane models. The calculated spectra using a sheet like structure which

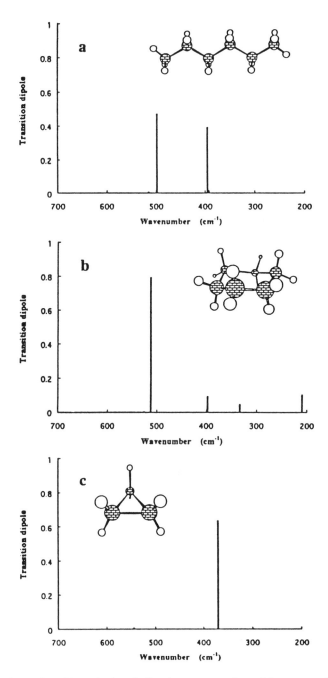

Figure 6.   The calculated vibration spectra of model compounds.
(a) hexasilane, (b) cyclohexasilane, (c) cyclotrisilane, (d) cyclotetrasilane, and (e) cyclopentasilane.

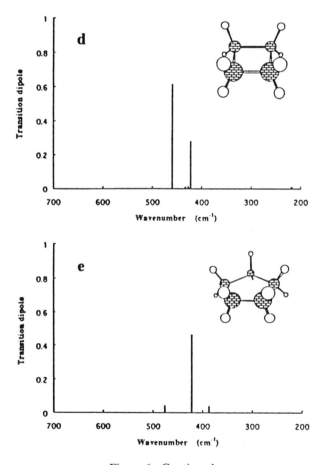

Figure 6. Continued.

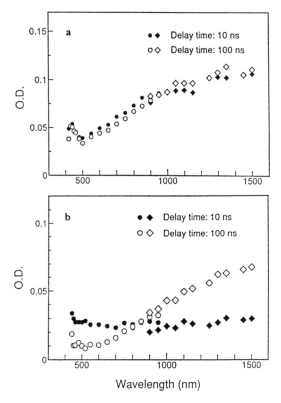

Figure 7.   Transient absorption spectra of 80 mM solution of poly(*n*-propylsilyne) obtained by pulse radiolysis.   (a) radical anion in THF, (b) radical cation in CH$_2$Cl$_2$.

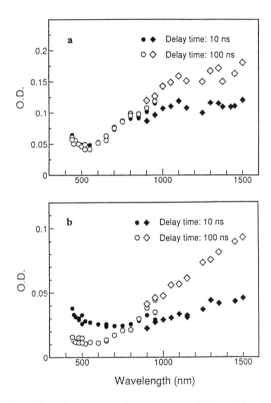

Figure 8. Transient absorption spectra of 80 mM solution of poly(*n*-hexylsilyne) (low molecular weight; Mw = 12,000) obtained by pulse radiolysis. (a) radical anion in THF and (b) radical cation in CH_2Cl_2.

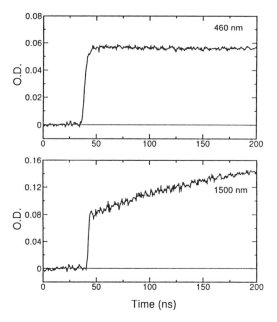

Figure 9.   Absorption-time profiles of radical anion of   poly(*n*-hexylsilyne) (low molecular weight; Mw = 12,000) in THF.

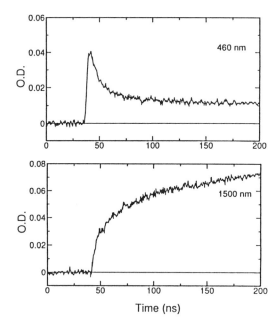

Figure 10.   Absorption-time profiles of radical cation of   poly(*n*-hexylsilyne) (low molecular weight; Mw = 12,000) in CH_2Cl_2.

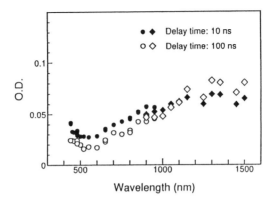

Figure 11. Transient absorption spectra of 80 mM poly(*n*-hexylsilyne) (high molecular weight; Mw = 119,000) in THF obtained by pulse radiolysis.

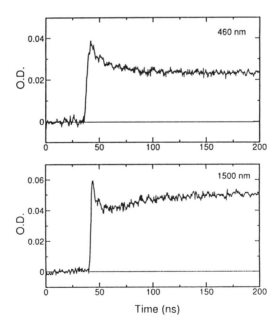

Figure 12. Absorption-time profiles of radical anion of poly(*n*-hexylsilyne) (high molecular weight; Mw = 119,000) in THF.

consists of several 6-membered rings became more broad one over wide wavenumber. These results suggest that polysilyne consists of a localized silicon ring structure connected by a linear silicon chain. As a main Si skeleton of poly(phenylsilyne), a silicon 6-membered ring can be expected.

The far-IR spectra of alkyl-substituted polysilynes show absorption in a lower energy region compared to poly(phenylsilyne). Judging from the comparison between the observed and calculated spectra, there are smaller silicon rings in the Si skeleton of the alkyl-substituted polysilynes. The formation of the small silicon ring is possible because the steric hindrance of the alkyl group is smaller than that of the phenyl group. The far-IR spectra of poly(n-hexylsilyne) with high molecular weight are quite broad as shown in Figure 4. A broad band at 393 cm^{-1} with shoulders at 496 and 463 cm^{-1} were observed. The calculated peak wave numbers of 5-memberd, 4-membered, and 3-menbered silicon rings correspond to the broad band. The absorption shoulder at 463 cm^{-1} is attributable to a linear Si-Si chain. The shoulder at 496 cm^{-1} can be assigned to a 6-membered ring. Many kinds of cyclic silicon rings connected by linear Si-Si chains are contained in the Si skeleton of poly(n-hexylsilyne).

Radical Ions of Polysilynes. There are two possible assignments for the radical ion of aryl-substituted polysilane. One is a main-chain type radical ion, and the other is a side-chain type radical ion. In the former, the radical ion is located in the Si-Si chain, and in the latter, it exists on the organic substituent. The radical ion may be stabilized by the aromatic side chain. The molecular orbital of the radical anion of poly(methylphenylsilylene) model compound was calculated by the MNDO method. The molecular orbital of the SOMO level is mainly a silicon orbital and the distribution to the phenyl group is not significant. The main chain type radical ion of polysilane was suggested.

The assignment of the radical ion site can be confirmed experimentally by investigation of the radical ion of the alkyl-substituted polysilane. An alkane radical ion has a lifetime shorter than 1 ns. The radical ion of alkyl-substituted polysilyne is not influenced by the alkyl side chain. The poly(n-propylsilyne) and poly(n-hexylsilyne) radical ions showed quite broad transient absorption spectra as shown in Figures 7 and 8. This confirms that the radical ions are the Si-Si main-chain type. The IR bands of the alkyl-substituted polysilynes appear at longer wavelength than poly(phenylsilyne). In the radical ions of poly(phenylsilyne), the radical ion site may be localized due to the stabilization of the phenyl group. The radical ions of the alkyl-substituted polysilynes are delocalized along the Si skeleton judging from the transient absorption spectra extended to the longer wavelength. The broadness of the transient absorption spectra is caused by the band structure of the polysilynes.

In a chemical sense, the anionic polaron and the cationic polaron can be interpreted as a radical anion and a radical cation, respectively, as shown in Figure 1. In the comparison between the band model and the molecular orbital theory, the energy level of the polaron corresponds to the SOMO level of radical ion. The radical anion shows the SOMO-LUMO transition with the lower transition energy and the HOMO-SOMO transition with the higher transition energy. The radical cation shows the HOMO-SOMO transition with the lower transition energy and the SOMO-LUMO transition with the higher transition energy. The IR-band of the polysilane radical ions may be explained by the SOMO-LUMO transition for the radical anion and the HOMO-SOMO transition for the radical cation. The polaron-like state may be formed immediately after the pulse radiolysis. However, the relaxation of the transient absorption spectra as shown in Figures 7 and 8 can not be explained by such energy diagram. The absorption-time profiles of the radical ions show quite different behavior. The IR band at 1500 nm increases whereas the visible band at 400 nm decreases. The spectral change suggests the structural change across the long-range of the Si chain and the formation of another electronic state.   As one of the possibilities, the formation of a charge resonance state accompanying the structural relaxation of the Si-Si chain can be considered.

Ushida et al. have reported the absorption spectra of the radical cation and anion of polysilylenes measured by γ-irradiated matrix isolation technique (*32*)   A sharp UV band and a broad IR band have been observed.   They observed the spectral change of the radical ion by photobleaching.   It was explained by the relaxation of the radical ion site accompanying the structural relaxation of the Si-Si chain.   Immediately after the irradiation, there are various types of charge traps.   By photobleaching, the charges migrate toward a more stable trap.   The UV band and the IR band were assigned to a local excitation (LE) band and a charge resonance (CR) band, respectively.   As a possible structure of the CR state, Ushida et al. proposed an intersegment CR model where the charge resonance state was formed among all-trans σ-conjugated segments (*32*).

The another noticeable feature on the transient absorption spectra is the different relaxation behavior between the radical anion and the radical cation.   The time-dependent spectral change of the radical cation is significant compared to the radical anion for polysilynes.   Ushida et al. also reported that almost no photochemical relaxation was observed for radical anions and full relaxed anions seemed to have excited immediately after the irradiation (*32*).   They pointed out that the stabilization energy of the negative charge is always smaller than that of the positive charge in the γ-irradiated matrix isolation technique and the negative charge can delocalize along the polysilane main chain.   The difference between a radical anion and a radical cation was exhibited even in a liquid phase by the pulse radiolysis technique. Therefore, another reason should be considered to explain the difference.   The SOMO of the radical cation is formed by the electron abstraction from bonding orbital HOMO.   This may weaken the Si-Si bonding and cause the geometric change around the radical cation site.   In the case of the σ-conjugated polysilanes, the energy level of the radical ions is the Si-Si σ-bond and the electronic and the geometric structures are affected seriously by the formation of the radical cation sites.   Such distortion near the charged site must be characteristic of low dimensional silicon compounds.

Conclusions

The far-IR spectra showed that the structure of polysilyne consists of a localized silicon ring connected by a linear Si-Si chain.   The silicon network structure of polysilyne is not so grown compared to amorphous silicon.   The broad absorption and emission spectra are due to the σ-conjugation along the cyclic structure with a branching silicon chain.   The radical ions of poly(*n*-propylsilyne) and poly(*n*-hexylsilyne) showed broad absorption bands in the visible and the IR region.   The radical ion of polysilane is a main chain type where the radical ion is located in the σ-conjugated Si-Si chain.   The broadness of the absorption band of the polysilyne radical ions must be caused by the band structure of a quasi-two dimensional silicon network structures.   The IR-band of the polysilyne radical ions can be explained by the SOMO-LUMO transition for the radical anion and the HOMO-SOMO transition for the radical cation.   The polaron-like state may be formed immediately after the pulse radiolysis.   A time-dependent spectral change of the radical cation suggests the relaxation of the polymer chain after the formation of the radical ion site.   The increase in the intensity of the IR band is due to the formation of the CR state where the radical ion site is stabilized after the geometric change of the σ-conjugated polysilyne chain.

Acknowledgments

The authors thank Dr. M. Suezawa and Professor K. Sumino of Tohoku University for the technical advice in the measurement of far-IR absorption spectra and their discussion.

424   POLYMERIC MATERIALS FOR MICROELECTRONIC APPLICATIONS

Literature Cited

1. Wesson, J. P.; Williams, T. C. *J. Polym. Sci., Polym. Chem. Ed.* **1980**, *18*, 959.
2. West, R; David, L. D.; Djurovich, P. I.; Stearley, K. L.; Srinivasan, K. S. V.; Yu, H. *J. Am. Chem. Soc*. **1981**, *103*, 7352.
3. Trujillo, R. E. *J. Organomet. Chem*. **1980**, *198*, C27.
4. Su, W. P.; Schrieffer, J. R.; Heeger, A. J. *Phys. Rev. Lett.* **1979**, *42*, 1698.
5. Scott, J. C.; Pfluger, P; Krounbi, M.; Street, G. B. *Phys. Rev.* **1983**, *B28*, 2140.
6. Kaneto, K.; Kohno, Y.; Yoshino, K. *Solid State Commun.* **1984**, *51*, 267.
7. Fichou, D.; Horowitz, G.; Xu, B.; Garnier, *Synthetic Metals* **1990**, *39*, 243.
8. Caspar, J. V.; Ramamurthy, V.; Corbin, D. R. *J. Am. Chem. Soc.* **1991**, *113*, 600.
9. Sun. Z. W.; Frank, A. J. *J. Chem. Phys.* **1991**, *94*, 4600.
10. Watanabe, A.; Mori, K.; Iwabuchi, A.; Iwasaki, Y.; Nakamura, Y.; Ito, O. *Macromolecules* **1989**, *22*, 3521.
11. Tagawa, S. In *Radiation Effects on Polymers*; Clough, R. L., Shalaby, S. W., Eds.; ACS Symposium Series 475, Am. Chem. Soc.: Washington, DC, 1991; Chapter 1.
12. Ban, H.; Sukegawa, K.; Tagawa, S. *Macromolecules* **1987**, *20*, 1775.
13. Ban, H.; Sukegawa, K.; Tagawa, S. *Macromolecules* **1988**, *21*, 45.
14. Irie, S.; Oka, K.; Irie., M. *Macromolecules* **1988**, *21*, 110.
15. Irie, S.; Irie, M., *Macromolecules* **1992**, *25*, 1766.
16. Tagawa, S.; Yoshida, Y. *Chem. Phys. Lett*, to be published.
17. Shida, T. *Electronic Absorption Spectra of Radical Ion*; Elsevier: 1988.
18. Bianconi, P. A.; Schilling, F. C.; Weidman, T. W. *Macromolecules* **1989**, *22*, 1697.
19. Bianconi, P. A.; Weidman, T. W. *J. Am. Chem. Soc*. **1988**, *110*, 2342
20. Furukawa, K.; Fujino, M.; Matsumoto, N. *Macromolecules* **1990**, *23*, 3423.
21. Wilson, W. L.; Weidaman, T. W. *J. Phys. Chem.* **1991**, *95*, 4568.
22. Watanabe, A; Matsuda, M. *Chem. Lett.* **1991**, 1101.
23. Kobayashi, H.; Ueda, T.; Kobayashi, T.; Washio, M.; Tabata, Y.; Tagawa, S. *Radiat. Phys. Chim.* **1983**, *21*, 13.
24. Kobayashi, H.; Ueda, T.; Kobayashi, T.; Tagawa, S.; Tabata, Y. *Nucl. Instrum. Meth*. **1981**, *179*, 223.
25. Stewart, J. J. *QCPE Bull.* **1989**, *9*, 10.
26. Shen, S. C.; Fang, C. J.; Cardona, M.; Genzel, L. *Phys. Rev. B* **1980**, *22*, 2913.
27. Alben, R.; Weaire, D.; Smith, J. E. Jr.; Brodsky, M. H. *Phys. Rev. B* **1975**, *11*, 2271.
28. Fong, C. Y.; Nichols, C. S. ;Guttman, L,; Kein, B. M. *Phys Rev B* **1986**, *34*, 2402.
29. Gilman, H.; Schwebke, G. L. *J. Am. Chem. Soc*. **1964**, *86*, 2693.
30. Watanabe, A.; Komatsubara, T.; Matsuda, M.; Yoshida, Y.; Tagawa, S. *J. Photopolym. Sci. Technol.* **1992**, *5*, 545.
31. Watanabe, A.; Nagai, Y.; Matsuda, M.; Suezawa, M.; Sumino, K. *Chem. Phys. Lett.* **1993**, *207*, 132.
32. Ushida, K.; Kira, A.; Tagawa, S.; Yoshida, Y.; Shibata, H. *Polym. Mat. Sci. Eng.* **1992**, *66*, 299.

RECEIVED September 13, 1994

Chapter 34

Optical Properties of Silicon-Based Polymers with Different Backbone Structures

Katsunori Suzuki[1], Yoshihiko Kanemitsu[1], Soichiro Kyushin[2], and Hideyuki Matsumoto[2]

[1]Institute of Physics, University of Tsukuba, Tsukuba, Ibaraki 305, Japan
[2]Department of Chemistry, Gunma University, Kiryu, Gunma 376, Japan

We have studied optical properties of quasi–one–dimensional Si backbone polymers and small Si clusters with eight Si atoms. The branch and ladder Si polymers whose backbones are constructed of the organosilicon units having three Si–Si bonds exhibit broad photoluminescence (PL) spectra. These PL characteristics are entirely different from those of the chain structure constructed of the organosilicon units having two Si–Si bonds only. Weak visible PL in Si based materials is caused by the introduction of Si atoms with three Si–Si bonds into the Si backbone. Moreover, even in small Si_8 clusters, the PL spectra strongly depend on the shape of the clusters. In the cubic structure, the weak and temperature-sensitive PL originates from the radiative recombination of triplet excitons. Silicon–based polymers and clusters exhibit a wealth of unique optical phenomena.

Optical and electronic properties of low–dimensional semiconductor nanostructures have attracted much attention, because they exhibit new quantum phenomena and have potentials for becoming novel and future optoelectronic devices (1,2). In exploring new optoelectronic materials and devices, a great deal of research effort is focused on reducing the dimensionality of the electronic structures. In this sense, chemically synthesized semiconducting polymers are regarded as natural quantum wires whose unique properties are primarily attributed to the quantum confinement effect on the conjugated electrons delocalized in the one-dimensional (1D) polymer backbone chains. The polysilanes, σ–conjugated polymers, are well–known as 1D silicon–based materials that have alkyl or aryl groups in their side chains (3).

Although the highly hydrogenated amorphous silicon (a-Si:H) also contains Si chains, there are many differences in optical properties between a-

0097–6156/94/0579–0425$08.00/0

Si:H and chain–like Si polymers. For example, the broad photoluminescence (PL) spectrum due to the band tail emission is observed in a–Si:H but a sharp PL band with a large quantum efficiency is observed in various chain–like Si polymers. The broad PL in a–Si:H rather resembles that in network Si polymers (4). Because the local Si structures can be controlled in chemically synthesized Si polymers, studies of optical properties of Si polymers with different backbone structures help to understand the electronic structures and the microscopic PL mechanisms in a variety of Si–based materials. Moreover, studies of optical properties of small Si clusters help to understand the dimension effects on electronic properties of Si–based nanostructure, and the mechanism of visible PL in nanometer–size Si crystallites such as porous Si (5).

In this paper, we report the optical properties of quasi–1D Si polymers with different backbone structures and small Si clusters. We discuss the luminescence properties in a variety of Si materials.

Experiment

Three types of quasi–1D Si polymers used in this work are (a) poly(methylphenylsilane), (b) poly[(dimethylphenylsilyl)phenylsilane] and (c) dodecaisopropyltetracyclodecasilane. The Si backbone structures are illustrated in Figure 1 and hereafter, we call these structures as the chain [Figure 1(a)], branch [Figure 1(b)], and ladder structures [Figure 1(c)]. In the chain structure, the organosilicon units on the polymer backbone have two Si–Si bonds only. In the branch and ladder structures, the polymer backbones are constructed of the organosilicon units having three Si–Si bonds.

Chain, ladder, and cubic structures of Si_8 clusters are also illustrated in Figure 2. Synthetic and purification methods were described in the literature (6). It is theoretically reported that the small clusters of *pure* Si form crystal structures that are neither diamond–like nor tetrahedral. This complication makes it difficult to understand the size dependence of electronic properties in *pure* Si clusters systematically, because the changes of size are accompanied by the changes of symmetry and the formation of dangling bonds. Therefore, saturatedly bonded Si clusters terminated by an organic substituent have no dangling bonds and become a new model system for small Si clusters.

Absorption spectra of Si polymers dissolved in tetrahydrofuran (THF) were measured. For PL spectrum measurements, thin solid films were prepared on a quartz substrate from THF solution. The PL spectra were measured by using 325 nm excitation light from a He–Cd laser. Picosecond PL decay under a 1–ps and 300–nm laser excitation was measured using a monochromator of subtractive dispersion and a synchroscan streak camera. The calibration of the spectral sensitivity of the whole measuring systems was performed by using a tungsten standard lamp.

Results and Discussion

Quasi–One–Dimensional Si Polymers.   Figure 3 shows PL and absorption spectra in quasi–1D Si polymers. In the chain structure [Figure 3(a)], sharp PL

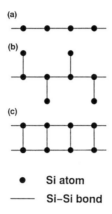

●   Si atom

──   Si-Si bond

Figure 1. Illustration of quasi–1D Si polymers with different backbones: (a) chain, (b) branch, and (c) ladder structures.

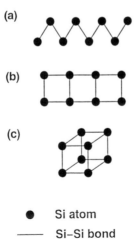

●   Si atom

──   Si-Si bond

Figure 2. Si_8 clusters: (a) chain, (b) ladder, and (c) cubic structures.

and sharp absorption bands are observed. The structures of chain–like Si polymers take a variety of conformations such as *trans*–planar, *trans*–*gauche*, 7/3 helical, and disordered forms, depending on organic substituents attached to the polymer backbones (7). However, the PL band is very sharp in all conformations including disordered forms. Therefore, the sharp PL band is the most important feature of the optical properties of the chain structure.

In the branch and ladder structures, broad PL spectra are observed. Moreover, the intensities of the lowest absorption peak per organosilicon unit in the branch and ladder structures are very small compared with that in the chain structure. In the ladder structure, the peak of the weak PL appears near the edge of the low energy tail of the lowest absorption band. This feature is often observed in an indirect gap semiconductors. These broad visible PL spectra were not observed in the chain structure at room temperature. The weak and sharp phosphorescence spectrum is observed near the strong fluorescence at low temperatures below 100 K. This phosphorescence spectrum is completely quenched at room temperature. The characteristics of the board PL spectra in the branch and ladder structures are different from those of phosphorescence in the chain structure.

Figure 4 shows the initial PL decay dynamics at peak energies in the chain, branch and ladder structures. The initial PL decay in the chain and ladder structures approximately exhibit a single exponential decay but the initial PL decay in the branch structure shows nonexponential decay. The nonexponential PL decay and the broad PL spectrum in the branch structure resemble the band–tail emission in a–Si:H.

To evaluate the radiative decay rate of excitons, we approximately determine the time constant τ_{PL} of the initial decay by using the single exponential function indicated by broken lines. In the branch and ladder structures, the quantum efficiency of PL, η, is very low ($\leq 10^{-3}$), but τ_{PL} is large (~0.8 ns), compared with those of the chain structure. Here, we can estimate the radiative decay rate of excitons τ_R^{-1} determined from the PL lifetime and the PL quantum yield, η/τ_{PL}. On the other hand, the radiative decay rate based on one unit in the polymer backbone τ_{abs}^{-1} can be evaluated from the integral of the lowest absorption band. The ratio τ_{abs}/τ_R roughly gives the delocalized region of excitons. These calculations show that in the branch and ladder structures excitons are strongly localized on two or three Si atoms (8), but in the chain structure excitons are delocalized over about thirty Si atoms (9). Excitons are strongly localized at the Si units with three Si–Si bonds.

We consider the origin of the broad PL spectra in the branch and ladder structures from a viewpoint of the electronic band structures of Si polymers. The band calculations of the chain, branch, and ladder structures were performed with the first–principle local density functional method (10). The characteristic features of these band structures can be summarized as follows:
(1) The band–gap energies of the branch and ladder structures are smaller than that of the chain structure.
(2) The E–k dispersions of the branch and ladder structures are flat compared with those of the chain structure.

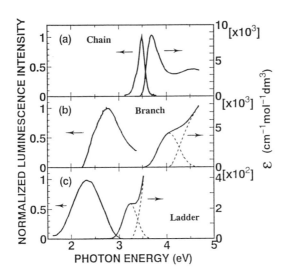

Figure 3. Optical absorption and normalized photoluminescence spectra in Si polymers: (a) chain, (b) branch, and (c) ladder structures. The broken lines in absorption spectra are optimum tails of two Gaussian functions.

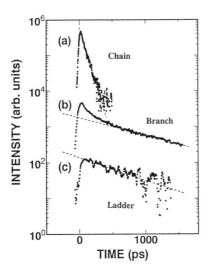

Figure 4. Picosecond PL decay of three Si polymers: (a) chain, (b) branch, and (c) ladder structures. The broken lines indicate the single exponential decay with a time constant of (a) 80 ps, (b) 880 ps, and (c) 700 ps.

(3) The density of states near the top of valence band and the bottom of the conduction band in the branch and ladder structures is larger than that of the chain structure.

(4) The energy of the indirect optical transition is close to that of the direct optical transition in the branch and ladder structures.

The above four characteristics of the band structure give an answer to why the branch and ladder structures exhibit broad PL spectra in the visible region [(1): PL appears in the visible region, (2) and (3): the PL spectrum is broad, and (4): the PL lifetime is long.]. A large part of experimental results are well explained by the electronic band structures of polymers.

However, it is considered that the band structure of the crystalline state is an oversimplified picture. The polymers have noncrystalline characteristics and are "soft" materials. Wilson and Weidman (4) and Fujiki (11) pointed out that in the network polymers, the branching points and the random network structures give rise to localized electronic states known as the band tail of the both conduction and valence bands and excitons are localized near the branching points. However, in the chain structure, the disorder does not critically affect the observed PL spectrum, because the sharp PL band is observed even in the disorder forms of the chain structure. Therefore, we consider that *broad PL bands observed in the branch and ladder structures and Si network polymers are closely related to the existence of Si atoms having three or four Si–Si bonds*. The disorder effects also play a very important role in the formation of localized states in the Si materials constructed of Si units with three or four Si–Si bonds.

Chemically Synthesized Small Si Clusters. Figure 5 shows the optical absorption and PL spectra of Si_8 clusters with different structures. In the chain structure, the dangling bonds are terminated by the methyl and phenyl groups. In the ladder structure, the dangling bonds are terminated by the isopropyl groups. In the cubic structure, the dangling bonds are terminated by the 1,1,2–trimethylpropyl groups. In the short Si chain, a sharp and strong PL band was observed. However, the PL efficiency and the absorption coefficient per Si unit are small compared with those of the long chains [see, Figure 3 (a)]. In the ladder structure, a broad PL band was observed in the visible spectral region. The PL spectra and efficiency of the small ladder structure are very similar to those of the long ladder structures, because excitons are localized on two or three Si atoms in the ladder structure, as discussed in the previous section.

In the cubic structure, the PL was not observed at room temperature but was observed at low temperatures below 80 K. The PL lifetime was about 3 ms. Our light–induced ESR experiment (12) suggests that the weak and temperature–sensitive PL is due to the recombination of the triplet exciton. These luminescence properties of the cubic structure are entirely different from those of the chain and ladder structure. Even in the very small Si materials, the optical and electronic structures are very sensitive to the Si backbone structures. Further experimental studies are needed to understand the nature of electronic structures of a cubic Si_8 cluster. The understanding of the luminescence properties of

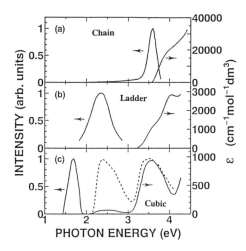

Figure 5. Optical absorption and normalized photoluminescence spectra in small Si clusters: (a) chain, (b) ladder, and (c) cubic structures. The broken line in the cubic structure is the excitation spectrum at the PL peak energy. The PL and PL excitation spectra of the cubic structure were measured at 20 K.

small clusters provide information about the mechanism of the visible luminescence from Si-based nanostructures such as porous Si. We hope that this work will stimulate a theoretical consideration of chemically synthesized small Si clusters.

Conclusions

We have studied the optical properties of quasi-1D Si polymers with different backbone structures and chemically synthesized Si clusters. The branch and ladder Si polymers constructed of the organosilicon units with three Si-Si bonds exhibit broad PL spectra with long decay times. Broad PL bands observed in Si materials are closely related to the existence of Si atoms having three or four Si-Si bonds. Moreover, the very weak PL from the cubic cluster is due to the radiative recombination of triplet excitons. The optical properties of Si polymers and small Si clusters are very sensitive to the Si backbone structures.

Acknowledgments

One of the authors (Y. K.) is grateful to the IKETANI Science and Technology Foundation for financial support.

Literature Cited

(1) Brus, L. *Appl. Phys.* **1991**, A53, 465.
(2) Hanamura, E. *Phys. Rev.* **1988**, B38, 1288.
(3) Miller, R. D.; Michl, J. *Chem. Rev.* **1989**, 89, 1359.
(4) Wilson, W. L.; Weidman, T. W. *J. Phys. Chem.* **1991**, 94, 4568.
(5) Kanemitsu, Y.; Suzuki, K.; Uto, H.; Masumoto, Y.; Matsumoto, T.; Kyushin, S.; Higuchi, K.; Matsumoto, H. *Appl. Phys. Lett.* **1992**, 61, 2446.
(6) Matsumto, H.; Higuchi, K.; Kyushin, S.; Goto, M. *Angre. Chem. Int. Ed. Engl.* **1992**, 31 1355.
(7) Tachibana, H.; Matsumoto, M.; Tokura, Y.; Moritomo, Y.; Yamaguchi, A.; Koshihara, S.; Miller, R. D.; Abe, S. *Phys. Rev.* **1993**, B47, 4363.
(8) Kanemitsu,Y.; Suzuki, K.; Masumoto, Y.; Komatsu, T.; Sato, Y.; Kyushin, S.; Matsumoto, H. *Solid State Commun.* **1993**, 86, 545.
(9) Kanemitsu, Y.; Suzuki, K.; Nakayoshi, Y.; Masumoto, Y. *Phys. Rev.* **1992**, B46, 3916.
(10) Takeda K. In *Light Emission from Novel Silicon Materials*, Kanemitsu, Y.; Kondo, M.; Takeda, K. Eds. Physical Society of Japan, Tokyo, 1994; pp.1-29.
(11) Fujiki, M. *Chem. Phys. Lett.* **1992**, 198, 177.
(12) Kanemitsu, Y.; Suzuki, K.; Kondo, M.; Matsumoto, H. *Solid State Commun.* **1994**, 89, 619.

RECEIVED September 13, 1994

Chapter 35

Synthesis and Properties of Polysilanes Prepared by Ring-Opening Polymerization

Eric Fossum and Krzysztof Matyjaszewski[1]

Department of Chemistry, Carnegie Mellon University, 4400 Fifth Avenue, Pittsburgh, PA 15213

A series of cyclotetrasilanes with varying numbers of methyl and phenyl substituents, $Me_nPh_{8-n}Si_4$, where n=3,4,5, and 6, have been prepared, characterized, and polymerized employing silyl cuprates. The polymerizations proceed with two inversions of configuration at both the attacked silicon atom and the newly formed reactive center and with some regioselectivety. The polymers have been analyzed by ^1H, ^{13}C, and ^{29}Si NMR spectroscopy, along with UV spectroscopy.

Polysilanes (polysilylenes) are a relatively new class of polymers with some unusual properties related to their intrinsic electronic and morphological structures.(1-3) The significant delocalization of electrons in the Si-Si backbone leads to both intense UV absorptions and fluorescence emissions. The optoelectronic properties depend strongly on the structure of the substituents attached to the backbone, the chain conformation, and potential defects, such as siloxane linkages or branching. In a similar manner the morphology of polysilanes depends on the structure and symmetry of the substituents and their configurations, or microstructure. These factors control the overall morphology of the polymer chains. The establishment of structure-property relationships for polysilanes requires well-defined polymers, as even small amounts of defects may strongly affect some properties and lead to erroneous correlations. Therefore, our continuous efforts are focused on the synthesis of well-defined polysilanes.

Polysilanes are typically prepared by the reductive coupling of disubstituted dichlorosilanes with molten sodium. The resulting polymers have broad and uneven polymodal molecular weight distributions and unpredictable molecular weights. We have previously improved the synthesis of poly(methylphenylsilylene) by reducing the reaction temperature below 60 °C in the presence of ultrasound(4). Monomodal high polymers ($M_n \approx 100,000$) with relatively narrow molecular weight distributions ($M_w/M_n < 1.3$) were prepared. The absence of the low molecular weight fraction ($M_n < 10,000$) was ascribed to the suppression of side processes limiting chain growth. On the other hand, the narrow molecular weight distributions were explained by the selective degradation of high molecular weight polymers by frictional forces accompanying the cavitation process. Only macromolecules above the chain entanglement limit could be degraded by mechanical forces.

---

[1]Corresponding author

0097–6156/94/0579–0433$08.00/0

Polysilanes are also prepared by the dehydrogenative coupling of primary silanes(5). However, this procedure usually results in low molecular weights (M_n<10,000), broad polydispersities, and branching points. There are two alternative methods for the synthesis of polysilanes with improved structural control, including the anionic polymerization of masked disilenes(6) and ring opening polymerization (ROP) of strained cyclosilanes(7). The ring-opening process will be discussed in detail in this article. Three aspects of ring-opening polymerization will be covered: 1) monomer synthesis, 2) stereoselectivity of the ring-opening polymerization of 1,2,3,4-tetramethyl-1,2,3,4-tetraphenylcyclotetrasilane, and 3) regioselectivity of the ring-opening polymerization of cyclotetrasilanes with variable numbers of methyl and phenyl substituents $Me_nPh_{8-n}Si_4$, where n=3,5, and 6.

Experimental. All experiments were performed in a VAC HE dry box under a nitrogen atmosphere with less than 1 ppm of moisture and oxygen. 1,2,3-Trimethylpentaphenyl cyclotetrasilane was prepared by treating octaphenyl cyclotetrasilane with three equivalents of trifluoromethanesulfonic acid followed by methylation with methylmagnesium bromide. 1,2,3,4-Tetramethyl-1,2,3,4-tetraphenylcyclotetrasilane was prepared by previously reported methods(7). 1,1,2,3,4-Pentamethyl-2,3,4-triphenylcyclo-tetrasilane, 1,1,2,2-3,4-hexamethyl-3,4-diphenylcyclotetrasilane, and 1,1,2,3,3,4-hexamethyl-3,4-diphenyl-cyclotratsilane were prepared by treating 1,2,3,4-tetramethyl-1,2,3,4-tetraphenyl-cyclotetrasilane with the respective equivalents of triflic acid, followed by methylation. The monomers were polymerized using the silyl cuprate $(PhMe_2Si)_2Cu(CN)Li_2$ in THF with reaction times of 1 hour.

Monomers

Most if not all known linear polysilanes are only kinetic products. Exposure to radiation, as well as to strong nucleophilic reagents, leads to complete polymer degradation(8) In solution, linear polysilanes degrade to cyclic products, usually to strainless cyclopenta- and cyclohexasilanes. The majority of known cyclosilanes, prepared by the reductive coupling process, are thermodynamic products and can not be converted to linear polymers. An equilibrium mixture of cyclic compounds is sometimes formed, like permethylated cyclopenta-, cyclohexa- and cycloheptasilanes(1). The preferred ring size depends on the bulkiness of substituents attached to Si atoms. For the dimethylsilyl group, the six membered ring is favored. For larger dialkyl and arylalkyl substituents a mixture of six- and five-membered rings is often observed(9). Diarylsilyl groups lead to a mixture of five- and four-membered rings. For very bulky groups, like di-t-butyl or dimesityl, stable three membered rings or even disilenes have been reported(10). This brief review of known cyclosilanes indicates that bulky substituents at each Si atom decrease the preferred ring size at the expense of the angular strain. Apparently, ring strain is tolerated much easier by atoms from the second than those from the first period.

Attempts to polymerize some of the cyclotetrasilanes formed in the reductive coupling process have been unsuccessful. Therefore, our strategy was based on preserving the ring structure and replacing the bulky substituents by smaller ones to increase the ring strain. The easily available octaphenylcyclotetrasilane was converted to 1,2,3-trimethyl-1,2,3,4,4-pentaphenylcyclotetrasilane and to 1,2,3,4-tetramethyl-1,2,3,4-tetraphenylcyclo-tetrasilane (n=3 and 4, respectively) in the following way:

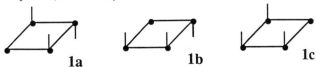

An NMR analysis of $Me_3Ph_5Si_4$ indicates that the predominant stereoisomer has an all-trans configuration of the methyl groups (**1a**). The ^1H NMR spectrum displays two peaks in the methyl region in a 2:1 ratio indicating the all-trans structure. The ^{29}Si and ^{13}C NMR spectra show similar patterns. The all-cis structure (**1c**) (also 2:1 pattern) is sterically more hindered and less likely to form.

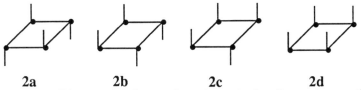

There are four possible isomers of $Me_4Ph_4Si_4$ formed in this reaction but, ^1H, ^{13}C, and ^{29}Si NMR spectra indicate that the all-cis isomer (**2d**) is absent and the other three isomers are formed in comparable amounts.

It was possible to enrich the reaction mixture in the all-trans isomer (**2a**) (>90%) by repetitive crystallization from n-hexane. The structure of this isomer was proven by detailed NMR studies of the compound containing ^{13}C enriched methyl groups and also, the synthesis of 1,1,2,3,4-pentamethyl-2,3,4-triphenylcyclotetrasilane(11). The all-trans isomer is an important starting product for the potential synthesis of stereoregular poly(methylphenylsilylene).

Attempts to prepare the pentamethyl derivative by the direct dearylation of octaphenylcyclotetrasilane with five equivalents of triflic acid were unsuccessful. The fifth equivalent of acid leads to ring cleavage rather than to the desired pentatriflate derivative. This can be ascribed to both electronic and steric effects of the bulky triflate groups. A successful approach was based on the dearylation of a tetramethyl derivative (**2**). Preparation of $Me_5Ph_3Si_4$ starting from only one isomer of $Me_4Ph_4Si_4$ (all-trans) results in the formation of only one stereoisomer. The reaction mixture also contains 5% $Me_4Ph_4Si_4$ and 5% $Me_6Ph_2Si_4$, which can be ascribed to the limited chemoselectivity of dearylation(9). The ^{29}Si NMR spectrum shows the expected pattern of peaks, 1:2:1, and indicates the selective formation of only one isomer, **3a**.

Preparation of $Me_6Ph_2Si_4$ from a mixture of stereoisomers of $Me_4Ph_4Si_4$, gives rise to a mixture of both geometrical and stereoisomers possessing either a 1,3 or 1,2 arrangement of the Me_2Si units. The possible isomers are shown below:

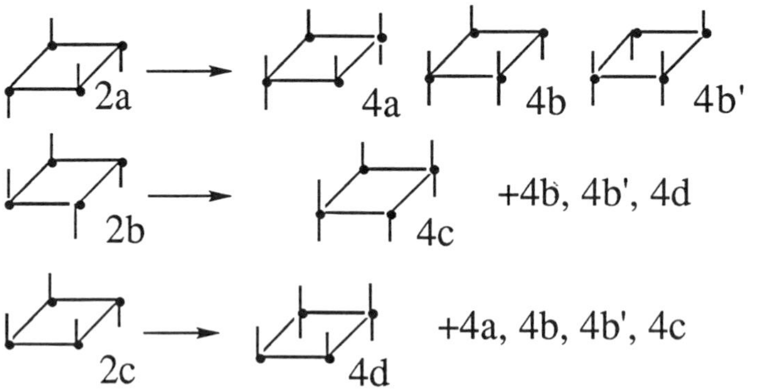

2a → 4a   4b   4b'

2b → 4c   +4b, 4b', 4d

2c → 4d   +4a, 4b, 4b', 4c

The ^{13}C NMR spectrum displays six peaks in the methyl region which result from the various methyl groups present in the isomers. Because there are only six major peaks present, the reaction may proceed preferentially by either the 1,2 or 1,3 arrangement of the triflate groups, but the dominating pathway is not yet established.

All of the cyclotetrasilanes, $Me_nPh_{8-n}Si_4$, with n=3,4,5,6 are sufficently strained and can be polymerized. The polymers are well soluble and can be characterized. When n=2, the solubility decreases strongly and it is difficult to dissolve the polymers precipitated from alcohol or hexane.

Polymerization

Cyclotetrasilanes with methyl and phenyl substituents can be easily polymerized by anionic initiators, such as alkyl lithium, silyl potassium, or silyl lithium compounds. The reaction is very slow in non polar media (benzene, toluene). Reactions are accelerated by the addition of small amounts of tetrahydrofuran (THF) (1%), or by complexing the alkali metal cations with crown ethers and cryptands. In a typical reaction, the monomer is first converted to a linear polymer which is subsequently degraded to strainless cyclosilanes. The proper choice of reaction conditions allows termination after the polymer is formed and before cyclic compounds start to appear. For example, reaction in benzene at 20 $^{\circ}$C with $[M]_0 = 0.5$ mol/L and $[(BuLi/cryptand[2.1.1] \approx 1:1)]_0 = 0.005$ mol /L provides a polymer with $M_n \approx 40,000$ after 20 min. and depolymerization starts after 3 hours. Polymerization is favored by the ring strain of the four-membered ring and the depolymerization by entropic factors (Scheme I).

Not all cyclotetrasilanes can be successfully polymerized. For example, octaphenylcyclotetrasilanes can only be isomerized to decaphenylcyclopentasilane, but can not be converted to high molecular weight poly(diphenylsilylene)(9). This can be ascribed to its lower strain and to the repulsive interactions between bulky phenyl groups in the hypothetical open chain polymer. Additionally, steric effects may be responsible for the slower propagation and the relative increase of the rate of degradation. Equilibrium polymerizations require that $[M]_0 > [M]_e$. The low solubility of octaphenylcyclotetrasilane may prohibit reaching concentrations above the equilibrium monomer concentration ($[M]_e$).

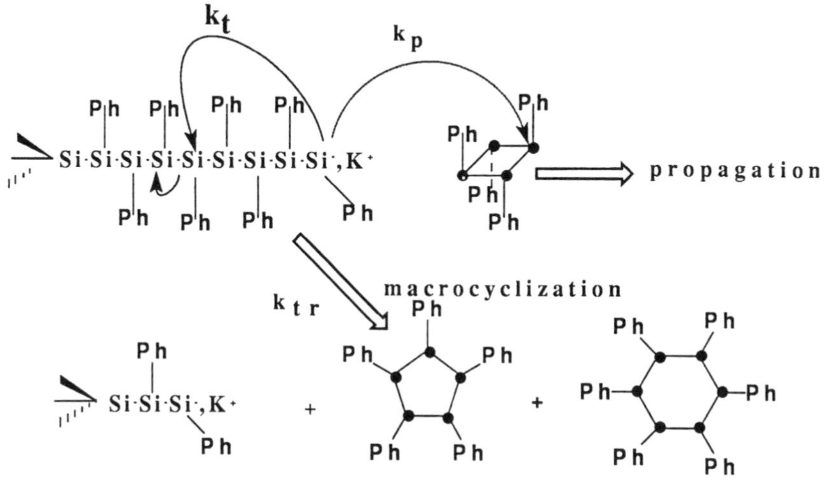

Scheme I

Microstructure/Stereoselectivity. A mixture of three isomers of 1,2,3,4-tetramethyl-1,2,3,4-tetraphenylcyclotetrasilane, as well as the enriched (>90%) all-trans isomer, with silyl lithium or potassium or alkyl lithium as the initiator, benzene as the solvent, and in the presence of THF, crown ethers, or cryptands, always provides polymers with a microstructure very similar to the polymer obtained by the reductive coupling process. Thus, although control of molecular weights (usually up to $M_n \approx 50,000$) and polydispersities (usually $M_w/M_n < 1.5$) can be significantly improved in the ROP process, in comparison with reductive coupling, the polymers are nearly stereorandom. It seems that the ring opening proceeds with at least racemization at either the attacked Si atom in the ring or at the resulting Si anion. Thus, only one out of four bonds in the cyclotetrasilane remains fixed and this is too little to observe a significant change in the microstructure.

We have attempted to use various, milder reagents for ROP in order to better control the stereochemistry at the two Si atoms which may invert or retain their configurations. Application of transition metals, such as derivatives of Pd or Pt led to oligomerization, and in the case of $Pd(PPh_3)_4$, resulted in high yields (>80%) of a dimer (1,2,3,4,5,6,7,8-octamethyl-1,2,3,4,5,6,7,8-octaphenylcyclo-octasilane), which, starting from **2a**, has presumably an all-trans configuration(12).

High molecular weight polymers have been prepared using silyl cuprates as initiators(12). Higher order cuprates are formed from two equivalents of $PhMe_2SiLi$ and one equivalent of CuCN(13). The aggregated structure $[PhMe_2SiLi]\cdot CuCN$ is a more sluggish initiator than silyl lithium and requires more polar THF as a solvent. To complete polymerization at $[I]_0 \approx 0.01$ mol/L in THF at 20 °C, nearly one hour is required for silyl cuprates, but only a few seconds for silyl lithium; after two minutes in the latter case only strainless cyclosilanes are formed as a result of depolymerization.

Additionally, silyl cuprates affect the microstructure of the resulting polymers. ^{29}Si NMR data (Figure 1) reveals that the polymer has 75% atactic (-38.5 ppm) and

25% (-41.0 ppm) isotactic triads. No syndiotactic triads at -39.5 ppm were formed. These assignments were based on a comparison of the ^{29}Si NMR spectra of polymers obtained from polymerization of **2a** and a mixture of **2a**, **2b**, and **2c**. One way of distinguishing between ring opening occuring by two inversions of two retentions of configuration would be the preparation of 1:1 adducts of the monomer with the initiator. The stereochemistry of the resulting pentamers would reveal the mechanism of polymerization. However, efforts to prepare oligomers by the reaction of one monomer unit per initiator were unsuccessful due to scrambling reactions, which led to the formation of di-, tri-, tetra-, and pentasilanes. The resulting microstructure indicates that ROP must proceed with two inversions of configuration at the attacked Si atom and at the new nucleophilic center. The "anionic" Si atom does not racemize, but retains the inverted configuration until a new monomer is attacked.

A plausible mechanism is shown in Scheme II. A pentacoordinated anionic intermediate is proposed where the largest group, the polymer chain, is in an apical position. The leaving group, ring skeleton, must also be in an apical position. This leads to an inversion of configuration at the attacked silicon atom. The bulky cuprate cation apparently prefers a backside attack, which leads to the second inversion of configuration at the newly formed reactive center. Apparently, the smaller lithium cation can approach the anionic center from both sides leading to racemization of the new active center.

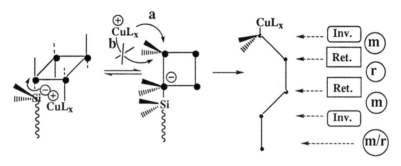

$$CuL_x \Rightarrow a >> b; \; Li^+/[1.1.2] \Rightarrow a \approx b$$

Scheme II

Microstructure/Regioselectivity. The polymerization of $Me_3Ph_5Si_4$ is expected to occur by attack of the anion on a less hindered PhMeSi unit and subsequent ring opening results in the formation of the more stable Ph_2Si anion. If each monomer unit is attacked in the same fashion, then an alternating copolymer, $[(MePhSi)_3-(Ph_2Si)]_n$ should result. Assuming the same mechanism of polymerization using silyl cuprates is applicable, the central three PhMeSi units would lose the all-trans configuration of substituents because of an inversion at the attacked silicon atom. Therefore, the signals for the PhMeSi appear predominantly as heterotactic triads near -38.5 ppm. The ^{29}Si NMR spectrum of the polymer prepared by polymerization of 1 is shown in Figure 2.

In this DEPT (Distortionless Enhancement by Polarization Transfer)-acquired spectrum the intensity of the peaks from the PhMeSi units is enhanced due to DEPT in comparison with the Ph_2Si units. If parameters optimized for phenyl protons are used the signals for the Ph_2Si units appear at -30 ppm. The sharp, downfield signal may be assigned to the central PhMeSi unit in the heterotactic triads (rm= mr), which are separated by Ph_2Si units. Because these triads can be coupled in equal probabilities,

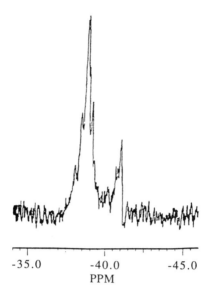

Figure 1. ^{29}Si NMR spectrum of the polymer obtained from polymerization of **2a** with $(PhMe_2Si)_2Cu(CN)Li_2$ in THF; $[M]_0$= 0.63 M, $[I]_0$ = 0.013 M, 1 h at RT.

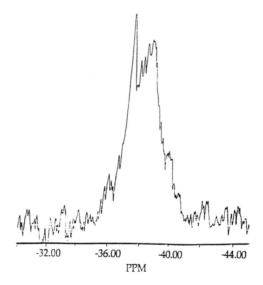

Figure 2. ^{29}Si NMR spectrum of the polymer prepared by ROP of $Me_3Ph_5Si_4$ with $(PhMe_2Si)_2Cu(CN)Li_2$ in THF; $[M]_0$= 0.505 M, $[I]_0$= 0.01 M 6 h.

m' and r' (Scheme III), the signals of the PhMeSi units adjacent to Ph_2Si units can absorb at slightly different chemical shifts leading to the broad, featureless peak observed.

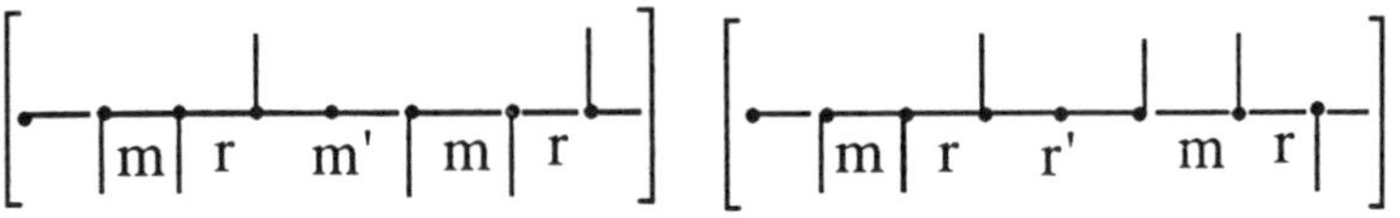

Scheme III

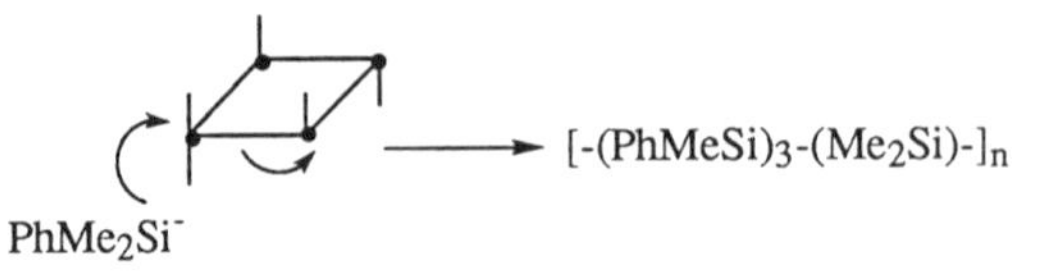

$PhMe_2Si^-$ $[-(PhMeSi)_3-(Me_2Si)-]_n$

Scheme IV

Ring opening of $Me_5Ph_3Si_4$ should occur via the pathway shown in Scheme IV. Attack of the initiator and the growing chain is expected to occur at the sterically less hindered Me_2Si unit resulting in the formation of a new reactive anion, $MePhSi^-$, whose configuration would be inverted. This would lead to heterotactic triads of PhMeSi units separated by a Me_2Si unit. The ^{29}Si NMR spectrum of the polymer obtained from ring opening polymerization of 3a is presented in Figure 3. The peaks between -35.0 and -37.0 ppm are assigned to the Me_2Si units.

The peak at -41.0 ppm, corresponding to isotactic polymer, accounts for less than 5% of the PhMeSi units. This indicates that some regioselectivety is lost in this system as the only route to form isotactic or syndiotactic triads would be for the propagating PhMeSi anion to attack a PhMeSi unit in the next monomer unit. However, because the percentage of isotactic triads is small, the preferential ring opening of this monomer appears to occur via the pathway shown in Scheme III.

Polymerization of the mixture of isomers of $Me_6Ph_2Si_4$ can occur via a number of pathways. However, ring opening of each individual isomer is expected to occur via attack of the initiator or growing chain on the sterically less hindered Me_2Si unit giving rise to the more stable anion $MePhSi^-$ as the new reactive site. The ^{29}Si NMR spectrum is shown in Figure 4 and contains several peaks. These can be separated into two groups: the peaks from -35.0 to -37.5 ppm correspond to Me_2Si groups and the peaks from -37.5 to -40.0 ppm are assigned to the MePhSi units. The peaks for the Me_2Si and PhMeSi units should be present in a 1:1 ratio, but the polarization transfer leads to more enhanced signals for the former.

Because the monomer mixture contained a variety of different isomers it is difficult to arrive at any firm conclusions about the regio- and stereoselectivety of the polymerization. However, the absence of any isotactic triads at -41.0 ppm indicates the reaction is at least partially regioselective; polymerization of 4b, 4b', and 4d could lead to the formation of some isotactic triads if the ring opening was not regioselective. A more detailed analysis requires the acquisition of ^{29}Si NMR spectra without polarization transfer and possibly separation of the individual isomers of 4.

Polymer Properties. Table 1 gives the relevant molecular weight data and also the absorption maxima for the above polymers. The λ_{max} of the materials shows the expected dependence on the percentage of aromatic groups present in the polymer. No thermal transitions were detected for the polymers by differential scanning calorimetry indicating a very low degree of crystallinity.

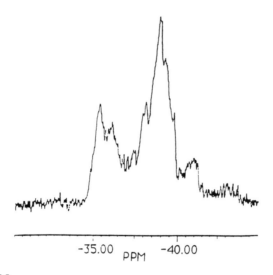

Figure 3. ^{29}Si NMR spectrum of the polymer obtained from ROP of Me$_5$Ph$_3$Si$_4$ with (PhMe$_2$Si)$_2$Cu(CN)Li$_2$ in THF; [M]$_O$= 0.324 M, [I]$_O$= 0.065 M, 1h at RT.

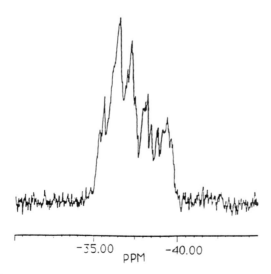

Figure 4. ^{29}Si NMR spectrum of the polymer prepared by ROP of Me$_6$Ph$_2$Si$_4$ with (PhMe$_2$Si)$_2$Cu(CN)Li$_2$ in THF, [M]$_O$= 0.397 M, [I]$_O$= 0.005 M, 1 h at RT.

Table 1. Absorbance data for Polymers 1-4

polymer	M_n	M_w/M_n	λ_{max}
1	23,300	1.5	353
2)	48,000	1.5	336
3	10,500	2.0	332
4	12,000	1.6	328

Conclusions

A series of cyclotetrasilanes with varying numbers of methyl and phenyl substituents have been prepared and polymerized using silyl cuprates. Silyl cuprates lead to the ring-opening process with two inversions of configuration. In the case of polymerization of all-trans isomer **2a**, the ROP results in poly(methylphenylsilylene) with a unique tacticity (75% atactic, 25% isotactic, and 0% syndiotactic triads). The polymerizations of $Me_3Ph_5Si_4$, $Me_5Ph_3Si_4$, and $Me_6Ph_2Si_4$ appear to occur with some regioselectively. The polymers have been analyzed using ^1H, ^{13}C, and ^{29}Si NMR spectroscopy, along with UV spectroscopy. Further work to determine the stereochemistry and a more detailed picture of the regiochemistry of the polymers is in progress.

Acknowledgements. Support of the Office of Naval Research for this work is appreciated. K.M. acknowledges support by the National Science Foundation, as well as from DuPont, Eastman Kodak, Xerox, and PPG Inc. within the Presidential Young Investigator Award.

Literature Cited

1. West, R. *J. Organomet. Chem.* **1986,** 300, 327.
2. Miller, R. D.; Michl, J. *Chem. Rev.* **1989,** 89, 1359.
3. Matyjaszewski, K.; Cypryk, M.; Frey, H.; Hrkach, J.; Kim, H. K.; Moeller, M.; Ruehl, K.; White, M. *J. Macromol. Sci., Chem..* **1991,** A28, 1151.
4. Kim, H. K.; Matyjaszewski, K. *J. Amer. Chem. Soc.* **1988,** 110, 3321.
5. Woo, H. G.; Waltzer, J. F.; Tilley, T. D. *Macromolecules,* **1991,** 24, 6863.
6. Sakamoto, K.; Obata, K.; Hirata, H.; Nakajima, M.; Sakurai, H. *J. Amer. Chem. Soc.* **1989,** 111, 7641.
7. Gupta, Y.; Cypryk, M.; Matyjaszewski, K. *J. Amer. Chem. Soc.* **1991,** 113, 1046.
8. Kim, H. K.; Matyjaszewski, K. *J. Polym. Sci., Chem.* **1993,** 31, 299.
9. Matyjaszewski, K. *Makromol. Chem. Macromol. Symp.* **1991,** 42/43, 269.
10. Masamune, S.; Hanezawa, Y.; Murakami, S.; Bally, T.; Blount, J. F. *J. Am. Chem.* **1982,** 104, 1150.
11. Fossum, E.; Gordon-Wylie, S. W.; Matyjaszewski, K. Organometallics **1994,** 13, 1695.
12. Cypryk, M.; Chrusciel, J.; Fossum, E.; Matyjaszewski, K. *Makromol. Chem., Macromol. Symp.* **1993,** 73, 167.
13. Sharma, S.; Oehlschlager, A. C. *Tetrahedron* **1989,** 45, 557.

RECEIVED September 13, 1994

Chapter 36

New Synthesis and Functionalization of Photosensitive Poly(silyl ether) by Addition Reaction of Bisepoxide with Dichlorosilane

Atsushi Kameyama, Nobuyuki Hayashi, and Tadatomi Nishikubo[1]

Faculty of Engineering, Kanagawa University, Rokkakubashi, Kanagawa-ku, Yokohama 221, Japan

A poly(silyl ether) **P-1** containing silicon-silicon bond as a new class of silicon-containing polymer was successfully synthesized by polyaddition of 1, 2-dichloro-tetramethyldisilane with bisphenol A diglycidyl ether using quaternary onium salts. The polymer was further modified readily with photo-crosslinkable compounds by the substitution reaction using 1,8-diazabicyclo-[5.4.0]-undecene-7 under mild conditions to give multifunctional photopolymers **P-2** having both a positive-working moiety in the main chain and a negative-working moiety in the side chain. Photochemical properties of **P-1** and **P-2** were investigated. **P-1** was decomposed smoothly in a solution by irradiation with UV light. It was, furthermore, found that the photochemical reaction of **P-2** was controlled easily by selecting the wavelength of the irradiation.

Silicon-containing polymers have attracted considerable attention in the field of microlithography, since they have characteristic advantages such as resistance to oxygen-reactive ion etching and adhesion to variety of materials. Poly(siloxane)s or poly(silane)s have been investigated as useful materials for photolithography. Particularly, polymers containing silicon-silicon bonds undergo photochemical reactions upon UV irradiation (*1-3*). Poly(siloxane)s are usually synthesized (*4*) by the catalytic ring opening polymerization of cyclic siloxanes, and poly(silane)s have been obtained (*5*) by the reaction of dichlorosilanes with sodium metal. Poly(silyl

---

[1]Corresponding author

0097–6156/94/0579–0443$08.00/0

ether)s are categorized as a new class of silicon-containing polymers with various structures in the main chains and poly(silyl ether)s have different characteristics from poly(siloxane)s. Concerning the synthesis of poly(silyl ether)s, Imai et al. reported (6) the synthesis and characterization of poly(silyl ether)s containing disilane moieties by the reaction of 1,2-bis(diethylamino)-tetramethyldisilane with various bisphenols. Although it is known that the addition reaction (7) of silyl chlorides with cyclic ethers give the corresponding silyl ethers, there were no reports on the synthesis of poly(silyl ether)s based on the silyl-ether formation reaction.

We recently reported (8) in a communication the new synthesis of poly(silyl ether)s containing a disilane unit by the reaction of dichlorosilanes with bisepoxides, which was catalyzed efficiently by quaternary onium salts similar to the polyaddition (9) of diacyl chlorides with bis(cyclic ether)s giving the polyesters carrying reactive pendant chloromethyl groups. These polymers having pendant chloromethyl groups are very interesting materials from the viewpoint of designing new type of functional polymers, since the pendant chloromethyl groups can be modified readily by conventional reactions.

Meanwhile photosensitive polymers having both negative-working and positive-working moieties in the side chains or in the main chains are interesting multifunctional photopolymers. We recently reported (10) for the first time the synthesis of new multifunctional copolyamides containing both cyclobutane rings and conjugated double bonds in the main chain by the polycondensation of various diamines with bis(p-nitrophenyl) β-truxinate and bis(p-nitrophenyl) esters of the unsaturated dicarboxylic acids. It was also found that the photochemistry of these copolyamides can be controlled by the selection of the exposing wavelength.

From this background, we designed new type of multifunctional poly(silyl ether) with silicon-silicon bond in the main chain and photo-crosslinkable moieties in the side chains. In this paper, we wish to report the successful synthesis of a new poly(silyl ether) containing silicon-silicon bond by the polyaddition of bisepoxides with dichlorosilane compounds. Functionalizations of the resulting poly(silyl ether) were performed by the substitution reaction with photo-crosslinkable compounds using 1,8-Diazabicyclo-[5.4.0]-undecene-7 (DBU). Photochemical properties of the multifunctional poly(silyl ether)s thus obtained were also investigated.

EXPERIMENTAL

Measurements.   ^{1}H NMR spectra were obtained on a JEOL EX-90 or FX-200 operating in the pulsed Fourier-transform (FT) modes, using tetramethylsilane (TMS) as an internal standard in chloroform-*d*. IR spectra were recorded on a JASCO IR-700. UV spectra were recorded on a Shimadzu model UV-2100 spectrophotometer. The $\overline{M}n$ and $\overline{M}w/\overline{M}n$ of polymers were measured with a TOSOH HLC-8020 GPC unit using TSK-Gel columns (eluent: N,N-dimethyl-formamide (DMF), calibration: polystyrene standards).

Materials.   1,2-Dichlorotetramethyldisilane (CMDS) was synthesized starting from hexamethyldisilane according to the reported procedure (*11*). 4-Dimethylamino-cinnamic acid (MAC), 4-dimethylamino-α-cyanocinnamic acid (MACC), cinnamylidenecyanoacetic acid (CCA), and 4-dimethylamino-4'-hydroxychalcone (MAHC) were prepared by the Knoevenagel condensation according to the literature (*12*).   Bisphenol A diglycidyl ether (BPGE) was recrystallized from methanol /ethyl methyl ketone (4/1, v/v).   Triphenylphosphine (TPP) was purified by recrystallization from methanol.   Tetrabutylammonium bromide (TBAB) was recrystallized from ethyl acetate.   Other quaternary ammonium halides and cesium fluoride, and 18-crown-6 were used as received.   Phenyl glycidyl ether (PGE) and DBU were purified by distillation.   Solvents such as toluene, nitrobenzene, tetrahydrofurane (THF), DMF, N,N-dimethylacetamide (DMAc), N-methyl-2-pyrolidone (NMP), and hexamethylphosphoroamide (HMPA) were dried and distilled before use.

Synthesis of model compound by the addition reaction of CMDS with PGE.   To a solution of PGE (1.502 g, 10 mmol) and TBAC (0.014 g, 0.05 mmol) in CHCl$_3$ (1 mL) was added dropwise a solution of CMDS (0.936 g, 5 mmol) in 2 mL of CHCl$_3$ at 0 °C, and then the reaction mixture was stirred at ambient temperature for 7 h.   The reaction mixture was evaporated and the reaction product isolated by column chromatography with silica gel using chloroform/carbon tetrachloride (1/2, v/v) as the eluent.   Yield; 85 %.   IR (neat, cm^{-1}): 1247 (νC-O-C), 1093, 1049 (νSi-O-C), and 770 (νC-Cl).   ^{1}H NMR (CDCl$_3$, TMS): δ 0.27 (s, 12 H, CH$_3$), 3.42-3.85 (m, 4 H, CH$_2$Cl), 3.90-4.05 (m, 4 H, CH$_2$), 4.10-4.35 (m, 2 H, CH), 6.80-7.40 (m, 10 H, Ar).

Synthesis of poly(silyl ether) by the polyaddition of CMDS with BPGE. To BPGE (0.8511 g, 2.5 mmol) in a 5 mL round-bottom flask was added dropwise the solution of CMDS (0.4680 g, 2.5 mmol) and 1 mol% of tetrabutylammonium chloride (TBAC, 0.025 mmol) in 2 mL of toluene at 0 °C. The reaction mixture was stirred at 0 °C for 1 h and at ambient temperature for 23 h. The reaction mixture was washed three times with water, and then poured into n-hexane. The polymer isolated was reprecipitated twice from chloroform in n-hexane and dried in vacuo at 60 °C to obtain the targeted polymer (**P-1**). Yield; 91 %. $\overline{M}$n; 24,000. IR (neat, cm^{-1}): 1247 (νC-O-C), 1093, 1046 (νSi-O-C), and 769 (νC-Cl). ^{1}H NMR (CDCl$_3$, TMS): δ 0.31 (s, 12 H, Si-CH$_3$), 1.61 (s, 6 H, CH$_3$), 3.52-3.80 (m, 4 H, CH$_2$Cl), 3.82-4.00 (m, 4 H, CH$_2$), 4.05-4.35 (m, 2 H, CH), 6.76-7.12 (m, 8 H, Ar).

Typical procedure for reaction of the poly(silyl ether) with photo-crosslinkable compounds. **P-1** (1.055 g, 2 mmol), MACC (0.714 g, 4.4 mmol), and DBU (0.670 g, 4.4 mmol) were dissolved in 5 mL of DMSO, which was stirred at 60 °C for 48 h. The reaction mixture was poured into methanol. The polymer isolated was purified by reprecipitation from chloroform in methanol to give the corresponding polymer (**P-2b**, 0.343 g). The degree of substitution of the resulting polymer was estimated by ^{1}H NMR to be 39 %. IR (film, cm^{-1}) 2210 (νCN), 1803, 1714 (νC=O), 1609 (νC=C), and 1246 (νC-O-C). ^{1}H NMR (CDCl$_3$, TMS) δ 0.15-0.31 (m, Si-CH$_3$), 1.61 (s, CH$_3$ of BPGE unit), 3.09 (s, N-CH$_3$), 3.50-4.55 (m, CH$_2$, CH), 6.55-6.90 (m, Ar), 7.12 (d, J = 8.57, Ar of BPGE), 7.92 (d, J = 9.01, Ar of MACC), and 8.62(s, CH =C).

Photochemical reaction of P-1 in THF solution. **P-1** (0.132 g, 0.25 mL) was dissolved in 100 mL of THF in a quartz cell reactor. The solution was bubbled with nitrogen for 1 h and then irradiated using a 500-W high-pressure mercury lamp (Ushio Electric Co,: USH 500D) under nitrogen. The change of the molecular weight was measured by GPC.

Photochemical reaction of P-2 in the film state. A THF solution of P-2 (0.5 g/10 mL) was cast on the inside wall of a quartz cell and dried. The absorbance at 230 nm was 0.35-0.4. The polymer film on the quartz cell was irradiated by a 500-W high-pressure mercury lamp (USH-500D) through a monochromator (JASCO

Model CT-10). Appearance or disappearance of absorption peaks was measured using a UV spectrophotometer.

RESULTS AND DISCUSSION

The addition reaction of CMDS with PGE was examined as a model reaction for the polyaddition of CMDS with BPGE. The reaction was carried out using TBAC as the catalyst at 0 °C for 1h and at ambient temperature for 7 h to give the corresponding silyl ether **1** as addition product in 85 % yield. The structure of **1** was ascertained by means of IR, ^{1}H NMR spectroscopy, and elemental analysis. The ^{1}H NMR proved that β-cleavage of the epoxy ring of PGE occurred selectively in the reaction to afford **1**. Thus it was found that the reaction of CMDS with epoxides proceeded very smoothly and regioselectively under mild conditions.

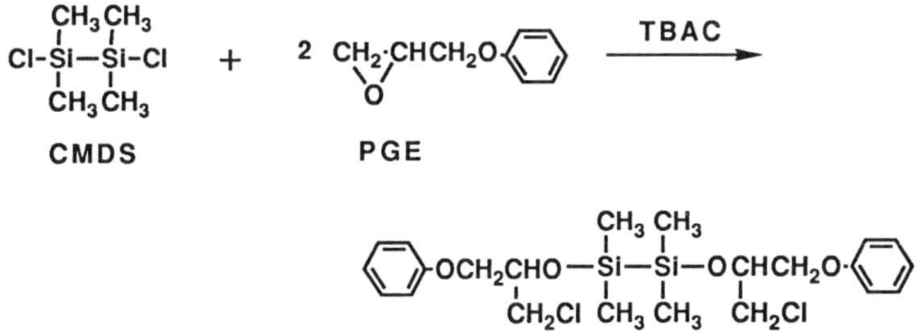

Scheme 1

Based on the result of the model reaction, the polyaddition reaction of CMDS with BPGE was carried out using 1 mol% of TBAC in various solvents at 0 °C for 1 h and at ambient temperature for 23 h. The results are summarized in Table I. The

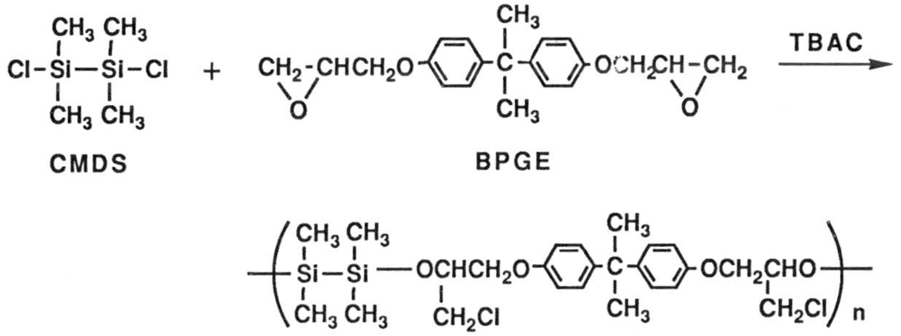

P-1

Scheme 2

polymerization proceeded very efficiently in THF, toluene, and DMAc to afford the corresponding poly(silyl ether) **P-1** with number-average molecular weight ($\overline{\mathrm{Mn}}$) of

Table I

Solvent Effect on the Polyaddition of CMDS with BPGE[a]

Run	Solvent	Yield/%	$\overline{\mathrm{Mn}} \times 10^{-3}$	$\overline{\mathrm{Mw}}/\overline{\mathrm{Mn}}$
1	THF	88	28.7	1.59
2	Toluene	91	25.4	1.54
3	Benzene	80	7.4	1.35
4	PhNO$_2$	80	19.2	1.38
5	DMAc	88	24.3	1.54
6	NMP	89	19.5	1.45
7	HMPA	79	5.1	1.18

a)The reaction was carried out with 2.5 mmol of CMDS and BPGE using 1 mol% of TBAC in 2 ml of solvent at 0 °C for 1 h and at r.t. for 23 h.

Table II

Polyaddition of BPGE with CMDS Using Various Catalysts[a]

Run	Catalyst	Yield (%)	$\overline{\mathrm{Mn}} \times 10^{-3}$	$\overline{\mathrm{Mw}}/\overline{\mathrm{Mn}}$
1	TPP	90	9.3	1.49
2	TPP/KI	79	9.1	1.47
3	DMAP	82	6.7	1.31
4	18-C-6/CsF	66	3.8	1.00
5	TMBAC	81	10.3	1.24
6	TBAB	88	29.0	1.62
7	TBAC	91	25.4	1.54
8	TBPB	85	27.7	1.60
9	TBPC	87	26.4	1.54

a)The reaction was carried out with 2.5 mmol of CMDS and BPGE using 1 mol% of various catalysts in 2 ml of toluene at 0 °C for 1h and at r.t. for 23 h.

24,000-29,000 in high yield. The molecular weight of the polymer obtained was relatively low, when the reaction was conducted in chloroform or benzene. The IR spectrum of **P-1** obtained showed characteristic absorptions at 2954 cm^{-1} due to νC-H, 1247 cm^{-1} due to νC-O-C, 1093 and 1046 cm^{-1} due to νSi-O-C, 769 cm^{-1} due to νC-Cl, respectively. ^{1}H NMR spectral data supported the structure of **P-1** as shown in Scheme 2, and the methine protons were observed at 4.05-4.35 ppm with the expected intensity ratio. This result means that the polyaddition of CMDS with BPGE proceeded regioselectively in the presence of TBAC. In the UV absorption spectrum of **P-1** measured in THF, an absorption maximum was observed at 230 nm, which was assigned to the silicon-silicon bond. It was proved that the reaction using quaternary onium salts as catalysts took place smoothly in various solvents including cyclic ether, aromatic solvents, and aprotic polar solvents, particularly, THF, toluene, and DMAc were suitable solvent to obtain **P-1** with high molecular weight.

The polyaddition of CMDS with BPGE was conducted using various catalysts in toluene (Table II). In the case of the reaction using TPP as the catalyst, the molecular weight of the polymer obtained was low. N,N-Dimethylaminopyridine (DMAP) and 18-crown-6/CsF complex also gave the relatively low molecular weight polymer. On the other hand, the corresponding polymer with high molecular weight was obtained when the reaction was conducted using quaternary onium salts such as TBAB, TBAC, and tetrabutylphosphonium bromide and chloride. It was, therefore, found that quaternary onium salts have higher catalytic activity than TPP or DMAP, although it is reported (*13*) that TPP and quaternary onium salts catalyzed efficiently the reaction of silyl chlorides with cyclic ethers.

Table III shows the effect of TBAB concentration on the polyaddition of CMDS with BPGE. The reaction was catalyzed effectively by even 0.5 mol% of TBAB to produce **P-1** with $\overline{M}n$ of 27,000. In the case of the reaction using 1 or 2 mol% of TBAB, the molecular weight of **P-1** obtained was about 30,000.   However, increasing TBAB concentration further to 4 mol% tended to decrease the molecular weight ($\overline{M}n$; 25,000). In the reaction using 4 mol% of TBAB side reactions seem to occur slightly owing to a small amount of water contained in TBAB catalyst.

Table III
Effect of TBAB Concentration[a]

Run	TBAB/mol%	Yeild/%	$\overline{M}n \times 10^{-3}$	$\overline{M}w/\overline{M}n$
1	0.5	88	27.0	1.66
2	1	88	29.0	1.62
3	2	87	31.0	1.54
4	4	84	25.0	1.48

a)The reaction was carried out with 2.5 mol of CMDS and BPGE in 2 ml of toluene and using TBAB as a catalyst at 0 °C for 1h and at r.t. for 23 h.

P-1 is insoluble in methanol and acetonitrile, but soluble in various organic solvents including ketones, halogenated hydrocarbons, aromatic solvents, and aprotic polar solvents.

As shown in Figure 1, the molecular weight of **P-1** decreased immediately when the irradiation was conducted in THF solution using a 500-W high pressure mercury lamp, particularly in the initial stage of the irradiation. On the contrary, the decrease in the molecular weight was not observed without irradiation. This result can be explained by the cleavage of the silicon-silicon bond in the polymer backbone upon UV light irradiation. It was thus demonstrated that **P-1** has a function as a positive type photopolymer.

Further chemical modification of **P-1** was studied with photo-crosslinkable compounds such as MAC and MACC, which have absorption bands in the visible region. The reaction was carried out using DBU as a base in DMSO at 60 °C for 48 h, and the results are shown in Table IV. The reaction of **P-1** with MAC proceeded readily under the reaction conditions to afford the targeted polymer **P-2a** with 73 % of degree of substitution, which was estimated by the ^{1}H NMR spectrum. In the case of the reaction with MACC, the expected polymer **P-2b** was also obtained with

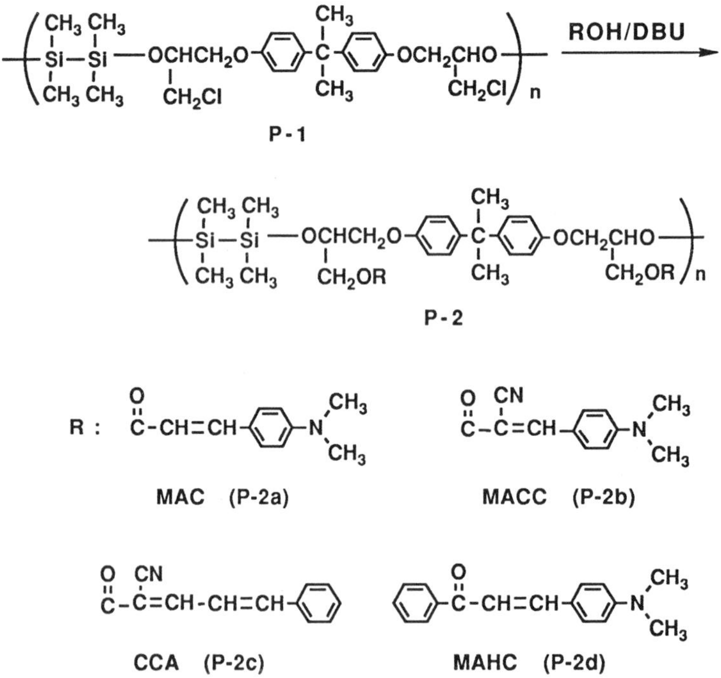

Scheme 3

39 % of degree of substitution. On the other hand, the modified polymers were not obtained when the reaction was conducted with CCA and MAHC under similar reaction condition.

Table IV

Chemical Modification of P-1 with Photo-crosslinkable Compounds[a]

Run	Polym.	Compound	Yield /g	D.S.[b] /%	$\overline{Mn}$ x10^{-3}	$\overline{Mw}/\overline{Mn}$	λmax /nm
1	P-2a	MAC	0.525	73	5.7	1.24	365.0
2	P-2b	MACC	0.343	39	7.0	1.16	420.0
3	P-2c[c]	CCA	0.896	---	7.2	1.20	---
4	P-2d	MAHC	0.154	---	4.3	1.04	---

a)The reaction was carried out with 2.0 mmol of poly(silyl ether) and 4.4 mmol of photo-crosslinkable compound and 4.4 mmol of DBU in 5 ml of DMSO at 60 °C for 48 h.
b)Degree of substitution estimated by ^{1}H-NMR.
c)Reaction time:96 h.

The photochemical reaction of **P-2a** containing pendant MAC moiety was performed in the film state by irradiation with 365 nm light through a monochromator. Figure 2 shows the change of the UV spectrum of **P-2a** film which was formed from THF solution. The absorption due to the MAC moiety at 365 nm decreased rapidly and two isosbestic points were observed at 245 nm and 300 nm. This result means that the photo-crosslinking reaction of **P-2a** proceeded selectively upon 365 nm light irradiation. Figure 3 shows the time-course of the conversion of the C = C bond of the pendant MAC moiety upon 365 nm light irradiation. The conversion was 71 % for 5 min and 83 % for 15 min. It was thus demonstrated that the photo-crosslinking reaction proceeded very rapidly in the initial stage of the irradiation. After the photoreaction the polymer film became insoluble in THF. It was, therefore, proved that **P-2a** has a function as a negative type photoresist.

When the **P-2a** film was irradiated with 230 nm light, the absorption at 230 nm based on silicon-silicon bond decreased immediately. This result suggested that the photo-decomposition of **P-2a** occurred selectively in the initial stage as shown in Figure 4. However, a prolonged irradiation caused the decrease in intensity of the absorption at 365 nm due to the MAC moiety; that is, photo-crosslinking reaction as a side reaction seemed to occur by long irradiation with 230 nm light. The polymer film irradiated with 230 nm light for 10 min became soluble in methanol, although **P-2a** film was insoluble in methanol before irradiation. It is clear from these results

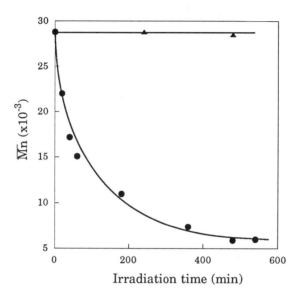

Figure 1. Effect of UV irradiation on the molecular weight change of **P-1** in THF. (▲): with irradiation, (●): without irradiation.

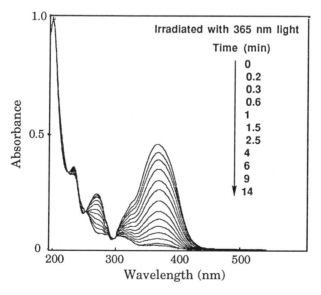

Figure 2. Change of UV spectrum of **P-2a** in the film state observed upon irradiation with 365 nm light.

that **P-2a** function as a positive-working resist.   **P-2b** containing MACC moiety in the side chain exhibited a behavior similar to **P-2a** toward the irradiation with 420 and 230 nm light and it was found **P-2b** also has both negative and positive capabilities.

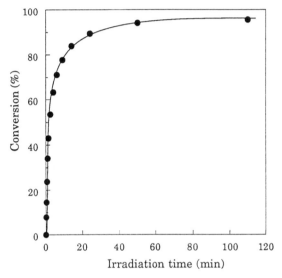

Figure 3. Photochemical reaction of MCA moiety in **P-2a** film by irradiation with 365 nm light.

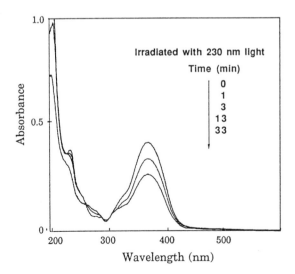

Figure 4. Change of UV spectrum of **P-2a** in the film state observed upon irradiation with 230 nm light.

CONCLUSION

The new poly(silyl ether) **P-1** containing photo-decomposable silicon-silicon bond was successfully synthesized by polyaddition of CMDS with BPGE using quaternary onium salts.   The poly(silyl ether) was modified easily with photo-crosslinkable compounds by the substitution reaction using DBU under mild conditions to give multifunctional photopolymers **P-2** having both positive-working moiety in the main chain and negative-working moiety in the side chain, respectively. It was demonstrated that the photochemical reaction of **P-2** was controlled easily by selecting the wavelength of the irradiation.   These multifunctional polymers are a new type of photopolymers.

REFERENCES

1. X. H. Zhang and R. West, *J. Polym. Chem., Polym. Chem. Ed.*, **22**, 225 (1984).
2. P. Trefonas III and R. West, *J. Am. Chem. Soc.*, **1985**, *107*, 2737.
3. M. Ishikawa, N. Hongzhi, K. Matsusaki, K. Nate, T. Inoue, and H. Yokono, *J. Polym. Sci., Polym. Lett.*, **1984**, *22*, 669.
4. For example: P. V. Wright, "Ring Opening Polymerization", Vol. 2, K. J. Ivin and T. Saegusa, Ed., Elsevier, London, 1984, pp 1055-1133.
5. For example: R. West, *J. Organo-metallic Chem.*, **1986**, *300*, 327.
6. M. Padmanaban, M. Kakimoto, and Y. Imai, *J. Polym. Sci., Part A: Polym. Chem.*, **1990**, *28*, 2997.
7. For example: W.P. Weber, "Silicon Reagents for organic Synthesis"; Springer, Berlin, 1983, pp 21-39.
8. T. Nishikubo, A. Kameyama, and N. Hayashi, *Polym. J.*, **1993**, *25*, 1003.
9. a) A. Kameyama, S. Watanabe, E. Kobayashi, and T. Nishikubo, *Macromolecules*, **1992**, *25*, 2307. b) A. Kameyama, Y. Yamamoto, and T. Nishikubo, *J. Polym. Sci., Part A: Polym. Chem.*, **1993**, *31*, 1639.
10. T. Nishikubo, T. Iizawa, Y. Shiozaki, and T. Koito, *J. Polym. Sci., Part A: Polym. Chem.*, **1992**, *30*, 449.
11. H. Sakurai, M. Tominaga, T. Watanabe, and M. Kumada, *Tetrahedron Lett.*, **1966**, 5493.
12. G. Jones, "Organic Reactions", Vol. 15, John Wiley and Sons, New York, 1967, pp 204.
13. G. C. Andrews, T. C. Crawford, and L. D. Contillo, Jr., *Tetrahedron Lett.*, **1981**, *22*, 3803.

RECEIVED September 13, 1994

Author Index

Affiliation Index

Subject Index

Production: Meg Marshall
Indexing: Deborah H. Steiner
Acquisition: Anne Wilson
Cover design: Amy Hayes

Printed and bound by Maple Press, York, PA

Highlights from ACS Books

Bestsellers from ACS Books

The ACS Style Guide: A Manual for Authors and Editors
Edited by Janet S. Dodd
264 pp; clothbound ISBN 0–8412–0917–0; paperback ISBN 0–8412–0943–X

Understanding Chemical Patents: A Guide for the Inventor
By John T. Maynard and Howard M. Peters
184 pp; clothbound ISBN 0–8412–1997–4; paperback ISBN 0–8412–1998–2

Chemical Activities (student and teacher editions)
By Christie L. Borgford and Lee R. Summerlin
330 pp; spiralbound ISBN 0–8412–1417–4; teacher ed. ISBN 0–8412–1416–6

Chemical Demonstrations: A Sourcebook for Teachers,
Volumes 1 and 2, Second Edition
Volume 1 by Lee R. Summerlin and James L. Ealy, Jr.;
Vol. 1, 198 pp; spiralbound ISBN 0–8412–1481–6;
Volume 2 by Lee R. Summerlin, Christie L. Borgford, and Julie B. Ealy
Vol. 2, 234 pp; spiralbound ISBN 0–8412–1535–9

Chemistry and Crime: From Sherlock Holmes to Today's Courtroom
Edited by Samuel M. Gerber
135 pp; clothbound ISBN 0–8412–0784–4; paperback ISBN 0–8412–0785–2

Writing the Laboratory Notebook
By Howard M. Kanare
145 pp; clothbound ISBN 0–8412–0906–5; paperback ISBN 0–8412–0933–2

Developing a Chemical Hygiene Plan
By Jay A. Young, Warren K. Kingsley, and George H. Wahl, Jr.
paperback ISBN 0–8412–1876–5

Introduction to Microwave Sample Preparation: Theory and Practice
Edited by H. M. Kingston and Lois B. Jassie
263 pp; clothbound ISBN 0–8412–1450–6

Principles of Environmental Sampling
Edited by Lawrence H. Keith
ACS Professional Reference Book; 458 pp;
clothbound ISBN 0–8412–1173–6; paperback ISBN 0–8412–1437–9

Biotechnology and Materials Science: Chemistry for the Future
Edited by Mary L. Good (Jacqueline K. Barton, Associate Editor)
135 pp; clothbound ISBN 0–8412–1472–7; paperback ISBN 0–8412–1473–5

For further information and a free catalog of ACS books, contact:
American Chemical Society
Distribution Office, Department 225
1155 16th Street, NW, Washington, DC 20036
Telephone 800–227–5558